D0153821

LINEAR MATHEMATICS

A PRACTICAL APPROACH

PATRICIA
CLARK
KENSCHAFT
Professor Emerita
Montclair State University

DOVER PUBLICATIONS, INC., Mineola, New York

Bibliographical Note

Linear Mathematics: A Practical Approach, first published by Dover Publications, Inc., in 2013, is an unabridged republication of the first edition of the work originally published by Worth Publishers, Inc., New York, in 1978.

International Standard Book Number

ISBN-13: 978-0-486-49719-8
ISBN-10: 0-486-49719-4

Manufactured in the United States by Courier Corporation
49719401 2013
www.doverpublications.com

To my spouse, Frederick Chichester, whose versatile skills provided me with the basic ideas of Section 6.4, with numerous delicious meals, and with continuous emotional support

and to my children, Lori and Edward Kenschaft, who have encouraged my career by proofreading parts of this book and making up some exercises, by helping (sometimes cheerfully) with the housework, and by patience through many long-distance telephone calls

Preface

Mathematics is both one of the great achievements of human creativity and a pervasive practical skill in our culture, vital to higher education in almost every discipline. This text is designed to help nontechnical majors learn enough challenging mathematics to experience both the beauty of abstract patterns and the excitement of discovering how they can model the "real" world.

Carefully chosen examples and exercises are the basis of the mathematical presentation. Short examples showing how the mathematics can be applied to the real world occur in most sections of the book; the numbers are kept small in these examples so that a student can follow them and do the corresponding exercises without groping through a maze of arithmetic. A few longer examples with realistic numbers are also included; these can be omitted or used as an entire lesson's discussion. The important topic of input–output analysis appears several times in the linear algebra half of the book to help students develop a genuine "feel" for this basic application.

Each section of the book corresponds to a day's lesson and is accompanied by two or three sets of problems. Either exercise set A or set B provides a complete homework assignment: together, the sets include plenty of problems for review and classroom demonstration. Set C provides supplementary problems, usually more advanced than those in sets A and B, but sometimes displaying different applications. At the end of each chapter is a sample test. The answers to all sample tests and to exercise sets A and C are included since I believe that instant feedback spurs students to continue the essential practice that exercises provide. Flexibility is added by including the answers to the B exercises only in the Instructor's Manual. These can be easily reproduced by teachers who want to share them with their students.

The only prerequisite for the course is two years of high school algebra—or one good year. No knowledge of geometry or calculus is needed to study this book. Although topics not usually covered in elementary algebra appear from the first sections to stimulate the students' interest, there is a considerable amount of review material included near the beginning of the book, since most students appreciate such help when starting a new math course. More thorough reviews of the elementary topics essential for reading this book are found in the appendixes, along with exercises and answers.

Covering one section a day with time out for reviews and tests, I can complete about 30 of the 44 sections of this text in a three-credit, one-semester course at Montclair State College. (Starred sections and chapters are optional; later material is not dependent on them.)

There are many ways that sections can be omitted to adapt the book to a semester course. I would suggest that the following sections are essential for an introductory course in linear algebra and linear programming: 1.1, 1.2, 1.3, 1.4, 1.5, 2.1, 2.2, 2.4, 3.1, 3.2, 3.4, 5.1, 5.2, 5.4, 6.1, and 6.2.

A fairly traditional course in linear algebra interwoven with modern applications can proceed directly through the book until Section 6.2, omitting Sections 1.6 and 2.6.

A linear mathematics course similar to those called "Finite Mathematics" can be given by covering Sections 1.1, 1.2, 1.3, 1.4, 1.5, 2.1, 2.2, 2.3, 2.4, 5.1, 5.2, 5.3, 5.4, 6.1, 6.2, 9.1, 9.2, and 9.3, and Chapter 10. If ample time is desired for probability, Chapter 9 can be inserted any time after Chapter 1.

A liberal arts survey course on the uses of mathematics in the contemporary world can be based on Chapters 1, 2, 5, 8, and 9. It is also possible to cover the first four sections of Chapter 2 before Chapter 1 if a professor prefers to introduce solving simultaneous equations before presenting matrices.

Most of these approaches have already been class-tested using earlier editions of the book; this diagram shows which chapters are required for those following and may be helpful to others planning their own course.

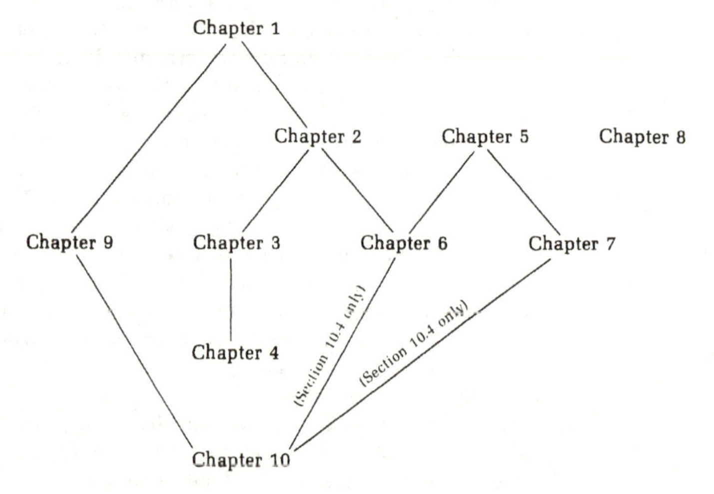

If the appendixes are to be used as lessons, Appendix 3 should be inserted before Section 1.5, Appendix 1 before Chapter 2, and Appendix 2 before Chapter 5.

Many thanks are hereby relayed to the students and to my colleagues at Montclair State College who have given a steady stream of suggestions and corrections, semester after semester, as this book went through its four preliminary versions. I have been repeatedly touched by the friendliness and openness with which these comments have been generously offered, always with an eye to improving the book's effectiveness. I am indeed fortunate to teach and write in such a supportive atmosphere.

I am grateful to Kenneth Bergstresser of Washington State University and to David A. Cox of Douglass College, Rutgers University, for offering suggestions after preliminary editions of this book were used in their classes. Others who have read large portions of the text and wrote detailed, painstaking comments include Robert Bixby, Northwestern University; Robert Canavan, Monmouth College; Robert Hughes, Boise State University; Kenneth Kalmanson, Montclair State College; Joseph Rosenstein, Rutgers University; M. A. Shiro, Bloomfield College: Neil Weiss, Arizona State University; and Frank Young, Knox College. Kenneth Bergstresser and Kusum Jain proofread the typeset pages and recalculated the answers to all the problems. To all of them I want to express deep appreciation.

I also wish to thank Peter Casazza, University of Alabama, Huntsville; Charles Jackson, Jamestown Community College; Sister Janice Marie, Our Lady of Angels College; Thomas Kearns, Northern Kentucky University; Calvin King, Tennessee State University; Russell Lee, Allan Hancock College; John Mack, University of Kentucky; Conrad McKnight, University of Southwestern Louisiana; James Osborn, Georgia Institute of Technology; Richard Panicucci, Fairleigh Dickinson University, Teaneck, New Jersey; Peter Rice, University of Georgia; James Russo, Roger Williams College: Roger Turcotte, Collège Brébeuf, Montréal; and Wayne Wallace, University of Wisconsin, Oshkosh, for their valuable suggestions for improving the manuscript.

Finally I want to thank the people at Worth Publishers, who not only recruited the impressive array of reviewers listed above, but have thoughtfully helped in every stage of the book's planning and production. Although there are a few names I sorely want to mention, I know that if I begin, I will not be able to find a sensible stopping place until I have mentioned almost everyone in the company.

Like a child, a book grows with the input of so many people that the only time a parent feels completely responsible for it is when it errs. Yet as this book-child of mine approaches adulthood, I find myself hoping that it conveys to its readers the excitement of discovery that I myself have felt while learning about the pervasive power of mathematics to improve our understanding of many other fields.

January, 1978 Patricia Clark Kenschaft
Upper Montclair, N.J.

Contents

Applications

BIOLOGY

Human blood types

Exercise 1.2.B *problem 7*

Bacteria surviving in a test tube

Examples 2.1.3, 3.1.10

Exercise 2.1.C *problem 1*

Animals grazing

Exercise 5.3.C *problem 4*

Sex of an expected baby

Example 9.1.10

Heredity and genes

Example 9.1.22

Exercises 9.1.A *problems 8, 9* B *problem 8*

Nesting habits of birds

Examples 9.2.16, 9.2.17

Number of species of pea plants

Exercise 9.2.A *problem 6*

Pine tree deaths

Example 10.1.9

Exercise 10.1.A *problem 1*

SOCIOLOGY

Demographic input–output matrix

Example 1.1.12

Hiring practices and discrimination

Example 9.2.18

Exercises 9.2.A *problem 10* B *problem 11*

Home owning and renting of American whites and nonwhites

Exercise 9.3.A *problem 6*

Death rates of American white and nonwhite babies

Exercises 9.3.B *problem 4* 10.1.A *problem 5*

Rumors

Example 9.4.9

Rural–urban migrations

Exercise 9.4.A *problems 3, 5*

Relationship between auto accidents and age in the United States

Exercise Sample Test 9 *problem 5*

ECOLOGY

Ecological systems, input–output matrices

Example 1.3.10

Exercises 1.2.A *problem 3* B *problem 7*

Stocking a lake with fish totaling maximum weight

Example 5.2.6

Exercise 5.2.B *problem 4*

Cleaning up a river

Example Section 6.4

Exercises 6.4.A, B

Enforcing antipollution laws

Exercise 10.1.B *problem 6*

TRANSPORTATION STUDIES

Mileage table

Example 1.2.6

Traffic flow

Example 3.3.1

Exercises 3.3.A, B Sample Test 3 *problem 7*

Toll booths

Examples 9.1.19, 10.1.2, 10.1.8

Exercise 9.1.A *problem 4*

Accidents and distance driven

Exercise 9.3.B *problem 1*

Enforcing parking regulations

Exercise 10.1.A *problem 6*

PSYCHOLOGY

College test scores

Exercise 1.2.A *problem 4*

Pecking order of chickens

Example 9.2.15

Verifying the skill of wine tasters

Exercise 9.2.A *problem 11*

Verifying claims of ESP

Exercise 9.2.B *problem 12*

Movements of mice in a partitioned cage

Exercise 9.4.B *problems 2b, 3, 4b*

Grading essay exams

Exercise Sample Test 9 *problem 3*

Intercepting an enemy or robber

Exercises 10.2.A *problem 2c* B *problem 2c*

Greek–Turkish Cypriot conflict

Example 10.3.5

Prisoner's dilemma problem

Exercise 10.3.A *problem 4*

COMMUNICATION STUDIES

Postage rates

Examples 1.1.6, 1.1.8, 1.3.8

Cryptography

Examples 1.6.1, 1.6.2
Exercises 1.6.A *problems 1, 2, 3* B *problems 1, 2, 3*

Distributing advertising money among the media

Exercises 2.2.A *problem 5* B *problem 5*
Example 6.3.1

TV exposure time

Exercises 8.4.A *problem 4* B *problem 4*

Telephone calls received by a switchboard

Exercises 9.1.B *problem 4* Sample Test 9 *problem 2*

Newspaper circulation

Exercise 9.4.B *problems 2a, 3*

FUND RAISING

Obtaining grants from foundations

Exercises 1.3.C *problems 1, 2* 10.1.B *problems 5*

Alumni fund trends

Example 9.4.8

DIETETICS

Filling specified requirements with given foods

Exercise 2.2.C *problems 3, 4*

Finding the minimum cost of a diet with given nutritional requirements

Examples 5.4.1, 6.3.3 Section 5.5
Exercises 5.4.A *problems 1, 2, 3, 4* B *problems 1, 2, 3, 4* 5.5.A *problems 2, 3, 4* B *problems 2*

BUSINESS AND ECONOMICS

Motorcycle tire prices

Exercise 1.1.A *problem 7*

Steel shelving prices

Exercise 1.1.B *problem 7*

Parts-listing problems

Examples 1.2.1, 1.3.9, 2.6.2
Exercises 1.2.A *problems 8, 9* B *problems 8, 9* 1.3.A *problems 6, 7* B *problems 6, 7* 2.6.A *problems 2, 3, 4* B *problems 2, 3, 4*

Leontief matrices

Example 1.2.7
Exercises 1.2.A *problems 1, 2* B *problems 1, 2*

Stadium ticket sales

Example 1.3.6
Exercises 1.3.A *problems 2, 3* B *problems 2, 3*

Computing time needed for a series of jobs

Exercises 1.3.A *problems 4, 5* B *problems 4, 5*

Open Leontief models

Examples 1.4.1, 2.5.4
Exercises 1.4.A *problem 3* B *problem 3* C *problem 1* Sample Test 1 *problem 8* 2.5.A, B, C

Filling orders

Examples 2.1.1, 2.1.2, 2.2.1

LINEAR MATHEMATICS

A PRACTICAL APPROACH

Chapter 1

MATRICES: BASIC SKILLS AND APPLICATIONS

1.1 Definitions, Addition, Scalar Multiplication, and Notation

Arithmetic is the study of numbers. Geometry is the study of shapes. Algebra is the study of equations. And linear algebra, the subject matter of the first four chapters of this book, is the study of matrices.

1.1.1 Definition

A matrix (plural: matrices) is a rectangular array of numbers.

1.1.2 Example

$$\begin{bmatrix} 1.22 & 1.64 & 1.97 \\ 0.90 & 1.56 & 1.89 \\ 0.45 & 0.88 & 1.08 \\ 0.21 & 0.21 & 0.30 \end{bmatrix}$$

is a matrix. You have already seen matrices in many contexts. The matrix above, for example, appears in a table of the cost of mailing books:

	8 oz	1 lb	2 lb
Airmail	1.22	1.64	1.97
First class	0.90	1.56	1.89
Printed matter	0.45	0.88	1.08
Book rate	0.21	0.21	0.30

The first chapter of this text reminds you of some facts you already know about matrices, introduces some unfamiliar applications, and explores some properties of matrices that mathematicians use. In many ways matrices are like numbers; they can be added and multiplied, for example. And they also have novel useful properties, as we shall see. The next three chapters explore these properties further.

In Chapters 5 through 8 of this book rectangular arrays of numbers will be used in a somewhat different subject, linear programming, which uses some of the same skills that you will develop in linear algebra. Linear programming problems involve maximizing or minimizing some variable (often profit or cost, respectively) given certain restrictions on other variables. A good course in first-year algebra is the only prerequisite for this text; if you need to review algebra, you can read the appendixes and do the exercises given there. No geometry or calculus will be used.

The words row and column in matrix theory follow ordinary English usage. The elements (numbers) in the first *row* of the matrix in Example 1.1.2 are 1.22, 1.64, and 1.97; the elements in the first *column* are 1.22, 0.90, 0.45, and 0.21. We say that this matrix is a 4 by 3 matrix because it has 4 rows and 3 columns.

1.1.3 Definition

A matrix with n rows and m columns is said to be of *dimension* n by m.

1.1.4 Definition

Two matrices are said to be *equal* if they have the same dimension and each element in one is equal to the corresponding element in the other.

Matrix addition is straightforward. Two matrices must have the same dimension if they are to be added; to add them, we merely add corresponding elements.

1.1.5 Example

$$\begin{bmatrix} 2 & 3 \\ 1 & 4 \end{bmatrix} + \begin{bmatrix} 7 & -1 \\ 0 & 5 \end{bmatrix} = \begin{bmatrix} 2+7 & 3-1 \\ 1+0 & 4+5 \end{bmatrix} = \begin{bmatrix} 9 & 2 \\ 1 & 9 \end{bmatrix}$$

1.1.6 Example

Suppose that the Postal Service decides to increase the rates across-the-board by \$0.02 over the rates given in Example 1.1.2. To find the new matrix describing the postal rates, we add to our original matrix one that has the same dimensions and in which each element equals 0.02.

$$\begin{bmatrix} 1.22 & 1.64 & 1.97 \\ 0.90 & 1.56 & 1.89 \\ 0.45 & 0.88 & 1.08 \\ 0.21 & 0.21 & 0.30 \end{bmatrix} + \begin{bmatrix} 0.02 & 0.02 & 0.02 \\ 0.02 & 0.02 & 0.02 \\ 0.02 & 0.02 & 0.02 \\ 0.02 & 0.02 & 0.02 \end{bmatrix} = \begin{bmatrix} 1.24 & 1.66 & 1.99 \\ 0.92 & 1.58 & 1.91 \\ 0.47 & 0.90 & 1.10 \\ 0.23 & 0.23 & 0.32 \end{bmatrix}$$

Subtraction of matrices is similar to addition; it is done element by element. Again, the matrices must have the same dimension.

1.1.7 Example

If the Postal Service decides to decrease (!) the rates across-the-board by $0.02 from those given in Example 1.1.2, we can find the new rates as follows:

$$\begin{bmatrix} 1.22 & 1.64 & 1.97 \\ 0.90 & 1.56 & 1.89 \\ 0.45 & 0.88 & 1.08 \\ 0.21 & 0.21 & 0.30 \end{bmatrix} - \begin{bmatrix} 0.02 & 0.02 & 0.02 \\ 0.02 & 0.02 & 0.02 \\ 0.02 & 0.02 & 0.02 \\ 0.02 & 0.02 & 0.02 \end{bmatrix} = \begin{bmatrix} 1.20 & 1.62 & 1.95 \\ 0.88 & 1.54 & 1.87 \\ 0.43 & 0.86 & 1.06 \\ 0.19 & 0.19 & 0.28 \end{bmatrix}$$

Suppose that the Postal Service decides instead to increase all the rates by 10 percent. This would be the same as multiplying all the rates by 1.1 (because an increase of 10 percent is the same as adding 0.1 times the original price to the original price and $0.1x + x = 1.1x$). Thus, to get the matrix describing the new rates, we would use scalar multiplication; that is, we would multiply every element of the matrix by the same number (which is called a "scalar" in this context).

1.1.8 Example

Use scalar multiplication to show how a 10 percent rate increase affects the postal rate matrix of Example 1.1.2. Round off the results to the nearest cent.

$$(1.1)\begin{bmatrix} 1.22 & 1.64 & 1.97 \\ 0.90 & 1.56 & 1.89 \\ 0.45 & 0.88 & 1.08 \\ 0.21 & 0.21 & 0.30 \end{bmatrix} = \begin{bmatrix} (1.1)1.22 & (1.1)1.64 & (1.1)1.97 \\ (1.1)0.90 & (1.1)1.56 & (1.1)1.89 \\ (1.1)0.45 & (1.1)0.88 & (1.1)1.08 \\ (1.1)0.21 & (1.1)0.21 & (1.1)0.30 \end{bmatrix} = \begin{bmatrix} 1.34 & 1.80 & 2.17 \\ 0.99 & 1.72 & 2.08 \\ 0.50 & 0.97 & 1.19 \\ 0.23 & 0.23 & 0.33 \end{bmatrix}$$

If the postal rates are doubled, the scalar multiplication of the matrix is easy indeed:

$$(2)\begin{bmatrix} 1.22 & 1.64 & 1.97 \\ 0.90 & 1.56 & 1.89 \\ 0.45 & 0.88 & 1.08 \\ 0.21 & 0.21 & 0.30 \end{bmatrix} = \begin{bmatrix} 2.44 & 3.28 & 3.94 \\ 1.80 & 3.12 & 3.78 \\ 0.90 & 1.76 & 2.16 \\ 0.42 & 0.42 & 0.60 \end{bmatrix}$$

Matrix Notation

Often we shall want to denote the elements of a matrix symbolically. The standard way of writing a general 2 by 3 matrix is

$$A = \begin{bmatrix} a_{11} & a_{12} & a_{13} \\ a_{21} & a_{22} & a_{23} \end{bmatrix}$$

The subscripts look like two-digit numbers occurring in the expected order, but each subscript actually consists of two different one-digit numbers, the first indicating the row in which the element appears and the second indicating the column. The upper right element in the preceding matrix is read "a one three." (In mathematics books it is usual to omit punctuation between such numbers because they are generally less than 10 and are written frequently. In applied books, where the matrices are often large, a comma may be used between the two subscripts.)

It is customary to use a capital letter to denote a whole matrix and the corresponding lowercase letter with subscripts to denote elements in that matrix.

1.1.9 Example

Find a_{13}, a_{31}, a_{24}, and a_{42} in the following 3 by 4 matrix:

$$A = \begin{bmatrix} 1 & 2 & 3 & 4 \\ 5 & 6 & 7 & 8 \\ 9 & 10 & 11 & 12 \end{bmatrix}$$

Solution:

a_{13} indicates the element in the first row and the third column, which is 3. Similarly, $a_{31} = 9$ and $a_{24} = 8$. There is no a_{42}, because there is no fourth row.

1.1.10 Example

Write a 2 by 3 matrix such that $a_{11} = 4$, $a_{12} = 5$, $a_{13} = -1$, $a_{21} = 3$, $a_{22} = -2$, and $a_{23} = 0$.

Solution:

$$A = \begin{bmatrix} 4 & 5 & -1 \\ 3 & -2 & 0 \end{bmatrix}$$

1.1.11 Example

(a) Standard notation for matrices is to use the subscript i to refer to the row and j to refer to the column. Using this notation, write an arbitrary element in the second row of an unknown matrix.

(b) Designate an arbitrary element in the third column of an unknown matrix.

Solution:

(a) a_{2j} designates an element in the second row and the jth column.
(b) a_{i3} designates an element in the ith row and the third column.

The notation described above is especially useful when it is necessary to refer to a specific element in a large matrix.

The next example is our first example of an input–output matrix, a recurring theme in this text.

1.1.12 Example in Sociology

Given any group of women between the ages of 20 and 30 (as our input) we can expect 10 years later to have (as our output) a slightly smaller number of women between the ages of 30 and 40 plus a much smaller number of daughters under the age of 10. (It is customary to ignore males in such discussions, and we follow the prevailing custom. This custom may change when couples choose the sex of their children in significant numbers—but so will other customs.)

Approximating the statistics for American women in 1970, we obtain the input–output table shown in Figure 1.1–1. Each number in this table represents a proportion. For example, the fact that $a_{12} = 0.973$ means that from a "typical" group of 1000 girls between the ages of 0 and 9, 973 will survive to be teenagers (by which we mean people between the ages of 10 and 19) 10 years later. And the fact that $a_{21} = 0.070$ means that a "typical" group of 1000 young women between the ages of 10 and 19 will produce 70 girl children during the next 10 years. (Notice that such statements would be very difficult to express if the notations a_{12} and a_{21} were not available.) Thus a number in this table represents the proportion of the input group (listed on the left of the matrix) which will produce a member of the output group (listed at the top of the matrix).

Figure 1.1–1

	Output									
Input	0–9	10–19	20–29	30–39	40–49	50–59	60–69	70–79	80–89	90–99
0–9	0	0.973	0	0	0	0	0	0	0	0
10–19	0.070	0	0.992	0	0	0	0	0	0	0
20–29	0.147	0	0	0.985	0	0	0	0	0	0
30–39	0.105	0	0	0	0.979	0	0	0	0	0
40–49	0.009	0	0	0	0	0.945	0	0	0	0
50–59	0	0	0	0	0	0	0.857	0	0	0
60–69	0	0	0	0	0	0	0	0.667	0	0
70–79	0	0	0	0	0	0	0	0	0.549	0
80–89	0	0	0	0	0	0	0	0	0	0.254
90–99	0	0	0	0	0	0	0	0	0	0

SUMMARY

This section defined the term "matrix" and by using examples showed how to add and subtract matrices and how to multiply them by a number (scalar multiplication). Standard matrix notation was introduced and used to explain an example in sociology.

Mathematics is not a spectator sport. To implant the ideas of each section in your mind, it is important to do either Exercises A or B of each section. Exercises C contain supplementary problems.

EXERCISES 1.1. A

1. In Figure 1.1–1 (page 5), find a_{12}, a_{43}, a_{78}, and a_{39}.
2. If $A = \begin{bmatrix} 1 & -1 \\ 2 & 4 \end{bmatrix}$ and $B = \begin{bmatrix} 0 & 3 \\ -2 & 5 \end{bmatrix}$, find $A + B$, $A - B$, and $3A$.
3. If $A = \begin{bmatrix} 1 & 0 & 2 \\ 0 & 1 & 3 \end{bmatrix}$ and $B = \begin{bmatrix} 0 & 1 & 4 \\ 1 & 0 & -1 \end{bmatrix}$, find $2A + 3B$.
4. $\begin{bmatrix} 2 & 3 \\ 4 & 5 \\ 4 & 7 \end{bmatrix}$ is a ______ by ______ matrix.
5. Write the 2 by 3 matrix with $a_{11} = 5$, $a_{22} = 0$, $a_{23} = -1$, $a_{13} = -6$, $a_{21} = 4$, and $a_{12} = 2$.
6. Write a matrix with $a_{22} = 4$, $a_{11} = 1$, $a_{21} = 0$, and $a_{12} = -3$.
7. (a) Write the following information in matrix form: a motorcycle tire of size 2.25/2.50 × 16″ costs $9.79; if it is size 3.25/3.50 × 17″, it costs $16.50; if it is size 3.25/3.50 × 18″, it costs $18.59; a tire of size 2.75/3.00 × 16″ costs $12.99; one of 2.25/2.50 × 18″ costs $11.50; a tire of size 3.25/3.50 × 16″ costs $14.89; one of size 2.75/3.00 × 18″ costs $15.50; one of size 2.75/3.00 × 17″ costs $13.50; and one of size 2.25/2.50 × 17″ costs $9.89.
   (b) Use matrix notation to show how a uniform price rise of $0.50 affects the matrix you have written.
   (c) Use scalar multiplication to show how a 10 percent tax affects the prices in your original matrix.

EXERCISES 1.1. B

1. In Figure 1.1–1 (page 5), find a_{31}, a_{52}, a_{89}, and a_{26}.
2. If $A = \begin{bmatrix} 2 & 0 \\ 3 & -1 \end{bmatrix}$ and $B = \begin{bmatrix} 1 & 3 \\ 5 & -4 \end{bmatrix}$, find $A + B$, $A - B$, and $2A$.
3. If $A = \begin{bmatrix} 3 & 0 & 1 \\ 2 & 1 & 0 \end{bmatrix}$ and $B = \begin{bmatrix} -2 & 1 & 0 \\ 4 & 0 & 1 \end{bmatrix}$, find $3A - B$.
4. $\begin{bmatrix} 2 & 4 & 6 & 8 & 7 \\ 1 & -2 & -5 & 0 & 1 \end{bmatrix}$ is a ______ by ______ matrix.

5. Write the 3 by 2 matrix with $a_{11} = 6$, $a_{31} = 0$, $a_{22} = -3$, $a_{12} = -2$, $a_{32} = 5$, and $a_{21} = 2$.
6. Write a matrix with $a_{21} = 3$, $a_{22} = 0$, $a_{11} = 7$, and $a_{12} = -2$.
7. (a) Steel shelving prices vary according to both the width and depth of the shelves. Compile the following information into a matrix. An assembly with 36-in.-wide shelves that are 12 in. deep costs $16.99. If the shelves are 36 in. wide and 18 in. deep, they cost $21.49; but if the depth is only 16 in., the shelves cost $19.99. A 16-in.-deep set of shelves that is 72 in. wide costs $35.99. If a set is 72 in. wide and 18 in. deep, it costs $39.99, and if it is 72 in. wide and 12 in. deep, it costs $29.99.
   (b) Use matrix addition to show how a price rise of $0.50 per assembly affects this matrix.
   (c) Use scalar multiplication to show how a 10 percent tax affects the prices in the original matrix.

ANSWERS 1.1. A

1. $a_{12} = 0.973$, $a_{43} = 0$, $a_{78} = 0.667$, $a_{39} = 0$

2. $A + B = \begin{bmatrix} 1 & 2 \\ 0 & 9 \end{bmatrix}$; $A - B = \begin{bmatrix} 1 & -4 \\ 4 & -1 \end{bmatrix}$; $3A = \begin{bmatrix} 3 & -3 \\ 6 & 12 \end{bmatrix}$

3. $\begin{bmatrix} 2 & 3 & 16 \\ 3 & 2 & 3 \end{bmatrix}$

4. 3 by 2

5. $\begin{bmatrix} 5 & 2 & -6 \\ 4 & 0 & -1 \end{bmatrix}$

6. $\begin{bmatrix} 1 & -3 \\ 0 & 4 \end{bmatrix}$

7. (a)

	16	17	18
2.25/2.50	9.79	9.89	11.50
2.75/3.00	12.99	13.50	15.50
3.25/3.50	14.89	16.50	18.59

(b)

	16	17	18
2.25/2.50	10.29	10.39	12.00
2.75/3.00	13.49	14.00	16.00
3.25/3.50	15.39	17.00	19.09

(c)

	16	17	18
2.25/2.50	10.77	10.88	12.65
2.75/3.00	14.29	14.85	17.05
3.25/3.50	16.38	18.15	20.45

(The rows and columns could be reversed in these matrices.)

1.2 Parts-Listing and Input–Output Matrices; Triangular, Diagonal, and Symmetric Matrices

This section explores more applications of matrices and describes some specific types of matrices. One of the early economic (Leontief) input–output matrices is presented in its original form. We begin with a simpler type of input–output matrix, which shows how the various ingredients used to make one product are assembled to make that product. This type of input–output matrix is often called a parts-listing matrix.

1.2.1 Example

Suppose that we are making tripods on which to rest a scope used by a surveyor (Figure 1.2–1). The tripod (T) consists of three legs (L) joined together by a bolt (B). Each leg consists of three rods (R) connected by two bolts. Write an *assembly graph* and a *quantity matrix.*

Solution:

An assembly graph is merely a diagram showing the quantity of each part required to make more complex parts. The assembly graph in Figure 1.2–2 exhibits diagrammatically the statements in the text of Example 1.2.1.

An assembly graph is useful in making a quantity matrix, another device for showing how parts contribute to a whole. To construct a quantity matrix, we must first order the items so that the letter at the

Figure 1.2–1

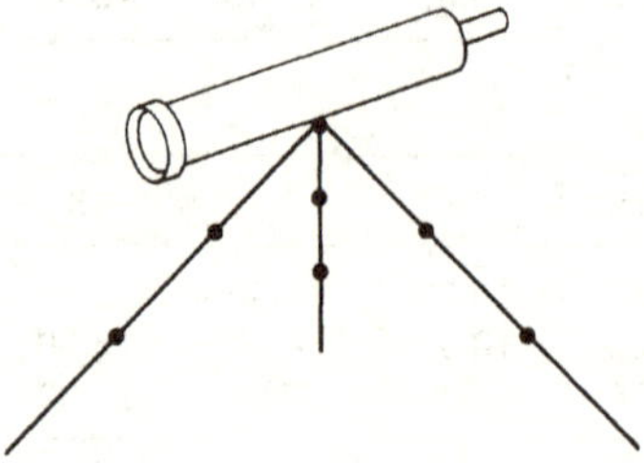

Figure 1.2–2

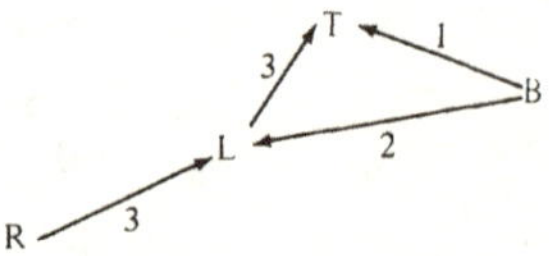

end of each arrow *follows* the letter at the arrow's point. Thus, in Example 1.2.1, both B and R must *follow* L in the ordering, and both B and L must *follow* T. Thus

$$T \quad L \quad B \quad R \qquad \text{and} \qquad T \quad L \quad R \quad B$$

are the only two lists that are in technological order for this tripod. We shall use the first list (for no particular reason). To construct the quantity matrix, we write the names of the items (actually, their abbreviations) in a technological order across the top from left to right and in the same technological order down the left side from top to bottom (Figure 1.2–3). To fill in the matrix, we ask ourselves how many of each of the items on the left of an element are needed to construct the item indicated above it. For example, we need three legs to make one tripod, so $a_{21} = 3$.

Figure 1.2–3

	T	L	B	R
T	0	0	0	0
L	3	0	0	0
B	1	2	0	0
R	0	3	0	0

Each number indicates only those components that contribute *immediately* to the item listed above it. For example, the number in the lower left corner is zero even though rods obviously appear in the finished tripods, because the rods are used only in the immediate construction of the legs, and when the tripod itself is assembled, it is the legs, not the rods, which are needed. In Sections 1.3 and 2.6 we shall continue our study of parts-listing matrices.

One of the most common and far-reaching uses of matrices in the social sciences is in input–output theory. If the large array of numbers in the next example dismays you, do not despair! A quick flip through the book should assure you that most of the matrices are much smaller; when student computations are anticipated, reality is abandoned in favor of easy numbers. But in this section we give due respect to the complexities of the real world.

In the late 1940s Wassily Leontief led a massive two-year study of how various industries in the United States economy interacted in the year 1947. The American economy was divided into 500 sectors, and a huge matrix was compiled showing how much each sector sold to the other 499 sectors and how much it reinvested in itself. Combining these 500 sectors into 42 major departments of production, a famous matrix (more palatable to the general public!) was drawn up and appeared in an article written by Leontief for *Scientific American* in October, 1951. We include a copy of this historical table on pages 10 and 11.

THIS TABLE SHOWS THE EXCHANGE OF GOODS

INDUSTRY

INDUSTRY PRODUCING	1 Agriculture and fisheries	2 Food and kindred products	3 Textile mill products	4 Apparel	5 Lumber and wood products	6 Furniture and fixtures	7 Paper and allied products	8 Printing and publishing	9 Chemicals	10 Products of petroleum and coal	11 Rubber products	12 Leather and leather products	13 Stone, clay and glass products	14 Primary metals	15 Fabricated metal products	16 Machinery (except electric)	17 Electrical machinery	18 Motor vehicles
1 AGRICULTURE AND FISHERIES	10.86	15.70	2.16	0.02	0.19	—	0.01	—	1.21	—	—	0.05	*	0.01	—	—	—	—
2 FOOD AND KINDRED PRODUCTS	2.38	5.75	0.06	0.01	*	*	0.03	*	0.79	*	—	0.44	*	*	*	*	*	—
3 TEXTILE MILL PRODUCTS	0.06	*	1.38	3.86	*	0.29	0.04	0.03	0.01	*	0.44	0.09	0.03	—	0.01	0.02	0.05	0.15
4 APPAREL	0.04	0.29	—	1.98	—	0.01	0.02	—	0.03	—	—	*	*	—	*	*	*	0.10
5 LUMBER AND WOOD PRODUCTS	0.15	0.10	0.02	*	1.08	0.39	0.27	*	0.04	0.01	—	0.02	0.02	0.06	0.06	0.09	0.05	0.05
6 FURNITURE AND FIXTURES	—	—	0.01	—	—	0.01	0.01	—	—	—	—	—	—	—	*	0.01	0.10	0.03
7 PAPER AND ALLIED PRODUCTS	*	0.52	0.06	0.02	*	0.02	2.08	1.08	0.33	0.11	0.02	0.05	0.18	*	0.09	0.04	0.07	0.03
8 PRINTING AND PUBLISHING	—	0.04	*	—	—	—	—	0.77	0.02	—	—	—	—	—	0.01	0.01	0.01	—
9 CHEMICALS	0.03	1.48	0.06	0.14	0.03	0.06	0.18	0.18	2.58	0.21	0.60	0.13	0.12	0.18	0.13	0.08	0.20	0.11
10 PRODUCTS OF PETROLEUM AND COAL	0.46	0.06	0.03	*	0.07	*	0.06	*	0.32	4.83	0.01	*	0.05	0.90	0.02	0.04	0.02	0.03
11 RUBBER PRODUCTS	0.12	0.01	0.01	0.02	0.01	0.01	0.01	*	*	*	0.04	0.05	0.01	*	0.01	0.13	0.03	0.50
12 LEATHER AND LEATHER PRODUCTS	—	—	*	0.05	*	0.01	—	*	—	—	—	1.04	—	—	*	0.02	*	0.01
13 STONE, CLAY AND GLASS PRODUCTS	0.06	0.25	*	*	0.01	0.03	0.03	—	0.26	0.05	0.01	0.01	0.43	0.21	0.07	0.07	0.12	0.19
14 PRIMARY METALS	0.01	*	—	*	0.01	0.11	—	0.01	0.19	0.01	0.01	*	0.04	6.90	2.53	2.02	1.05	1.28
15 FABRICATED METAL PRODUCTS	0.06	0.61	*	0.01	0.04	0.14	0.02	*	0.13	0.08	0.01	0.02	*	0.05	0.43	0.62	0.34	0.97
16 MACHINERY (EXCEPT ELECTRIC)	0.06	0.01	0.04	0.02	0.01	0.01	0.01	0.04	*	0.01	—	—	0.01	0.07	0.28	1.15	0.17	0.63
17 ELECTRICAL MACHINERY	—	—	—	—	—	—	—	—	*	—	—	—	0.01	0.05	0.24	0.58	0.06	0.62
18 MOTOR VEHICLES	0.11	*	—	—	*	—	—	—	—	*	—	—	*	*	0.03	0.03	0.01	4.40
19 OTHER TRANSPORTATION EQUIPMENT	0.01	—	—	—	—	—	*	—	*	*	*	—	*	*	—	—	*	0.01
20 PROFESSIONAL AND SCIENTIFIC EQUIPMENT	—	—	—	—	—	*	0.01	0.03	0.01	—	—	—	*	*	0.04	0.04	0.01	0.07
21 MISCELLANEOUS MANUFACTURING INDUSTRIES	*	0.01	*	0.26	*	0.02	0.01	—	0.03	—	*	0.02	0.01	*	0.02	0.05	0.11	0.02
22 COAL, GAS AND ELECTRIC POWER	0.06	0.20	0.11	0.04	0.02	0.02	0.12	0.03	0.19	0.56	0.04	0.02	0.20	0.35	0.08	0.10	0.05	0.06
23 RAILROAD TRANSPORTATION	0.44	0.57	0.08	0.06	0.14	0.05	0.22	0.07	0.29	0.27	0.04	0.04	0.15	0.52	0.13	0.16	0.07	0.23
24 OCEAN TRANSPORTATION	0.07	0.13	0.01	0.01	0.01	*	0.02	*	0.04	0.09	*	*	0.01	0.08	*	*	*	*
25 OTHER TRANSPORTATION	0.55	0.38	0.08	0.03	0.14	0.04	0.12	0.03	0.10	0.47	0.01	0.02	0.07	0.18	0.03	0.04	0.03	0.07
26 TRADE	1.36	0.46	0.23	0.37	0.06	0.06	0.18	0.03	0.17	0.02	0.05	0.08	0.05	0.38	0.29	0.26	0.14	0.06
27 COMMUNICATIONS	*	0.04	0.01	0.02	0.01	0.01	0.01	0.04	0.02	0.01	0.01	*	0.01	0.02	0.02	0.03	0.02	0.02
28 FINANCE AND INSURANCE	0.24	0.15	0.02	0.02	0.08	0.02	0.02	0.02	0.02	0.13	0.01	0.01	0.05	0.08	0.04	0.05	0.04	0.02
29 REAL ESTATE AND RENTALS	2.39	0.09	0.03	0.10	0.02	0.02	0.03	0.06	0.03	—	0.01	0.02	0.02	0.06	0.03	0.04	0.03	0.02
30 BUSINESS SERVICES	0.01	0.83	0.07	0.10	0.02	0.06	0.02	0.06	0.42	0.04	0.02	0.05	0.01	0.03	0.05	0.08	0.06	0.08
31 PERSONAL AND REPAIR SERVICES	0.37	0.12	*	*	0.04	*	*	0.02	0.01	0.01	*	*	0.03	0.01	0.01	0.01	*	*
32 NON-PROFIT ORGANIZATIONS	—	—	—	—	—	—	—	—	—	—	—	—	—	—	—	—	—	—
33 AMUSEMENTS	—	—	—	—	—	—	—	—	—	—	—	—	—	—	—	—	—	—
34 SCRAP AND MISCELLANEOUS INDUSTRIES	—	—	0.02	—	—	—	0.25	—	0.01	—	0.01	—	0.01	1.11	0.02	0.05	*	—
35 EATING AND DRINKING PLACES	—	—	—	—	—	—	—	*	—	—	—	—	—	—	—	—	—	—
36 NEW CONSTRUCTION AND MAINTENANCE	0.20	0.12	0.04	0.02	0.01	0.01	0.04	0.01	0.04	0.03	0.01	0.02	0.03	0.10	0.03	0.05	0.02	0.04
37 UNDISTRIBUTED	—	1.87	0.30	1.06	0.73	0.27	0.17	0.56	1.49	0.65	0.27	0.27	0.47	0.32	1.14	1.71	0.89	0.41
38 INVENTORY CHANGE (DEPLETIONS)	2.66	0.40	0.12	0.19	*	0.01	0.09	0.03	0.14	0.01	*	0.03	*	0.11	*	*	*	0.01
39 FOREIGN COUNTRIES (IMPORTS FROM)	0.89	2.11	0.21	0.20	0.10	0.01	0.62	0.01	0.59	0.26	*	0.04	0.14	0.62	0.01	0.05	*	0.02
40 GOVERNMENT	0.61	1.24	0.04	0.38	0.34	0.11	0.50	0.34	0.76	0.78	0.11	0.14	0.32	0.82	0.48	0.77	0.40	0.66
41 PRIVATE CAPITAL FORMATION (GROSS)	DEPRECIATION AND OTHER CAPITAL CONSUMPTION ALLOWANCES ARE INCLUDED IN HOUSEHOLD ROW																	
42 HOUSEHOLDS	19.17	7.05	3.34	4.24	2.72	1.12	2.29	3.14	3.75	5.04	1.08	1.20	2.35	5.53	4.14	6.80	3.41	3.39
TOTAL GROSS OUTLAYS	44.26	40.30	9.04	13.32	6.00	2.09	7.90	6.45	14.05	13.67	2.82	3.81	4.84	18.69	10.40	15.22	8.38	14.27

INTERINDUSTRY TABLE summarizes the transactions of the U. S. economy in 1947, for which preliminary data have just been compiled by the Bureau of Labor Statistics. Each number in the body of the table represents billions of 1947 dollars. In the vertical column at left the entire economy is broken down into sectors: in the

The original 1947 matrix was computed in billions of 1947 dollars. For example, reading across the top of the matrix, "agriculture and fisheries" reinvested about $10.86 billion into itself, sold $15.70 billion to "food and kindred products," about $2.16 billion to "textile mill products," only $0.02 billion to "apparel," and so on. Similarly, reading down the left column, we see that in 1947 "agriculture and fisheries" bought $2.38 billion from "food and kindred products," $0.06 billion from "textile mill products," and so on.

AND SERVICES IN THE U. S. FOR THE YEAR 1947

PURCHASING — FINAL DEMAND

Equipment · Manufacturing industries · Gas and electric power · Railroad transportation · Ocean transportation · Other transportation · Trade · Communications · Finance and insurance · Real estate and rentals · Business services · Personal and repair services · Non-profit organizations · Amusements · Scrap and miscellaneous industries · Eating and drinking places · New construction and maintenance · Undistributed · Inventory change (additions) · Foreign countries (exports to) · Government · Private capital formation (gross) · Households · Total gross output

24	25	26	27	28	29	30	31	32	33	34	35	36	37	38	39	40	41	42						Total gross output
—	*	*	—	*	*	0.01	—	*	—	—	—	—	0.12	—	—	0.87	0.09	0.17	1.01	1.28	0.57	0.02	9.92	44.26
—	0.01	0.02	*	0.08	0.01	0.03	0.07	0.01	—	—	—	*	0.25	*	0.02	3.47	*	0.42	0.88	1.80	0.73	—	23.03	40.30
0.01	0.05	0.08	0.07	—	0.01	0.01	0.03	*	—	—	*	0.03	*	—	0.01	—	0.05	0.52	0.06	0.92	0.10	0.02	1.47	9.84
0.01	*	*	*	*	*	*	0.02	*	—	—	—	0.02	0.02	*	0.01	0.02	*	0.15	0.21	0.30	0.28	*	9.90	13.32
0.03	*	0.06	0.06	—	0.01	*	0.03	*	—	0.14	*	*	*	—	0.11	0.01	2.33	0.35	0.17	0.17	0.01	0.04	0.07	6.00
0.02	*	—	*	—	—	*	—	*	0.04	0.08	—	—	*	—	—	—	0.20	0.20	0.08	0.03	0.05	0.57	1.46	2.89
0.02	0.08	0.07	*	*	—	*	0.57	*	*	—	*	0.06	0.03	—	0.68	0.06	0.17	0.31	0.04	0.15	0.06	—	0.34	7.90
—	*	—	*	0.04	*	0.02	0.18	0.03	0.21	—	2.45	0.03	0.17	0.01	0.01	0.03	—	0.68	*	0.07	0.16	0.09	1.49	6.45
0.02	0.05	0.17	0.06	0.03	0.01	0.02	0.07	*	*	—	0.01	0.20	0.22	*	0.03	0.04	0.64	1.25	0.30	0.81	0.19	—	1.96	14.05
0.01	*	0.01	0.47	0.27	0.09	0.45	0.20	*	0.01	0.78	*	0.06	0.06	*	0.01	0.01	0.62	0.36	0.06	0.68	0.18	*	2.44	13.67
0.01	*	0.04	*	*	—	0.13	0.06	*	0.01	*	—	0.07	*	—	*	*	0.06	0.47	0.09	0.17	0.02	0.01	0.71	2.82
*	0.01	0.01	*	—	—	*	*	—	—	—	—	0.03	0.01	—	0.01	—	*	0.29	0.11	0.06	0.03	0.02	2.03	3.81
0.01	0.03	0.06	0.02	0.01	*	*	0.04	*	—	—	—	0.02	0.01	—	*	0.06	1.74	0.36	0.10	0.21	0.02	0.01	0.34	4.84
0.43	0.07	0.20	0.05	0.20	—	0.01	—	*	—	—	—	—	*	—	0.15	*	1.19	1.24	0.16	0.77	0.02	—	0.02	18.69
0.10	0.07	0.04	*	0.03	*	0.01	0.06	*	—	—	*	0.03	0.01	—	0.06	0.02	3.09	1.44	0.21	0.39	0.05	0.28	0.95	10.40
0.22	0.03	*	0.03	0.06	—	0.01	0.01	—	0.02	—	—	0.15	*	—	0.07	—	0.51	2.24	0.37	1.76	0.18	5.82	1.22	15.22
0.12	0.03	0.02	0.02	0.04	—	0.01	0.01	0.05	—	—	0.01	0.08	*	—	0.04	—	0.77	1.27	0.25	0.44	0.17	1.75	0.93	8.38
*	—	—	0.01	*	—	0.13	0.02	*	—	*	—	1.05	*	—	0.07	*	0.04	0.67	0.40	1.02	0.15	2.98	3.13	14.27
0.30	—	—	*	0.04	0.08	0.13	—	—	—	—	—	*	—	—	0.01	—	*	0.46	0.02	0.32	1.25	1.20	0.17	4.00
0.02	0.18	0.02	*	—	—	*	—	*	—	—	0.01	0.05	0.18	—	0.01	—	0.02	0.24	0.03	0.18	0.08	0.26	0.62	2.12
*	0.03	0.16	*	*	*	*	0.01	*	—	—	0.15	0.16	0.05	0.05	0.11	0.02	0.03	0.68	0.04	0.19	0.08	0.51	1.89	4.76
0.03	0.01	0.03	1.27	0.44	*	0.09	0.49	0.01	0.06	3.15	*	0.31	0.16	0.05	—	0.22	0.03	0.02	0.03	0.35	0.20	—	—	9.21
6.04	0.01	0.03	0.15	0.41	*	0.06	0.08	*	0.01	0.42	0.03	0.03	0.05	*	0.03	0.25	0.71	0.30	0.08	0.59	0.33	0.27	2.53	9.85
*	*	0.01	*	—	0.22	—	—	—	—	—	—	—	—	—	*	—	—	*	—	1.16	0.31	—	0.10	2.29
0.01	0.01	0.01	0.03	0.19	0.04	0.25	0.31	*	*	0.13	0.03	0.01	0.02	*	0.02	0.10	0.57	0.17	0.04	0.32	0.35	0.10	4.77	9.86
0.07	0.04	0.05	0.05	0.03	0.01	0.42	0.20	0.01	0.04	0.75	0.14	0.37	0.29	0.01	0.09	1.06	2.52	1.01	0.20	1.00	0.05	2.34	26.82	41.66
0.01	0.01	0.01	0.02	0.02	*	0.04	0.33	0.06	0.09	0.06	0.43	0.12	0.07	0.01	—	0.01	0.04	0.08	—	0.04	0.15	—	1.27	3.17
0.02	0.01	0.02	0.05	0.02	0.12	0.30	1.00	*	1.85	0.56	0.02	0.12	0.09	0.03	—	0.07	0.40	—	—	0.14	0.03	—	6.99	12.81
0.02	0.01	0.03	0.05	0.02	0.01	0.15	1.96	0.05	0.21	0.21	0.06	0.71	0.40	0.18	—	0.39	0.08	—	—	—	0.22	0.80	20.29	28.86
0.01	0.05	0.06	0.01	0.02	*	0.03	1.71	0.09	0.14	0.04	0.06	0.12	0.02	0.10	—	0.06	0.13	0.42	—	*	0.04	—	0.18	5.10
*	*	*	0.02	0.11	0.01	0.26	1.42	0.02	0.11	0.03	0.07	0.56	0.08	0.02	0.03	0.23	0.82	1.17	—	—	0.08	0.27	8.35	14.30
—	—	—	—	—	*	*	—	—	0.02	—	—	—	0.09	—	—	—	—	0.16	—	—	5.08	—	8.04	13.39
—	—	—	—	—	—	—	—	—	—	—	—	—	0.01	0.39	—	—	—	0.01	—	0.13	—	—	2.40	2.84
—	*	—	—	—	—	0.04	0.39	0.01	0.11	0.03	0.02	*	*	0.01	—	—	*	0.01	—	0.03	*	—	—	2.13
—	—	—	—	—	—	0.01	—	—	—	—	—	—	0.15	—	—	—	—	—	—	—	—	—	13.11	13.27
0.02	0.01	0.02	0.27	1.12	*	0.13	0.18	0.18	0.03	4.08	*	0.06	0.34	0.02	—	0.07	0.01	—	—	—	5.26	15.70	0.15	28.49
0.34	0.19	0.87	0.25	0.10	0.04	0.03	2.59	0.01	0.71	0.36	0.31	1.13	0.91	0.22	—	0.59	0.43	—	—	—	—	—	—	21.60
0.01	0.06	0.16	*	—	—	—	—	—	—	—	—	—	—	—	0.40	—	—	—	—	0.02	—	—	—	4.43
0.01	0.05	0.14	0.01	0.04	0.50	0.08	—	0.03	0.10	—	—	—	—	*	0.07	—	—	0.01	—	—	1.31	—	1.32	9.52
0.12	0.13	0.19	1.14	0.91	0.26	0.77	3.30	0.44	1.11	4.00	0.21	0.50	0.17	0.32	0.07	1.41	0.47	2.19	0.34	0.83	3.46	0.22	31.55	63.69
1.95	0.90	2.17	5.11	5.70	0.90	6.20	26.42	2.15	7.93	14.06	1.08	8.20	9.41	1.50	—	4.20	10.73	2.27	—	0.85	30.06	—	2.12	223.58
4.00	2.12	4.76	9.21	9.95	2.29	9.86	41.66	3.17	12.81	28.86	5.10	14.30	13.39	2.94	2.13	13.27	28.49	21.60	5.28	17.21	51.29	33.29	194.12	

horizontal row at the top the same breakdown is repeated. When a sector is read horizontally, the numbers indicate what it ships to other sectors. When a sector is read vertically, the numbers show what it consumes from other sectors. The asterisks stand for sums less than $5 million. Totals may not check due to rounding.

Economists find it useful to show how the purchases of each industry relate to the total budget of that industry. Thus we have adapted Leontief's matrix to the matrix in Figure 1.2–4 by combining still more industries and by dividing each element within the matrix by the total budget of the industry indicated at the top of that element's *column*. Since, when all possible inputs and outputs are considered, the total budget for each industry equals both the total inputs and the total outputs for that industry, each element of the matrix in Figure 1.2–4 is that fraction of the total output (and input) of the industry shown at the top of its column which is bought from the industry shown to the left of its row.

Thus, reading down the left column of Figure 1.2–4, we see that 0.398 of the budget of "agriculture and food" was bought from itself, 0.007 of its budget was bought from "textiles and clothing," and 0.010 was bought from "coal and power." Rephrasing this, for each dollar of output, "agriculture and food" spent about $0.398 within its own industry, $0.007 on "textiles and clothing," and $0.01 from "coal and power."

Figure 1.2–4

that is bought from these industries:	Fraction of the total value of the budget of these industries									
	Purchasers (output)*									
Suppliers (input)	1	2	3	4	5	6	7	8	9	10
1. Agriculture and food	0.398	0.142	0	0.007	0.119	0.003	0	0	0	0.002
2. Textiles and clothing	0.007	0.116	0.003	0.010	0.074	0.006	0	0.011	0.011	0.001
3. Coal and power	0.010	0.007	0.307	0.021	0.033	0.015	0.067	0.006	0.004	0.016
4. Building and building materials	0.004	0.005	0.016	0.106	0.005	0.023	0.009	0.006	0.016	0.003
5. Chemicals and rubber	0.025	0.016	0.009	0.022	0.038	0.020	0.010	0.043	0.036	0.005
6. Paper and printing	0.007	0.006	0.005	0.005	0.023	0.310	0	0.002	0.009	0.028
7. Primary metals	0	0	0.003	0.035	0.012	0.001	0.369	0.090	0.214	0.002
8. Motor vehicles	0.001	0	0	0.001	0	0	0	0.308	0.002	0.009
9. Other metal goods	0.009	0.003	0.006	0.122	0.008	0.005	0.009	0.156	0.291	0.007
10. Services	0.106	0.073	0.061	0.161	0.075	0.067	0.070	0.040	0.051	0.138

* The numbers across the top of the matrix indicate the corresponding sector listed at the left side.

The sum of the numbers in each column must be less than 1 (unity), since they are the portion of an industry supplied by each other industry, and not all industries are included. The numbers across one row, however, are not so related to each other, since each is a proportion of the total for that *column*. For example, $a_{15} = 0.119$ says that the "chemicals and rubber" sector buys 11.9 percent of the value of its budget from the "agriculture and food" industry, but it does not tell what proportion of the "agriculture and food" sector is bought; that would require another matrix.

Notice that in input–output tables the industries on the left are putting *into* the industries at the top, which receive the *output*. The flow of goods comes *in at the left* and goes *out at the top*. (Ignore the fact that the output *of* the industries at the left is the input *of* the industries at the top; this interpretation reverses the conventional language.)

1.2.2 Example

(a) For each $1000 of output, how much did the "chemicals and rubber" sector buy from "agriculture and food"? How much did it spend on "primary metals"?
(b) On what two sectors is the building sector most dependent? (That is, from which two industries did it purchase the largest fraction of its budget?)
(c) "Motor vehicles" is highly dependent on its own products for producing and selling more. What is the industry on which it is next most dependent?
(d) "Coal and power" uses a significant proportion of its own output for reinvestment. If its prices rise, what other two industries will be most affected?

Solution:

(a) Looking at the top of the fifth column, we see that "chemicals and rubber" bought $119 worth of goods from "agriculture and foods" for each $1000 of output. It bought $12 per $1000 of output from "primary metals."
(b) Since the building sector buys 16.1 percent of the value of its output from "services," it is most dependent on the services sector. Its second most significant supplier is "other metal goods," from which it buys 12.2 percent of the value of its output.
(c) Since "motor vehicles" buys 15.6 percent of the value of its output from "other metal goods," this is the other sector on which it is next most dependent.
(d) The industries next most affected will be "primary metals" and "chemicals and rubber," since they purchase 6.7 and 3.3 percent, respectively, of their output from "coal and power."

Looking at Figure 1.2–4, we can extend part (d) of Example 1.2.2 by noting that a rise in "coal and power" causes a rise in "primary metals," which in turn causes a rise in "other metal goods," which will in turn increase the price of "motor vehicles." Such analysis is becoming increasingly important.

In 1973 Wassily Leontief received the Nobel Prize in Economics in recognition of the impact that his input–output analysis has had on economic planning and forecasting in most industrialized countries, both Communist and capitalist. It has been useful not only in judging the effects of an anticipated change in prices, but also in predicting the results of a shift in government spending policies and in determining how the waste produced in various sectors might be converted into useful products.

Input–output matrices are useful in many fields besides economics. In biology they can be used to show at what rate each type of organism consumes other types of organisms. In water pollution control, input–output matrices can be used to trace the entry and exit of pollutants from a river or other body of water. They can also show how populations (human, animal, or plant) grow, as in Example 1.1.12.

Notice that the proportions in the Leontief matrix of Figure 1.2–4 are not quite analogous to those in the demographic matrix given in Example 1.1.12. In demography it is usual to ask what proportion of a certain (input) group will survive or give birth, while in economics it is typical to ask what proportion of a purchasing (output) industry's budget has been obtained from each other industry. In this text we choose to conform to normal usage within each discipline rather than to force consistency within this survey of many fields.

You probably noticed that there were a lot of zeros in the parts-listing matrix, the Leontief matrix, and Example 1.1.12; many useful matrices are amply supplied with zeros. We now turn to a brief discussion of some specific forms of matrices.

1.2.3 Definition

In a square matrix the numbers on the line connecting the upper left corner to the lower right corner are called the *major diagonal.*

1.2.4 Definition

A square matrix that has only zeros except perhaps on the major diagonal is called a diagonal matrix.

$$\begin{bmatrix} -1 & 0 & 0 \\ 0 & 3 & 0 \\ 0 & 0 & 6 \end{bmatrix}$$

1.2.5 Definition

A square matrix with the numbers above the major diagonal all zero is called a lower triangular matrix.

$$\begin{bmatrix} 1 & 0 & 0 \\ 2 & 3 & 0 \\ 4 & 5 & -2 \end{bmatrix}$$

Because the elements are listed in technological order, a quantity matrix for a parts-listing problem is always a lower-triangular matrix. Notice that using mathematical notation we could have defined a lower triangular matrix to be a square matrix such that $a_{ij} = 0$ whenever $i < j$.

1.2.6 Definition

If a square matrix is symmetric around the major diagonal, it is called symmetric. In other words, a square matrix is symmetric if and only if $a_{ij} = a_{ji}$ for every i and j.

1.2.7 Example

A mileage table is a common application of a symmetric matrix:

	Albany, NY	Augusta, ME	Boston, MA	Hartford, CO	Montreal. Que.	New York, NY
Albany, NY	0	367	174	103	229	149
Augusta, ME	367	0	165	266	267	376
Boston, MA	174	165	0	103	323	213
Hartford, CO	103	266	103	0	321	113
Montreal, Que.	229	267	323	321	0	378
New York, NY	149	376	213	113	378	0

SUMMARY

The parts-listing problem was introduced as an example of both a lower triangular matrix and an input–output matrix. Leontief's own matrix, published in 1951, was shown to be a genuine application of matrices in economics. In Figure 1.2–4 this matrix was adapted to the form in which we shall use input–output matrices. This Leontief matrix is, of course, quite complicated, but simpler fictitious examples will be used when student computations are expected. In all examples inputs are listed down the left column and outputs are listed at the top. The section concluded with a brief survey of the form that matrices can take.

EXERCISES 1.2. A

1. What kind of matrix is each of the following?

(a) $\begin{bmatrix} 2 & 5 & -1 \\ 5 & 7 & 0 \\ -1 & 0 & 1 \end{bmatrix}$ (b) $\begin{bmatrix} 1 & 0 & 0 \\ -5 & 0 & 0 \\ 7 & 6 & 2 \end{bmatrix}$

(c) $\begin{bmatrix} 3 & 0 & 0 \\ 0 & -2 & 0 \\ 0 & 0 & 6 \end{bmatrix}$ (d) $\begin{bmatrix} 3 & -4 & 8 \\ 0 & 2 & 6 \\ 0 & 0 & -3 \end{bmatrix}$

(*Hint:* The last type was not defined in the text—but guess!)

2. Give two definitions of "upper triangular matrix," one patterned after Definition 1.2.5 and one using mathematical symbols, as appears just before Definition 1.2.6.

3. Complete:
   (a) The elements a_{ii} of a square matrix are called its ______________ ______________.
   (b) If a square matrix is such that $a_{ij} = a_{ji}$ for every i and j, it is called a ______________ matrix.
   (c) A matrix that is both a lower triangular matrix and a symmetric matrix is also a ______________ matrix.
4. The following matrix shows how the test scores of 1046 college sophomores were correlated. What kind of matrix is this?

	Spelling	Punctuation	Grammar	Vocabulary	Literature	Foreign literature
Spelling	1	0.621	0.564	0.476	0.394	0.389
Punctuation	0.621	1	0.742	0.503	0.461	0.411
Grammar	0.564	0.742	1	0.577	0.472	0.429
Vocabulary	0.476	0.503	0.577	1	0.688	0.548
Literature	0.394	0.461	0.472	0.688	1	0.639
Foreign literature	0.389	0.411	0.429	0.548	0.639	1

5. Using Figure 1.2–4, answer the following questions.
   (a) For each $1000 of output, how much did the textile and clothing industry spend for agricultural products? How much did it spend on coal and power?
   (b) How much did the agriculture sector spend for textiles and clothing per $1000 output?
   (c) On what sector is "chemicals and rubber" most dependent? Was the dependence reciprocal?
   (d) From what other sector did "building and building materials" buy the most?
   (e) Other than agriculture itself, which industry was most dependent on agriculture?
6. Suppose that a hypothetical economy contains only "agriculture" and "manufacturing" as its two sectors. Agriculture buys 25 percent of the value of its gross output from itself and 65 percent from manufacturing. Manufacturing buys 55 percent of the value of its total output from itself and 40 percent from agriculture. Write an input–output matrix that describes this economy.
7. An ecological system contains plants, herbivores (plant-eating animals), carnivores (animal-eating animals), and omnivores (animals that eat both plants and animals). Let us abbreviate these "sectors" of the ecological system by P, H, C, and O. The plants maintain themselves on inorganic substances, such as soil, water, and light, outside these four groups. The herbivores get all their food from plants. Suppose that in a hypothetical ecological system the diet of the carnivorous animals is 50 percent herbivores, 20 percent carnivores, and 30 percent omnivores. And the diet of omnivores consists of 40 percent plants, 35 percent herbivores, 10 percent carnivores, and 15 percent omnivores. Write an input–output matrix that describes this ecological system.
8. A holiday apron (A) consists of a frontpiece (F), two ties (T), and two pompoms (P). The frontpiece is made of one rectangle (R) and five pom-

poms. Write an assembly graph and quantity matrix that describe the situation.

9. A coat-hanger stand (C) consists of a large floor base plate (P); a 5-foot rod (R), connected to the base plate by three bolts (B); and one hook assembly (H), fastened to the top of the rod by a single bolt (B). The hook assembly consists of a tiny top plate (T); and six metal hooks (M), *each* attached to the top plate by a single bolt (B). Write an assembly graph and the quantity matrix that describe the stand.

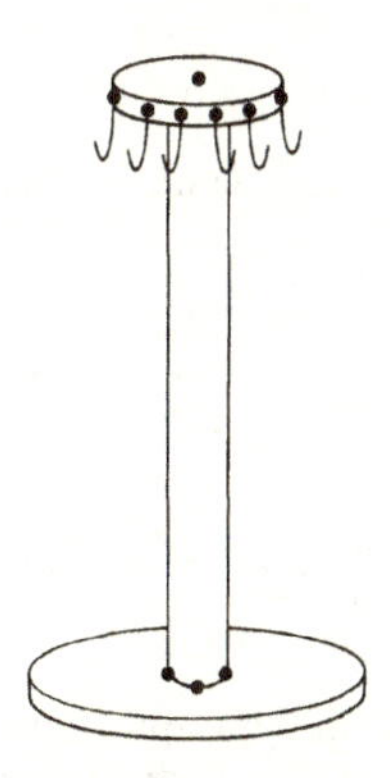

EXERCISES 1.2. B

1. What kind of matrix is each of the following?

(a) $\begin{bmatrix} 1 & 0 & 0 \\ -2 & 7 & 0 \\ 8 & -3 & 0 \end{bmatrix}$ (b) $\begin{bmatrix} -4 & 0 & 0 \\ 0 & 1 & 0 \\ 0 & 0 & 5 \end{bmatrix}$

(c) $\begin{bmatrix} 0 & 4 & -9 \\ 4 & 1 & 5 \\ -9 & 5 & 3 \end{bmatrix}$ (d) $\begin{bmatrix} 5 & -4 & 9 \\ 0 & 1 & -2 \\ 0 & 0 & 6 \end{bmatrix}$

(*Hint:* The last type was not defined in the text—but guess!)

2. Give two definitions of "upper triangular matrix," one patterned after Definition 1.2.5 and one using mathematical symbols, as appears just before Definition 1.2.6.

3. Complete:

(a) The elements a_{ii} of a square matrix are called its ____________ ____________.

(b) If a square matrix is such that $a_{ij} = a_{ji}$ for every i and j, it is called a __________ matrix.

(c) A matrix that is both a lower triangular matrix and a symmetric matrix is also a __________ matrix.

4. Human blood can be classified into eight types according to which (if any) of the three antigens (A, B, and Rh) it contains. A person may donate blood safely only to someone who already possesses the antigens of the donor. Thus a person who is A positive (meaning that he or she possesses the A and Rh antigens but not B) can donate to someone who is also A positive, or to someone who is AB positive (that is, has all three of the antigens), but to no one else. Similarly, a person who is AB positive can receive from anyone, but can give only to someone of the same blood type. The usual notation for blood lacking both the A and the B antigen is O. Thus a person who is O negative can donate to anyone but can receive only from someone of the same type.

   The following matrix has $a_{ij} = 1$ if the ith blood type can safely donate to the jth blood type and $a_{ij} = 0$ otherwise. What kind of a matrix is thereby formed?

	Recipient							
Donor	O neg.	A neg.	B neg.	AB neg.	O pos.	A pos.	B pos.	AB pos.
O neg.	1	1	1	1	1	1	1	1
A neg.	0	1	0	1	0	1	0	1
B neg.	0	0	1	1	0	0	1	1
AB neg.	0	0	0	1	0	0	0	1
O pos.	0	0	0	0	1	1	1	1
A pos.	0	0	0	0	0	1	0	1
B pos.	0	0	0	0	0	0	1	1
AB pos.	0	0	0	0	0	0	0	1

5. Use Figure 1.2–4 to answer the following questions.
   (a) For each $1000 of output, how much did the motor vehicles sector spend on textiles and clothing? How much did it spend on "other metal goods"?
   (b) How much did the primary metals sector spend on coal and power per $1000 output?
   (c) On which sector was textiles and clothing most dependent? Was the dependence reciprocal?
   (d) Other than itself, from what sector did "services" buy the most?
   (e) Which other sector was most dependent on the service sector?
6. Suppose that a hypothetical economy contains only two sectors, "agriculture" and "manufacturing." Agriculture buys 30 percent of the value of its gross output from itself and 60 percent from manufacturing. Manufacturing buys 50 percent of the value of its total output from itself and 35 percent from agriculture. Write an input–output matrix that describes this economy.
7. An ecological system contains plants, herbivores (plant-eating animals), carnivores (animal-eating animals), and omnivores (animals that eat both plants and animals). Let us abbreviate these "sectors" of the ecological

system by *P*, *H*, *C*, and *O*. The plants maintain themselves on inorganic substances, such as soil, water, and light, outside these four groups. The herbivores get all their food from plants. Suppose in a hypothetical ecological system that the diet of the carnivorous animals is 40 percent herbivores, 25 percent carnivores, and 35 percent omnivores. And the diet of omnivores consists of 35 percent plants, 30 percent herbivores, 25 percent carnivores, and 10 percent omnivores. Write an input–output matrix that describes this ecological system.

8. A salt-and-pepper holder (*Z*) consists of one leg piece (*L*), a holder piece (*H*), and a circle (*C*) for a handle. The leg piece is assembled from four

rods (*R*), and the holder piece is assembled from one rod (*R*) and two circles (*C*). Write an assembly graph and a quantity matrix that describes the situation.

9. A photo album (*A*) is assembled from a looseleaf folder (*F*), one blank page (*B*), and five picture pages (*P*). If each picture page requires one blank page (*B*) and 16 picture slots (*S*), write an assembly graph and a quantity matrix that describe the album.

ANSWERS 1.2. A

1. (a) symmetric (b) lower triangular (c) diagonal (d) upper triangular
2. (a) An upper triangular matrix is a square matrix such that all the numbers below the major diagonal are zero.
   (b) An upper triangular matrix is a square matrix such that $a_{ij} = 0$ whenever $i > j$.
3. (a) major diagonal (b) symmetric (c) Either "diagonal" or "upper triangular" is correct.
4. symmetric
5. (a) \$142, \$7 (b) \$7 (c) "agriculture"; no, "agriculture" depends more on both itself and "services" than on "chemicals and rubber," although this ranks third (d) "services" (e) "textiles and clothing"
6.

	Agriculture	Manufacturing
Agriculture	0.25	0.40
Manufacturing	0.65	0.55

7.

	P	H	C	O
P	0	1	0	0.40
H	0	0	0.50	0.35
C	0	0	0.20	0.10
O	0	0	0.30	0.15

8. There are many correct ways to pick the technological order. This is one:

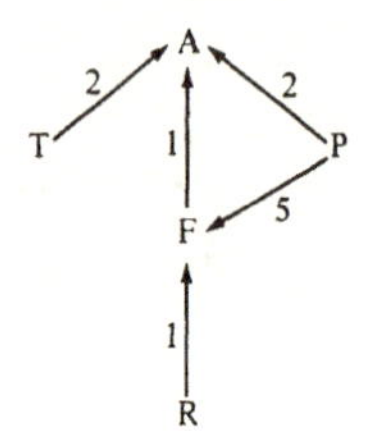

	A	F	R	P	T
A	0	0	0	0	0
F	1	0	0	0	0
R	0	1	0	0	0
P	2	5	0	0	0
T	2	0	0	0	0

9.

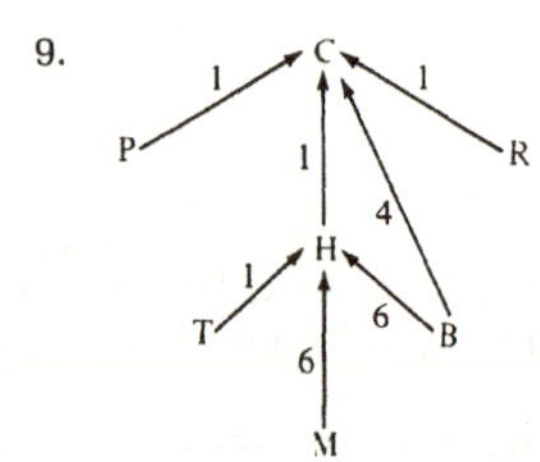

	C	H	R	P	T	M	B
C	0	0	0	0	0	0	0
H	1	0	0	0	0	0	0
R	1	0	0	0	0	0	0
P	1	0	0	0	0	0	0
T	0	1	0	0	0	0	0
M	0	6	0	0	0	0	0
B	4	6	0	0	0	0	0

1.3 Matrix Multiplication and Vector Inner Products

There are two types of multiplication involving matrices—scalar multiplication and matrix multiplication. The previous section explained how scalar multiplication involves multiplying each element of a matrix by a specific number, or scalar. Such multiplication proved to be easy.

Multiplying two matrices by each other, on the other hand, is like the old game of rubbing your stomach while patting your head. Your left hand moves from left to right while your right goes from top to bottom! To find an element a_{ij} in the product matrix, you run your left hand across the corresponding row i in the left matrix and your right hand down the corresponding column j in the right matrix, multiplying each entry in the left matrix by the matching number in the right matrix and adding these products.

1.3.1 Example

$$\begin{bmatrix} 2 & 3 \\ 4 & 5 \end{bmatrix}\begin{bmatrix} 6 & 7 \\ 8 & 9 \end{bmatrix} = \begin{bmatrix} 2\cdot 6 + 3\cdot 8 & 2\cdot 7 + 3\cdot 9 \\ 4\cdot 6 + 5\cdot 8 & 4\cdot 7 + 5\cdot 9 \end{bmatrix} = \begin{bmatrix} 36 & 41 \\ 64 & 73 \end{bmatrix}$$

The upper *left* number in the product matrix is determined by the *upper* row in the left matrix and the *left* column of the right matrix. The *upper right* number in the product matrix is determined by the *upper* row of the left matrix and the *right* column of the right matrix. And so forth.

1.3.2 Example

$$\begin{bmatrix} 6 & 7 \\ 8 & 9 \end{bmatrix}\begin{bmatrix} 2 & 3 \\ 4 & 5 \end{bmatrix} = \begin{bmatrix} 6\cdot 2 + 7\cdot 4 & 6\cdot 3 + 7\cdot 5 \\ 8\cdot 2 + 9\cdot 4 & 8\cdot 3 + 9\cdot 5 \end{bmatrix} = \begin{bmatrix} 40 & 53 \\ 52 & 69 \end{bmatrix}$$

The multiplication of matrices is not commutative. In other words, if you reverse the order of two matrices you are multiplying, you will not necessarily get the same answer. Notice that Examples 1.3.1 and 1.3.2 use the same matrices. But the answers are different because the order is reversed.

The lack of commutativity in multiplication is one important way that matrices differ from numbers. You know, of course, that any two numbers, x and y, are such that $x \cdot y = y \cdot x$; this is what we mean when we say that *the multiplication of numbers is commutative.*

Two matrices can be multiplied only if the left one has the same number of columns as the right one has rows. Thus there is no way to perform the multiplication $\begin{bmatrix} 1 & 2 & 3 \\ 4 & 5 & 6 \end{bmatrix}\begin{bmatrix} 7 & 8 \\ 9 & 0 \end{bmatrix}$. You will run out of numbers as you go down a column in the right-hand matrix—when you try to write $1 \cdot 7 + 2 \cdot 9 + 3 \cdot ?$, for example. But as long as the number of columns on the left equals the number of rows on the right, the multiplication can be performed, even if the matrices have otherwise different dimensions. You can tell whether two matrices can be multiplied together by writing their dimensions under them; they can be multiplied if and only if the inner two numbers are the same. The dimension of the product will be indicated by the outer two numbers.

1.3.3 Example

$$\underset{3 \text{ by } 2}{\begin{bmatrix} 3 & 1 \\ 2 & -1 \\ 0 & 3 \end{bmatrix}}\underset{2 \text{ by } 4}{\begin{bmatrix} 2 & 0 & 4 & 1 \\ 1 & 6 & 3 & -2 \end{bmatrix}} = \underset{3 \text{ by } 4}{\begin{bmatrix} 7 & 6 & 15 & 1 \\ 3 & -6 & 5 & 4 \\ 3 & 18 & 9 & -6 \end{bmatrix}}$$

1.3.4 Example

If $A = \begin{bmatrix} 1 & 2 & 3 \\ 4 & 0 & -1 \end{bmatrix}$ and $B = \begin{bmatrix} -1 & 0 \\ -2 & 1 \\ 3 & 0 \end{bmatrix}$, find AB and BA.

Solution:

$$AB = \begin{bmatrix} 1 & 2 & 3 \\ 4 & 0 & -1 \end{bmatrix} \begin{bmatrix} -1 & 0 \\ -2 & 1 \\ 3 & 0 \end{bmatrix} = \begin{bmatrix} 4 & 2 \\ -7 & 0 \end{bmatrix}$$

2 by (3 3) by 2 2 by 2

$$BA = \begin{bmatrix} -1 & 0 \\ -2 & 1 \\ 3 & 0 \end{bmatrix} \begin{bmatrix} 1 & 2 & 3 \\ 4 & 0 & -1 \end{bmatrix} = \begin{bmatrix} -1 & -2 & -3 \\ 2 & -4 & -7 \\ 3 & 6 & 9 \end{bmatrix}$$

3 by (2 2) by 3 3 by 3

1.3.5 Definition

A matrix that has only one column or only one row is called a <u>vector</u>.

1.3.6 Example

Suppose that a stadium sells four prices of tickets: adult ($3), student ($2), senior citizen ($1.50), and child ($1). This information might be expressed in vector form as [3, 2, 1.5, 1]; commas are often used to separate the elements of a row vector. Suppose also that at a certain game there are 1000 adults, 2000 students, 500 senior citizens, and 500 children. This might be expressed by the vector

$$\begin{bmatrix} 1000 \\ 2000 \\ 500 \\ 500 \end{bmatrix}$$

Then to find the total receipts of the stadium at this game, we need only take the matrix product of these two vectors:

$$[3, 2, 1.5, 1] \begin{bmatrix} 1000 \\ 2000 \\ 500 \\ 500 \end{bmatrix} = (3)1000 + (2)2000 + (1.5)500 + (1)500 = 8250$$

The total receipts are $8250. You may wonder what the advantage is of making up the vectors first. One answer is that for long lists of numbers, it is easier to keep one's own thoughts, and a computer's processes, organized.

The matrix product of two vectors has a special name.

1.3.7 Definition

The inner product of two vectors with the same number of elements is obtained by multiplying together the corresponding elements of the two vectors and then adding the products.

Taking vector inner products is a special form of matrix multiplication; to keep the notation consistent, the first vector "should" be a row vector and the second a column vector. Some books, however, simplify the typesetting by printing them either as two row vectors or as two column vectors.

In applications, matrix multiplication is often used to "pick off" just the right number to be multiplied in a mechanical way—ideally suited for a computer to "understand."

1.3.8 Example

Suppose that we have 5 packages of books weighing 8 oz each, 10 packages weighing 1 lb, and 20 packages weighing 2 lb. We want to compare the prices of sending these through the various services offered by the U.S. Postal Service. If we multiply the matrix of Example 1.1.2 by the vector suggested above, the resulting vector will list the total prices of sending this order for each of the types of mail services included in the original matrix. Study the matrix multiplication below to see how this works.

$$\begin{bmatrix} 1.22 & 1.64 & 1.97 \\ 0.90 & 1.56 & 1.89 \\ 0.45 & 0.88 & 1.08 \\ 0.21 & 0.21 & 0.30 \end{bmatrix} \begin{bmatrix} 5 \\ 10 \\ 20 \end{bmatrix} = \begin{bmatrix} 5(1.22) + 10(1.64) + 20(1.97) \\ 5(0.90) + 10(1.56) + 20(1.89) \\ 5(0.45) + 10(0.88) + 20(1.08) \\ 5(0.21) + 10(0.21) + 20(0.30) \end{bmatrix} = \begin{bmatrix} 61.90 \\ 57.90 \\ 32.65 \\ 9.15 \end{bmatrix}$$

1.3.9 Example

Refer back to the quantity matrix in Example 1.2.1. This matrix can be used to compute order sizes needed for the production of the items mentioned in that example. Suppose, for an easy first example, that we want to produce 1 tripod. Then we know we need 3 legs and 1 bolt. The matrix method for calculating this is below. Check this carefully and see why it works; before long the problems will become more difficult.

$$\begin{bmatrix} 0 & 0 & 0 & 0 \\ 3 & 0 & 0 & 0 \\ 1 & 2 & 0 & 0 \\ 0 & 3 & 0 & 0 \end{bmatrix} \begin{bmatrix} 1 \\ 0 \\ 0 \\ 0 \end{bmatrix} = \begin{bmatrix} 0 \\ 3 \\ 1 \\ 0 \end{bmatrix}$$

The vector at the left of the equal sign says that we want to produce 1 tripod and the vector on the right says we will need 3 legs and 1 bolt. Notice how the 1 in the upper part of the vector "peels off" exactly what is needed.

Now suppose that we want to make 7 tripods. Using matrices to compute how many items are needed, we write

$$\begin{bmatrix} 0 & 0 & 0 & 0 \\ 3 & 0 & 0 & 0 \\ 1 & 2 & 0 & 0 \\ 0 & 3 & 0 & 0 \end{bmatrix} \begin{bmatrix} 7 \\ 0 \\ 0 \\ 0 \end{bmatrix} = \begin{bmatrix} 0 \\ 21 \\ 7 \\ 0 \end{bmatrix}$$

and see that we need 21 legs and 7 bolts.

If we want to make 5 legs, we put the 5 in the second row of the vector and compute

$$\begin{bmatrix} 0 & 0 & 0 & 0 \\ 3 & 0 & 0 & 0 \\ 1 & 2 & 0 & 0 \\ 0 & 3 & 0 & 0 \end{bmatrix} \begin{bmatrix} 0 \\ 5 \\ 0 \\ 0 \end{bmatrix} = \begin{bmatrix} 0 \\ 0 \\ 10 \\ 15 \end{bmatrix}$$

We conclude that we need 10 bolts and 15 rods.

Suppose now that we want to make both 7 tripods and 5 legs. We can either add the answers of the previous two questions or we can do the entire computation using matrix multiplication:

$$\begin{bmatrix} 0 & 0 & 0 & 0 \\ 3 & 0 & 0 & 0 \\ 1 & 2 & 0 & 0 \\ 0 & 3 & 0 & 0 \end{bmatrix} \begin{bmatrix} 7 \\ 5 \\ 0 \\ 0 \end{bmatrix} = \begin{bmatrix} 0 \\ 21 \\ 17 \\ 15 \end{bmatrix}$$

Perhaps you can see how the matrix method, once the system is established, might be easier to use than brute force if the numbers are many and large and the items being produced are complicated.

Notice that if we add some rods and/or bolts in Example 1.3.9 but still require 7 tripods and 5 legs, the answer would not change. This is because we do not use any "ingredients" to make rods or bolts; they are assumed to be abundantly available in order to make more complex items.

1.3.10 Example in Biology

If we know what proportion of each kind of plant grown in a year in a certain ecological system is eaten by each kind of herbivore and what proportion of each kind of herbivore gets eaten by each kind of carnivore, we can use matrix multiplication to discover what proportion of each kind of plant eventually gets into each kind of carnivore. (The common unit of measurement is *biomass*, the product of the number of each species by the average mass of the individuals in that species.)

Suppose that a simple hypothetical ecological system has four categories of plants, p_1, p_2, p_3, and p_4; three types of herbivores, h_1, h_2, and h_3; and two carnivores, c_1 and c_2. Suppose also that the fraction of the total poundage of each category of plant produced in a season that is consumed by each kind of herbivore is given by the input–output matrix on the left below and the fraction of each category of herbivore that is consumed by each kind of carnivore is given by the matrix on the right.

	Herbivore		
Plant	h_1	h_2	h_3
p_1	0.1	0.2	0.3
p_2	0.2	0.2	0.1
p_3	0.2	0.3	0.1
p_4	0.3	0.1	0.1

	Carnivore	
Herbivore	c_1	c_2
h_1	0.1	0.2
h_2	0.2	0.3
h_3	0.3	0.1

To compute what proportion of each plant is consumed by each carnivore, we can multiply these two matrices:

$$\begin{bmatrix} 0.1 & 0.2 & 0.3 \\ 0.2 & 0.2 & 0.1 \\ 0.2 & 0.3 & 0.1 \\ 0.3 & 0.1 & 0.1 \end{bmatrix} \begin{bmatrix} 0.1 & 0.2 \\ 0.2 & 0.3 \\ 0.3 & 0.1 \end{bmatrix}$$

$$= \begin{bmatrix} (0.1)(0.1) + (0.2)(0.2) + (0.3)^2 & 0.02 + 0.06 + 0.03 \\ (0.2)(0.1) + (0.2)^2 + (0.1)(0.3) & 0.04 + 0.06 + 0.01 \\ 0.02 + 0.06 + 0.03 & 0.04 + 0.09 + 0.01 \\ 0.03 + 0.02 + 0.03 & 0.06 + 0.03 + 0.01 \end{bmatrix}$$

This last matrix is the basis for the following table:

	Carnivore	
Plant	c_1	c_2
p_1	0.14	0.11
p_2	0.09	0.11
p_3	0.11	0.14
p_4	0.08	0.10

The fact that $a_{11} = 0.14$ means that 14 percent of the total annual production of plant group p_1 winds up in the c_1 carnivores. Of this 14 percent, 1 percent comes via the h_1 herbivores [$0.01 = (0.1)(0.1)$], 4 percent via the h_2 herbivores [$0.04 = (0.2)(0.2)$], and 9 percent via h_3 herbivores [$0.09 = (0.3)(0.3)$].

SUMMARY

This section introduced vectors (as a particular type of matrix) and explained matrix multiplication and vector inner products. These were used to find total shipping prices, to compute the ingredients needed to make several items, and to explore a consumption matrix.

EXERCISES 1.3. A

1. For each pair of matrices, find the product, if it is defined.

(a) $\begin{bmatrix} 1 & -1 \\ 2 & 3 \end{bmatrix}\begin{bmatrix} -2 & 1 \\ 3 & 4 \end{bmatrix} =$ (b) $\begin{bmatrix} -2 & 1 \\ 3 & 4 \end{bmatrix}\begin{bmatrix} 1 & -1 \\ 2 & 3 \end{bmatrix} =$

(c) $\begin{bmatrix} 1 & -1 \\ 2 & 3 \end{bmatrix}\begin{bmatrix} 1 & 0 \\ 0 & 1 \end{bmatrix} =$ (d) $\begin{bmatrix} 1 & 0 \\ 0 & 1 \end{bmatrix}\begin{bmatrix} 1 & -1 \\ 2 & 3 \end{bmatrix} =$

(e) $\begin{bmatrix} 1 & -1 \\ 2 & 3 \end{bmatrix}\begin{bmatrix} 1 & 0 \\ 0 & 0 \end{bmatrix} =$ (f) $\begin{bmatrix} 1 & -1 \\ 2 & 3 \end{bmatrix}\begin{bmatrix} 0 & 0 \\ 0 & 1 \end{bmatrix} =$

(g) $\begin{bmatrix} 1 & 0 \\ 0 & 0 \end{bmatrix}\begin{bmatrix} 1 & -1 \\ 2 & 3 \end{bmatrix} =$ (h) $\begin{bmatrix} 0 & 0 \\ 0 & 1 \end{bmatrix}\begin{bmatrix} 1 & -1 \\ 2 & 3 \end{bmatrix} =$

(i) $\begin{bmatrix} 1 & 0 \\ 0 & 0 \end{bmatrix}\begin{bmatrix} 0 & 0 \\ 0 & 1 \end{bmatrix} =$ (j) $\begin{bmatrix} 2 & 3 & 4 \\ 5 & 6 & 7 \\ 8 & 9 & 1 \end{bmatrix}\begin{bmatrix} 1 & 0 & 0 \\ 0 & 1 & 0 \\ 0 & 0 & 1 \end{bmatrix} =$

(k) $\begin{bmatrix} 1 & 0 & 0 \\ 0 & 1 & 0 \\ 0 & 0 & 1 \end{bmatrix}\begin{bmatrix} 2 & 3 & 4 \\ 5 & 6 & 7 \\ 8 & 9 & 1 \end{bmatrix} =$

(l) $\begin{bmatrix} 2 & 3 & 4 \\ 5 & 6 & 7 \\ 8 & 9 & 1 \end{bmatrix}\begin{bmatrix} 1 & 0 & 0 \\ 0 & 1 & 0 \\ 0 & 0 & 0 \end{bmatrix} =$

(m) $\begin{bmatrix} 1 & 0 & 0 \\ 0 & 1 & 0 \\ 0 & 0 & 0 \end{bmatrix}\begin{bmatrix} 2 & 3 & 4 \\ 5 & 6 & 7 \\ 8 & 9 & 1 \end{bmatrix} =$

(n) $\begin{bmatrix} 2 & 3 & 4 \\ 5 & 6 & 7 \\ 8 & 9 & 1 \end{bmatrix}\begin{bmatrix} -1 & 3 & 0 \\ 2 & -4 & 6 \\ 0 & -2 & 1 \end{bmatrix} =$

(o) $\begin{bmatrix} 2 & 3 & -1 & 4 \\ 1 & -2 & 0 & 1 \end{bmatrix}\begin{bmatrix} 0 & 1 & 3 \\ 2 & -2 & 4 \\ 3 & -1 & 0 \\ 4 & 0 & -4 \end{bmatrix} =$

(p) $\begin{bmatrix} 0 & 3 \\ -1 & 2 \\ 4 & 0 \\ 1 & -2 \end{bmatrix}\begin{bmatrix} 3 & 2 & 0 \\ 4 & -1 & 2 \end{bmatrix} =$

(q) $\begin{bmatrix} 4 & 2 \\ 6 & 5 \\ 0 & 1 \\ 3 & 6 \end{bmatrix}\begin{bmatrix} 2 & 0 & 1 \\ -4 & 2 & -5 \\ 7 & -1 & 0 \end{bmatrix} =$

2. Suppose that at another game the stadium of Example 1.3.6 has 300 adults, 1000 students, 700 senior citizens, and 200 children. Use an inner product to find the total receipts.
3. Suppose that the same stadium at another game admits 300 students, 1000 senior citizens, 700 adults, and 200 children. Write this (carefully!) in vector form consistent with Example 1.3.6 and then use an inner product to find the total revenue.
4. Suppose that a person has three jobs to do, the first of which takes 5 minutes, the second of which takes 12 minutes, and the third of which takes 10 minutes. Write this information as a row vector. Suppose that he knows he will have to perform the first job five times, the second job twice, and the third job four times. Write this as a column vector. Find how long he must allow for all three jobs by taking an inner product.
5. In problem 4, suppose that he knows he must do the first job 10 times, the second 5 times, and the third 8 times. Compute the total time needed by taking an inner product.
6. Refer to problem 8 of Exercises 1.2.A (page 16). Suppose that you want to produce 8 aprons and 10 frontpieces. Use matrix multiplication to find out how much of each item is needed.
7. Referring to problem 9 of Exercises 1.2.A, use matrix multiplication to find how much of each item is needed to get ready to produce 12 coat-hanger stands and 8 hook assemblies.

EXERCISES 1.3. B

1. For each pair of matrices, find the product, if it is defined.

(a) $\begin{bmatrix} 2 & -1 \\ 4 & -3 \end{bmatrix}\begin{bmatrix} 4 & 1 \\ 3 & -2 \end{bmatrix} =$ (b) $\begin{bmatrix} 4 & 1 \\ 3 & -3 \end{bmatrix}\begin{bmatrix} 2 & -1 \\ 4 & -3 \end{bmatrix} =$

(c) $\begin{bmatrix} 2 & -1 \\ 4 & -3 \end{bmatrix}\begin{bmatrix} 1 & 0 \\ 0 & 1 \end{bmatrix} =$ (d) $\begin{bmatrix} 1 & 0 \\ 0 & 1 \end{bmatrix}\begin{bmatrix} 2 & -1 \\ 4 & -3 \end{bmatrix} =$

(e) $\begin{bmatrix} 2 & -1 \\ 4 & -3 \end{bmatrix}\begin{bmatrix} 1 & 0 \\ 0 & 0 \end{bmatrix} =$ (f) $\begin{bmatrix} 2 & -1 \\ 4 & -3 \end{bmatrix}\begin{bmatrix} 0 & 0 \\ 0 & 1 \end{bmatrix} =$

(g) $\begin{bmatrix} 1 & 0 \\ 0 & 0 \end{bmatrix}\begin{bmatrix} 2 & -1 \\ 4 & -3 \end{bmatrix} =$ (h) $\begin{bmatrix} 0 & 0 \\ 0 & 1 \end{bmatrix}\begin{bmatrix} 2 & -1 \\ 4 & -3 \end{bmatrix} =$.

(i) $\begin{bmatrix} 1 & 0 \\ 0 & 0 \end{bmatrix}\begin{bmatrix} 0 & 0 \\ 0 & 1 \end{bmatrix} =$ (j) $\begin{bmatrix} a & b & c \\ d & e & f \\ g & h & i \end{bmatrix}\begin{bmatrix} 1 & 0 & 0 \\ 0 & 1 & 0 \\ 0 & 0 & 1 \end{bmatrix} =$

(k) $\begin{bmatrix} 1 & 0 & 0 \\ 0 & 1 & 0 \\ 0 & 0 & 1 \end{bmatrix}\begin{bmatrix} a & b & c \\ d & e & f \\ g & h & i \end{bmatrix} =$

(l) $\begin{bmatrix} a & b & c \\ d & e & f \\ g & h & i \end{bmatrix}\begin{bmatrix} 1 & 0 & 0 \\ 0 & 1 & 0 \\ 0 & 0 & 0 \end{bmatrix} =$

(m) $\begin{bmatrix} 1 & 0 & 0 \\ 0 & 1 & 0 \\ 0 & 0 & 0 \end{bmatrix} \begin{bmatrix} a & b & c \\ d & e & f \\ g & h & i \end{bmatrix} =$

(n) $\begin{bmatrix} 3 & -2 & 4 \\ 0 & 1 & -3 \\ -1 & 2 & 0 \end{bmatrix} \begin{bmatrix} 4 & 0 & -1 \\ 3 & -2 & 1 \\ 0 & 3 & 2 \end{bmatrix} =$

(o) $\begin{bmatrix} 2 & 3 & -1 \\ 0 & 4 & 2 \end{bmatrix} \begin{bmatrix} 3 & 4 & 1 \\ 0 & -1 & 2 \\ 2 & 0 & -3 \end{bmatrix} =$

(p) $\begin{bmatrix} 0 & 1 \\ -2 & 3 \\ 4 & -5 \end{bmatrix} \begin{bmatrix} 1 & 3 & -4 \\ 2 & 0 & -1 \end{bmatrix} =$

(q) $\begin{bmatrix} 2 & 3 \\ 6 & 0 \\ -1 & 2 \end{bmatrix} \begin{bmatrix} 3 & 0 & 1 \\ -2 & 1 & 4 \\ 6 & 7 & 0 \end{bmatrix} =$

2. Suppose that at another game the stadium of Example 1.3.6 has 500 adults, 900 students, 600 senior citizens, and 300 children. Take an inner product to find the total receipts.
3. Suppose that the same stadium at another game admits 500 students, 900 senior citizens, 600 adults, and 300 children. Write this (carefully!) in vector form consistent with Example 1.3.6 and then take an inner product to find the total revenue.
4. Suppose that a person has three jobs to do, the first of which takes 8 minutes, the second of which takes 10 minutes, and the third of which takes 6 minutes. Write this information as a row vector. Suppose he knows that he will have to perform the first job 3 times, the second job 5 times, and the third job twice. Write this as a column vector. Find how much time should be allowed for all three jobs by taking an inner product.
5. In problem 4, suppose that he knows he must do the first job 4 times, the second 10 times, and the third 4 times. Compute the total time needed by taking an inner product.
6. Refer to problem 8 of Exercises 1.2.B (page 19). Suppose you want to produce 3 salt-and-pepper holders and 5 holder pieces. Use matrix multiplication to find out how much of each item is needed.
7. Referring to problem 9 of Exercises 1.2.B, use matrix multiplication to find how much of each item is needed to get ready to produce 8 photo albums and 10 picture pages.

EXERCISES 1.3. C

In making decisions as to which foundation is most likely to grant money for a specific project, assigning numerical values in vector form can help a professional fund raiser remember his past value judgments after interruptions. Robert Semple recommends the weighting for each factor in the total decision as shown below. By rating each factor on a scale of 1 to 5—the more emphatic the "yes" answer, the higher the rating—and taking an inner product, the likelihood of obtaining a grant from any particular foundation for a particular project can be measured.

Is your project in accord with the foundation's philosophy?	0.20
Is its focus in line with the foundation's primary interest?	0.20
Is the foundation free of incompatible restrictions as to geography, religious affiliation, and the like?	0.10
Is the amount you want compatible with its typical grant size?	0.10
Are the foundation's total assets large?	0.15
Can you meet the application deadline?	0.05
Do you have good rapport with the foundation?	0.20

For example, if foundation I rates 4 in "philosophy," 5 in "primary interest," 3 in "restrictions," 4 in "grant size," 5 in "total assets," 2 in "application deadline," and 2 in "rapport," its payoff rating would be 3.75, as follows:

$$[4, 5, 3, 4, 5, 2, 2]\begin{bmatrix} 0.20 \\ 0.20 \\ 0.10 \\ 0.10 \\ 0.15 \\ 0.05 \\ 0.20 \end{bmatrix} = 0.8 + 1 + 0.3 + 0.4 + 0.75 + 0.1 + 0.4 = 3.75$$

1. Find the payoff rating of a foundation that rates 4, 3, 5, 2, 5, 5, and 1 on the respective factors.
2. Find the payoff rating of a foundation that rates 4, 3, 5, 4, 5, 5, and 3 on the respective factors.

ANSWERS 1.3. A

1. (a) $\begin{bmatrix} -5 & -3 \\ 5 & 14 \end{bmatrix}$ (b) $\begin{bmatrix} 0 & 5 \\ 11 & 9 \end{bmatrix}$ (c) $\begin{bmatrix} 1 & -1 \\ 2 & 3 \end{bmatrix}$ (d) $\begin{bmatrix} 1 & -1 \\ 2 & 3 \end{bmatrix}$

   (e) $\begin{bmatrix} 1 & 0 \\ 2 & 0 \end{bmatrix}$ (f) $\begin{bmatrix} 0 & -1 \\ 0 & 3 \end{bmatrix}$ (g) $\begin{bmatrix} 1 & -1 \\ 0 & 0 \end{bmatrix}$ (h) $\begin{bmatrix} 0 & 0 \\ 2 & 3 \end{bmatrix}$

   (i) $\begin{bmatrix} 0 & 0 \\ 0 & 0 \end{bmatrix}$ (j) $\begin{bmatrix} 2 & 3 & 4 \\ 5 & 6 & 7 \\ 8 & 9 & 1 \end{bmatrix}$ (k) $\begin{bmatrix} 2 & 3 & 4 \\ 5 & 6 & 7 \\ 8 & 9 & 1 \end{bmatrix}$

   (l) $\begin{bmatrix} 2 & 3 & 0 \\ 5 & 6 & 0 \\ 8 & 9 & 0 \end{bmatrix}$ (m) $\begin{bmatrix} 2 & 3 & 4 \\ 5 & 6 & 7 \\ 0 & 0 & 0 \end{bmatrix}$ (n) $\begin{bmatrix} 4 & -14 & 22 \\ 7 & -23 & 43 \\ 10 & -14 & 55 \end{bmatrix}$

(o) $\begin{bmatrix} 19 & -3 & 2 \\ 0 & 5 & -9 \end{bmatrix}$ (p) $\begin{bmatrix} 12 & -3 & 6 \\ 5 & -4 & 4 \\ 12 & 8 & 0 \\ -5 & 4 & -4 \end{bmatrix}$ (q) cannot be done

2. $[3, 2, 1.5, 1]\begin{bmatrix} 300 \\ 1000 \\ 700 \\ 200 \end{bmatrix} = 4150$

3. $[3, 2, 1.5, 1]\begin{bmatrix} 700 \\ 300 \\ 1000 \\ 200 \end{bmatrix} = 4400$

4. $[5, 12, 10]\begin{bmatrix} 5 \\ 2 \\ 4 \end{bmatrix} = 89$ min

5. 190 min

6. 8 frontpieces, 10 rectangles, 66 pompoms, and 16 ties
7. 12 hook assemblies, 12 rods, 12 base plates, 8 top plates, 48 metal hooks, and 96 bolts

ANSWERS 1.3. C

1. 3.3
2. 3.9

1.4 Input–Output Models and Compact Notation

When a mathematical object, such as a matrix, equation, or graph, is used to describe a real-world relationship, we say that it is a mathematical model. A mathematical model (like a model of an airplane or a building) is only approximately accurate, but a good model based on data gathered in the past may forecast the future with remarkable accuracy.

We have already seen in the parts-listing problem how matrices may be used to model a production process within one company; now we shall explore how they can be used to model an entire economy. The Leontief matrix in Section 1.2 indicated how complicated such an analysis can be if it is realistic. Thus, in order to examine input–output theory without becoming distracted by overwhelming arithmetic, we shall study only simple hypothetical models.

1.4.1 Example

The following table is a model of a simple hypothetical economy with only two industries. The numbers in the table indicate billions of dollars. Primary inputs include land, labor, and any other inputs not currently produced by specific industries. Final demands include those outputs that are not immediately plowed back into the production cycle, such as goods produced for households (consumers), government, and export. The table says, for example, that there was \$100 worth of total agricultural output and that to obtain this there was \$30 worth of agricultural investment, $60 worth of manufacturing investment, and $10 worth of primary inputs. It also says that of the $100 worth of agricultural output, $30 is plowed back into agriculture, $48 worth is devoted to manufacturing, and $22 is available for households, government, and export.

	Agriculture	Manufacturing	Final demands	Gross outputs
Agriculture	30	48	22	100
Manufacturing	60	24	36	120
Primary inputs	10	48		

Such an analysis becomes especially useful when the assumption is made (as it often is) that the *proportions* between various parts of the economy remain constant even if the economy changes a bit. With this assumption we can use the preceding table to make a coefficient or technology matrix that may be useful to forecast possible alterations in the economy.

To obtain the technology matrix, we divide each element of that part of the table enclosed in the inner rectangle by the total of the column in which that particular element lies. Thus the technology matrix for the above economy is

$$\begin{bmatrix} \frac{30}{100} & \frac{48}{120} \\ \frac{60}{100} & \frac{24}{120} \end{bmatrix} = \begin{bmatrix} 0.3 & 0.4 \\ 0.6 & 0.2 \end{bmatrix}$$

(Figure 1.2–4 on page 12 is a real technology matrix.) Because of the way the technology matrix was obtained, we can now write the following equation:

$$\begin{bmatrix} 0.3 & 0.4 \\ 0.6 & 0.2 \end{bmatrix}\begin{bmatrix} 100 \\ 120 \end{bmatrix} + \begin{bmatrix} 22 \\ 36 \end{bmatrix} = \begin{bmatrix} 100 \\ 120 \end{bmatrix}$$

You should check to see that this matrix equation is true. (We shall define it to be the "technological equation.") Try to understand why it must be true; it may help to use the unreduced fractions given in the technology matrix above.

1.4.2 Definition

The technology matrix is that matrix such that

$$\begin{bmatrix} \text{technology} \\ \text{matrix} \end{bmatrix}\begin{bmatrix} \text{gross} \\ \text{outputs} \end{bmatrix} + \begin{bmatrix} \text{final} \\ \text{demands} \end{bmatrix} = \begin{bmatrix} \text{gross} \\ \text{outputs} \end{bmatrix}$$

where the gross outputs and final demands are written in column vector form. We shall refer to this equation as the technological equation.

The technological equation is our first example of a matrix equation. It is of the form $AX + D = X$, where A is the technology matrix (or Leontief matrix), X is the gross outputs vector, and D is the final demands vector.

By writing whole sets of equations as a simple matrix equation, we can often make the computations easier in the sense that they are more mechanical and a computer can do them. The technological equation given in Example 1.4.1 shows how a matrix equation can summarize linear numerical equations. The numerical equations that it summarizes are

$$(0.3)100 + (0.4)120 + 22 = 100$$
$$(0.6)100 + (0.2)120 + 36 = 120$$

It is important to be able to convert a matrix equation to a set of linear equations, and vice versa, especially equations of the form $AX = B$. In the following examples the matrix equation on the left is an abbreviated way of writing the set of equations on the right. Study each pair so you see the pattern and can change the notation in either direction when you do the exercises. You should be able to multiply the matrices on the left to get the set of equations on the right in each example.

1.4.3 Example

$$\begin{bmatrix} 2 & 4 \\ 5 & 3 \end{bmatrix} \begin{bmatrix} x_1 \\ x_2 \end{bmatrix} = \begin{bmatrix} 7 \\ 8 \end{bmatrix} \qquad \begin{aligned} 2x_1 + 4x_2 &= 7 \\ 5x_1 + 3x_2 &= 8 \end{aligned}$$

In the matrix equations above, $A = \begin{bmatrix} 2 & 4 \\ 5 & 3 \end{bmatrix}$ is called the coefficient matrix, $X = \begin{bmatrix} x_1 \\ x_2 \end{bmatrix}$ is called the solution vector, and $B = \begin{bmatrix} 7 \\ 8 \end{bmatrix}$ is called the constant vector or right-hand vector.

1.4.4 Example

$$\begin{bmatrix} a_{11} & a_{12} \\ a_{21} & a_{22} \end{bmatrix} \begin{bmatrix} x_1 \\ x_2 \end{bmatrix} = \begin{bmatrix} b_1 \\ b_2 \end{bmatrix} \qquad \begin{aligned} a_{11}x_1 + a_{12}x_2 &= b_1 \\ a_{21}x_1 + a_{22}x_2 &= b_2 \end{aligned}$$

1.4.5 Example

$$\begin{bmatrix} 1 & 1 & 0 \\ 2 & 1 & 1 \\ 3 & 0 & 3 \end{bmatrix} \begin{bmatrix} x_1 \\ x_2 \\ x_3 \end{bmatrix} = \begin{bmatrix} 2 \\ 3 \\ 4 \end{bmatrix} \qquad \begin{aligned} x_1 + x_2 \qquad\quad &= 2 \\ 2x_1 + x_2 + x_3 &= 3 \\ 3x_1 \qquad + 3x_3 &= 4 \end{aligned}$$

1.4.6 Example

$$\begin{bmatrix} a_{11} & a_{12} & a_{13} \\ a_{21} & a_{22} & a_{23} \\ a_{31} & a_{32} & a_{33} \end{bmatrix} \begin{bmatrix} x_1 \\ x_2 \\ x_3 \end{bmatrix} = \begin{bmatrix} b_1 \\ b_2 \\ b_3 \end{bmatrix} \qquad \begin{aligned} a_{11}x_1 + a_{12}x_2 + a_{13}x_3 &= b_1 \\ a_{21}x_1 + a_{22}x_2 + a_{23}x_3 &= b_2 \\ a_{31}x_1 + a_{32}x_2 + a_{33}x_3 &= b_3 \end{aligned}$$

SUMMARY

This section showed how one model of a simple economy can be written in matrix form by using a technology matrix in a technological equation. Then it gave other examples of how sets of linear equations can be written as matrices. Now it is your turn! Practice changing the following sets of linear equations into matrix equations.

(Notice that when the total dollars of an economy is given, as on pages 10 and 11 and page 31, the sum of the column for any one industry is the same as the sum of its row. But when a technology matrix is formed—as in Figure 1.2–4 or in this section—dividing each number by the sum of its column completely changes the additive relationship between the rows and columns.)

EXERCISES 1.4. A

1. Write the following sets of equations in the form $AX = B$, where A is a matrix and X and B are vectors. (*Hint*: Some of the coefficients may be zero.)

   (a) $\begin{aligned} x_1 + 4x_2 &= 7 \\ 3x_1 + 5x_2 &= 8 \end{aligned}$ (b) $\begin{aligned} 5x_1 - 3x_2 &= 9 \\ 4x_1 + 6x_2 &= 1 \end{aligned}$ (c) $\begin{aligned} 4x + 2y &= 8 \\ 8x - 7y &= 5 \end{aligned}$

   (d) $\begin{aligned} x_1 + 2x_2 + 3x_3 &= 4 \\ 5x_1 + 6x_2 + 7x_3 &= 8 \\ 9x_1 + 2x_2 - 3x_3 &= 0 \end{aligned}$ (e) $\begin{aligned} x_1 - x_2 &= 8 \\ x_2 - x_3 &= 6 \\ x_3 - x_1 &= 2 \end{aligned}$ (f) $\begin{aligned} 2x_2 + x_3 &= 7 \\ x_1 - x_2 + x_3 &= 4 \\ x_1 + 4x_3 &= -6 \end{aligned}$

2. In Example 1.4.4, on page 32, give the coefficient matrix, the solution vector, and the right-hand vector.
3. For each of the following models of hypothetical economies, find the technology matrix, write the technological equation, and check the equation by matrix multiplication.

   (a)

	Agriculture	Manufacturing	Final demands	Gross outputs
Agriculture	60	27	13	100
Manufacturing	20	54	16	90
Primary inputs	20	9		

   (b)

	Agriculture	Manufacturing	Final demands	Gross outputs
Agriculture	30	57	13	100
Manufacturing	50	19	26	95
Primary inputs	20	19		

4. An industry can be modeled using a mathematical pattern similar to that used to model an economy. For example, the metal industry can be divided into "primary metals" and "other metal goods." Suppose that

the amounts sold (in millions of dollars) of a fictitious metal industry are given in the table below. Write the "technology matrix" for this fictitious industry and write its "technological equation."

	Primary metals	Other metal goods	Sales to nonmetal industries	Total budget of metal industry
Primary metals	240	380	100	720
Other metal goods	420	190	150	760
Nonmetal inputs	60	190		

EXERCISES 1.4. B

1. Write the following sets of equations in the form $AX = B$, where A is a matrix and X and B are vectors. (*Hint:* Some of the coefficients may be zero.)

   (a) $\begin{aligned} 2x_1 + 3x_2 &= 5 \\ x_1 + 4x_2 &= 8 \end{aligned}$ (b) $\begin{aligned} 4x_1 - 2x_2 &= 7 \\ 5x_1 + x_2 &= 9 \end{aligned}$ (c) $\begin{aligned} 3x - 2y &= 6 \\ -x + 4y &= 9 \end{aligned}$

   (d) $\begin{aligned} x_1 + 5x_2 - 2x_3 &= 7 \\ 3x_1 + 2x_2 - 4x_3 &= 5 \\ 10x_1 - x_2 + 5x_3 &= 4 \end{aligned}$ (e) $\begin{aligned} x_2 + x_3 &= 4 \\ x_1 - x_2 &= 6 \\ x_1 + x_3 &= 9 \end{aligned}$ (f) $\begin{aligned} 2x_1 - 3x_2 &= 7 \\ 4x_2 + 6x_3 &= -8 \\ x_1 + x_2 - x_3 &= 3 \end{aligned}$

2. In Example 1.4.5 on page 32, give the coefficient matrix, the solution vector, and the right-hand vector.
3. For each of the following models of hypothetical economies, find the technology matrix, write the technological equation, and check the equation by matrix multiplication.

   (a)

	Agriculture	Manufacturing	Final demands	Gross outputs
Agriculture	25	12	13	50
Manufacturing	20	16	4	40
Primary inputs	5	12		

   (b)

	Agriculture	Manufacturing	Final demands	Gross outputs
Agriculture	18	18	9	45
Manufacturing	9	36	15	60
Primary inputs	18	6		

4. A company with two or more plants can be modeled using a mathematical pattern similar to that used to model an economy. Suppose that the transactions (in thousands of dollars) of a fictitious company are given

in the table below. Write the "technology matrix" and "technological equation" for this fictitious company.

	Plant I	Plant II	Sold outside the company	Total budget of the company
Plant I	70	30	20	120
Plant II	40	30	20	90
Primary inputs	10	30		

EXERCISES 1.4. C

In the following model of a hypothetical economy, find the technology matrix, write the technological equation, and check the equation by matrix multiplication.

	Agriculture	Manufacturing	Services	Final demands	Gross outputs
Agriculture	30	27	19	24	100
Manufacturing	20	36	9.5	24.5	90
Services	20	18	38	19	95
Primary inputs	30	9	28.5		

ANSWERS 1.4. A

1. (a) $\begin{bmatrix} 1 & 4 \\ 3 & 5 \end{bmatrix}\begin{bmatrix} x_1 \\ x_2 \end{bmatrix} = \begin{bmatrix} 7 \\ 8 \end{bmatrix}$ (b) $\begin{bmatrix} 5 & -3 \\ 4 & 6 \end{bmatrix}\begin{bmatrix} x_1 \\ x_2 \end{bmatrix} = \begin{bmatrix} 9 \\ 1 \end{bmatrix}$

   (c) $\begin{bmatrix} 4 & 2 \\ 8 & -7 \end{bmatrix}\begin{bmatrix} x \\ y \end{bmatrix} = \begin{bmatrix} 8 \\ 5 \end{bmatrix}$ (d) $\begin{bmatrix} 1 & 2 & 3 \\ 5 & 6 & 7 \\ 9 & 2 & -3 \end{bmatrix}\begin{bmatrix} x_1 \\ x_2 \\ x_3 \end{bmatrix} = \begin{bmatrix} 4 \\ 8 \\ 0 \end{bmatrix}$

   (e) $\begin{bmatrix} 1 & -1 & 0 \\ 0 & 1 & -1 \\ -1 & 0 & 1 \end{bmatrix}\begin{bmatrix} x_1 \\ x_2 \\ x_3 \end{bmatrix} = \begin{bmatrix} 8 \\ 6 \\ 2 \end{bmatrix}$

   (f) $\begin{bmatrix} 0 & 2 & 1 \\ 1 & -1 & 1 \\ 1 & 0 & 4 \end{bmatrix}\begin{bmatrix} x_1 \\ x_2 \\ x_3 \end{bmatrix} = \begin{bmatrix} 7 \\ 4 \\ -6 \end{bmatrix}$

2. Coefficient matrix: $\begin{bmatrix} a_{11} & a_{12} \\ a_{21} & a_{22} \end{bmatrix}$; solution vector: $\begin{bmatrix} x_1 \\ x_2 \end{bmatrix}$; right-hand vector: $\begin{bmatrix} b_1 \\ b_2 \end{bmatrix}$

3. (a) $\begin{bmatrix} 0.6 & 0.3 \\ 0.2 & 0.6 \end{bmatrix}\begin{bmatrix} 100 \\ 90 \end{bmatrix} + \begin{bmatrix} 13 \\ 16 \end{bmatrix} = \begin{bmatrix} 100 \\ 90 \end{bmatrix}$

(b) $\begin{bmatrix} 0.3 & 0.6 \\ 0.5 & 0.2 \end{bmatrix} \begin{bmatrix} 100 \\ 95 \end{bmatrix} + \begin{bmatrix} 13 \\ 26 \end{bmatrix} = \begin{bmatrix} 100 \\ 95 \end{bmatrix}$

4. $\begin{bmatrix} \frac{1}{3} & \frac{1}{2} \\ \frac{7}{12} & \frac{1}{4} \end{bmatrix} \begin{bmatrix} 720 \\ 760 \end{bmatrix} + \begin{bmatrix} 100 \\ 150 \end{bmatrix} = \begin{bmatrix} 720 \\ 760 \end{bmatrix}$

ANSWERS 1.4. C

$$\begin{bmatrix} 0.3 & 0.3 & 0.2 \\ 0.2 & 0.4 & 0.1 \\ 0.2 & 0.2 & 0.4 \end{bmatrix} \begin{bmatrix} 100 \\ 90 \\ 95 \end{bmatrix} + \begin{bmatrix} 24 \\ 24.5 \\ 19 \end{bmatrix} = \begin{bmatrix} 100 \\ 90 \\ 95 \end{bmatrix}$$

1.5 Identities and Inverses

We have seen that matrices are mathematical objects similar to numbers. In this section we shall investigate other ways in which matrices resemble and differ from numbers. We shall confine our attention to square matrices—those that have an equal number of rows and columns.

We recall that any two matrices of the same dimension can be added and subtracted. It is clear from the definition that the usual axioms of addition hold true.

Addition of matrices is associative: $(A + B) + C = A + (B + C)$

Addition of matrices is commutative: $A + B = B + A$

Furthermore, for each dimension of matrices there is a unique additive identity. This means that there is a matrix O which when added to any matrix A always gives A as an answer:

$$O + A = A \qquad \text{for every matrix } A$$

What is O? Convince yourself that the matrix of the same dimension as A consisting of all zeros has this property.

You might wonder if some other matrix also has this property—whether the additive identity is really unique. It is unique, and the proof is easy. Suppose that O and E were two additive identities for the same dimension of matrices. Then

$$O \underset{\substack{\text{because } E \\ \text{is an identity}}}{=} O + E \underset{\substack{\text{because } O \\ \text{is an identity}}}{=} E$$

We conclude that $O = E$, so there is only one identity.

Every matrix A has an additive inverse, or negative. This means for every matrix A there is a matrix $-A$ such that $A + (-A) = O$. What is $-A$? Just replace every element of A with its negative; in other words, multiply every element by -1.

To subtract B from A is the same as to add $-B$ to A, just as with numbers.

Multiplication of matrices differs from multiplication of numbers in three ways that we have already indicated.

1. There are two kinds of multiplication:
   (a) Multiplication of a matrix by a scalar (Example 1.1.8).
   (b) Multiplication of a matrix by a matrix (Example 1.3.8).
2. Multiplication of matrices is not commutative. Example 1.3.2:

$$\begin{bmatrix} 2 & 3 \\ 4 & 5 \end{bmatrix}\begin{bmatrix} 6 & 7 \\ 8 & 9 \end{bmatrix} \neq \begin{bmatrix} 6 & 7 \\ 8 & 9 \end{bmatrix}\begin{bmatrix} 2 & 3 \\ 4 & 5 \end{bmatrix}$$

3. The product of two nonzero matrices may be zero. Problem 1(i) of Exercises 1.3.A and 1.3.B:

$$\begin{bmatrix} 1 & 0 \\ 0 & 0 \end{bmatrix}\begin{bmatrix} 0 & 0 \\ 0 & 1 \end{bmatrix} = \begin{bmatrix} 0 & 0 \\ 0 & 0 \end{bmatrix}$$

These last two surprising properties are filled with implications for matrix theory. Before we explore these peculiarities, we point out some less surprising facts.

Multiplication of matrices is distributive. This parallels a similar fact for numbers and means that for any three matrices A, B, and C, such that the indicated multiplications can be performed, it is true that

$$A(B + C) = AB + AC \qquad \text{and} \qquad (A + B)C = AC + BC$$

While doing problem 1 of Exercises 1.3.A or 1.3.B, you may have guessed (correctly) that each set of square matrices of a given dimension has a multiplicative identity—that is, a matrix I which, when multiplied by any matrix A of that dimension, always gives the matrix A as an answer. Check the following computations.

$$\begin{bmatrix} 1 & 0 \\ 0 & 1 \end{bmatrix}\begin{bmatrix} a & b \\ c & d \end{bmatrix} = \begin{bmatrix} a & b \\ c & d \end{bmatrix} = \begin{bmatrix} a & b \\ c & d \end{bmatrix}\begin{bmatrix} 1 & 0 \\ 0 & 1 \end{bmatrix}$$

$$\begin{bmatrix} 1 & 0 & 0 \\ 0 & 1 & 0 \\ 0 & 0 & 1 \end{bmatrix}\begin{bmatrix} a & b & c \\ d & e & f \\ g & h & i \end{bmatrix} = \begin{bmatrix} a & b & c \\ d & e & f \\ g & h & i \end{bmatrix}$$

$$= \begin{bmatrix} a & b & c \\ d & e & f \\ g & h & i \end{bmatrix}\begin{bmatrix} 1 & 0 & 0 \\ 0 & 1 & 0 \\ 0 & 0 & 1 \end{bmatrix}$$

We conclude from these computations that the multiplicative identity for 2 by 2 matrices is $\begin{bmatrix} 1 & 0 \\ 0 & 1 \end{bmatrix}$ and the multiplicative identity for 3 by 3 matrices is $\begin{bmatrix} 1 & 0 & 0 \\ 0 & 1 & 0 \\ 0 & 0 & 1 \end{bmatrix}$ Can you guess the multiplicative identity for the 4 by 4 matrices? Each dimension of matrices "needs" its own identity because we cannot multiply square matrices of different dimensions together.

Bad news is just around the corner. It is *not* always possible to divide square matrices of the same dimension. Sometimes you can, but sometimes you cannot. This can be a big nuisance, but then perhaps life was not meant to be easy.

Division is not always possible, because multiplicative inverses do not always exist. If you do not see the connection, think for a moment about the division of numbers. To divide x by y is the same as to multiply x by y^{-1}, the reciprocal or multiplicative inverse of y. If y^{-1} does not exist, then we cannot compute $x/y = x \cdot y^{-1}$.

1.5.1 Definition

The multiplicative inverse, A^{-1}, of a square matrix, A, is a matrix such that $A \cdot A^{-1} = I$, where I is the multiplicative identity with the same dimension as A.

It can be shown that if A is a square matrix and $A \cdot A^{-1} = I$, then $A^{-1} \cdot A = I$; in other words, a square matrix always commutes with its inverse. Matrices that are not square never have two-sided inverses, and we shall ignore them in our discussion of inverses.

Many square matrices do not have inverses either, and it is not immediately clear whether a given matrix has a multiplicative inverse. Chapter 2 shows how to find the inverse when it exists. Meanwhile test your understanding of the concepts involved in the following example. You will understand matrices faster and better if you attempt to solve each problem yourself before looking at the answers. (Whenever the word "inverse" is used without a modifier, it means "multiplicative inverse.")

1.5.2 Example

Find the multiplicative inverse of each of the following matrices if it exists.

(a) $\begin{bmatrix} 1 & 0 \\ 0 & 1 \end{bmatrix}$ (b) $\begin{bmatrix} 2 & 0 \\ 0 & \frac{1}{3} \end{bmatrix}$ (c) $\begin{bmatrix} 1 & 0 \\ 0 & 0 \end{bmatrix}$ (d) $\begin{bmatrix} 0 & 1 \\ 1 & 0 \end{bmatrix}$

(e) $\begin{bmatrix} 5 & 7 \\ 2 & 4 \end{bmatrix}$

Solution:

(a) Since this is the multiplicative identity, its inverse is itself: $\begin{bmatrix} 1 & 0 \\ 0 & 1 \end{bmatrix}$.

(b) The obvious thing works: $\begin{bmatrix} \frac{1}{2} & 0 \\ 0 & 3 \end{bmatrix}$.

(c) This is tricky. There is no inverse! To verify this, assume that there were an inverse; we can call it $\begin{bmatrix} a & b \\ c & d \end{bmatrix}$. Then $\begin{bmatrix} 1 & 0 \\ 0 & 0 \end{bmatrix}\begin{bmatrix} a & b \\ c & d \end{bmatrix} = \begin{bmatrix} a & b \\ 0 & 0 \end{bmatrix}$, which will never have a 1 in the lower right corner. This contradiction shows that $\begin{bmatrix} a & b \\ c & d \end{bmatrix}$ cannot be the inverse for any a, b, c, and d.

(d) Again, its inverse is itself: $\begin{bmatrix} 0 & 1 \\ 1 & 0 \end{bmatrix}$. There are other matrices with this property. Can you find some of them?

(e) This one is hard. The answer is $\begin{bmatrix} \frac{2}{3} & -\frac{7}{6} \\ -\frac{1}{3} & \frac{5}{6} \end{bmatrix}$. You are not likely to have guessed this! In Chapter 2 we shall develop some efficient methods for finding multiplicative inverses that are not obvious. Meanwhile, you could solve such problems for 2 by 2 matrices by grinding out the algebra needed to solve simultaneous equations:

Assuming that the given matrix does have an inverse, we can write $\begin{bmatrix} 5 & 7 \\ 2 & 4 \end{bmatrix}\begin{bmatrix} a & b \\ c & d \end{bmatrix} = \begin{bmatrix} 1 & 0 \\ 0 & 1 \end{bmatrix}$, where a, b, c, and d are to be determined. We use matrix multiplication to obtain the following linear equations:

$$5a + 7c = 1 \qquad 5b + 7d = 0 \qquad 2a + 4c = 0 \qquad 2b + 4d = 1$$

The first and third of these give:

$10a + 14c = 2$	(multiplying the first by 2)
$10a + 20c = 0$	(multiplying the third by 5)
$-6c = 2$	(subtracting)
$\boxed{c = -\frac{1}{3}}$	(dividing)
$2a + 4(-\frac{1}{3}) = 0$	(substituting)
$\boxed{a = \frac{2}{3}}$	

The second and fourth of these give:

$10b + 14d = 0$	(multiplying the second by 2)
$10b + 20d = 5$	(multiplying the fourth by 5)
$6d = 5$	(subtracting)
$\boxed{d = \frac{5}{6}}$	(dividing)
$5b + 7(\frac{5}{6}) = 0$	(substituting)
$\boxed{b = -\frac{7}{6}}$	

Combining these four results, we obtain the inverse as stated.

You can see that this would be time-consuming for a 3 by 3 matrix. The Gauss–Jordan method, explained in Chapter 2, is much simpler. Meanwhile, the exercises in this chapter will be carefully chosen to be easy for you to do with your present skills; they are designed to help you get a "feel" for the ideas involved.

SUMMARY

Addition of matrices is associative and commutative. Each dimension of matrices contains a unique additive identity, the matrix of all zeros, and every matrix has an additive inverse called its negative.

Matrix multiplication (not to be confused with scalar multiplication) is *not* commutative, and the product of two nonzero matrices may be zero. Although each set of n by n square matrices has a multiplicative identity, not every square matrix has a multiplicative inverse. We postpone the detailed study of inverses until Chapter 2, but here we provide some relatively easy matrices for which you can find (multiplicative) inverses. (We follow the conventional usage in referring to a "multiplicative inverse" merely as an "inverse"; an additive inverse is a "negative.")

EXERCISES 1.5. A

1. Write the additive identity for:
   (a) The 2 by 2 matrices.
   (b) The 2 by 3 matrices.
   (c) The 5 by 5 matrices.
2. Write the additive inverse of:
   (a) $\begin{bmatrix} 2 & 3 \\ 107 & -6 \end{bmatrix}$ (b) $\begin{bmatrix} 6 & 7 & -9 \\ 0 & -2 & 1 \end{bmatrix}$
3. Matrices that are not square have different multiplicative identities for the left and right sides. Show that for the set of 2 by 3 matrices, $\begin{bmatrix} 1 & 0 \\ 0 & 1 \end{bmatrix}$ is a left identity (that is, $IA = A$, where A is any 2 by 3 matrix) and that $\begin{bmatrix} 1 & 0 & 0 \\ 0 & 1 & 0 \\ 0 & 0 & 1 \end{bmatrix}$ is a right identity ($AI = A$). (*Hint:* Let $A = \begin{bmatrix} a & b & c \\ d & e & f \end{bmatrix}$ and just substitute and compute.)
4. What is the multiplicative identity for the 4 by 4 matrices? For the 5 by 5 matrices?
5. Find the multiplicative inverse, if it exists, of:
   (a) $\begin{bmatrix} 3 & 0 \\ 0 & 4 \end{bmatrix}$ (b) $\begin{bmatrix} -1 & 0 & 0 \\ 0 & 2 & 0 \\ 0 & 0 & 3 \end{bmatrix}$ (c) $\begin{bmatrix} 1 & 0 & 0 \\ 0 & 1 & 0 \\ 0 & 0 & 1 \end{bmatrix}$

(d) $\begin{bmatrix} 0 & 1 \\ 0 & 0 \end{bmatrix}$ (e) $\begin{bmatrix} 0 & 1 & 0 \\ 1 & 0 & 0 \\ 0 & 0 & 1 \end{bmatrix}$ (f) $\begin{bmatrix} 0 & 1 & 0 & 0 \\ 1 & 0 & 0 & 0 \\ 0 & 0 & 0 & 1 \\ 0 & 0 & 1 & 0 \end{bmatrix}$

(g) $\begin{bmatrix} 3 & 0 & 0 & 0 \\ 0 & -4 & 0 & 0 \\ 0 & 0 & \frac{2}{3} & 0 \\ 0 & 0 & 0 & 8 \end{bmatrix}$ (h) $\begin{bmatrix} 2 & 5 \\ 3 & 8 \end{bmatrix}$

6. Perform the following multiplications. The purpose of this exercise is not only to give you further practice in multiplying matrices, but also to help you see patterns that will be useful in Chapter 2. If you feel an impulse to guess an answer before computing, yield to that impulse! Take time to explore patterns.

(a) $\begin{bmatrix} 0 & 1 & 0 \\ 1 & 0 & 0 \\ 0 & 0 & 1 \end{bmatrix}\begin{bmatrix} a & b & c \\ d & e & f \\ g & h & i \end{bmatrix} =$

(b) $\begin{bmatrix} 1 & 0 & 0 \\ 0 & 0 & 1 \\ 0 & 1 & 0 \end{bmatrix}\begin{bmatrix} a & b & c \\ d & e & f \\ g & h & i \end{bmatrix} =$

(c) $\begin{bmatrix} 0 & 0 & 1 \\ 0 & 1 & 0 \\ 1 & 0 & 0 \end{bmatrix}\begin{bmatrix} a & b & c \\ d & e & f \\ g & h & i \end{bmatrix} =$

(d) $\begin{bmatrix} 2 & 0 & 0 \\ 0 & 1 & 0 \\ 0 & 0 & 1 \end{bmatrix}\begin{bmatrix} a & b & c \\ d & e & f \\ g & h & i \end{bmatrix} =$

(e) $\begin{bmatrix} 1 & 0 & 0 \\ 0 & 1 & 0 \\ 0 & 0 & 4 \end{bmatrix}\begin{bmatrix} a & b & c \\ d & e & f \\ g & h & i \end{bmatrix} =$

(f) $\begin{bmatrix} 1 & 0 & 0 \\ 0 & 3 & 0 \\ 0 & 0 & 1 \end{bmatrix}\begin{bmatrix} a & b & c \\ d & e & f \\ g & h & i \end{bmatrix} =$

(g) $\begin{bmatrix} 1 & 2 & 0 \\ 0 & 1 & 0 \\ 0 & 0 & 1 \end{bmatrix}\begin{bmatrix} a & b & c \\ d & e & f \\ g & h & i \end{bmatrix} =$

(h) $\begin{bmatrix} 1 & 0 & 0 \\ 2 & 1 & 0 \\ 0 & 0 & 1 \end{bmatrix}\begin{bmatrix} a & b & c \\ d & e & f \\ g & h & i \end{bmatrix} =$

EXERCISES 1.5. B

1. Write the additive identity for:
   (a) The 3 by 3 matrices. (b) The 3 by 2 matrices.
   (c) The 4 by 4 matrices.

2. Write the additive inverse of:

(a) $\begin{bmatrix} 4 & 6 \\ 416 & -8 \end{bmatrix}$ (b) $\begin{bmatrix} \frac{1}{2} & 3 & 0 & 17 \\ -19 & 6 & -2 & -\frac{1}{4} \end{bmatrix}$

3. Matrices that are not square have different multiplicative identities for the left and right sides. Show that for the set of 3 by 2 matrices, $\begin{bmatrix} 1 & 0 & 0 \\ 0 & 1 & 0 \\ 0 & 0 & 1 \end{bmatrix}$ is a left identity (that is, $IA = A$, where A is any 2 by 3 matrix) and that $\begin{bmatrix} 1 & 0 \\ 0 & 1 \end{bmatrix}$ is a right identity ($AI = A$). (*Hint*: Let $A = \begin{bmatrix} a & b \\ c & d \\ e & f \end{bmatrix}$ and just substitute and compute.)

4. What is the multiplicative identity for the 4 by 4 matrices? For the 5 by 5 matrices?

5. Find the multiplicative inverse, if it exists, of:

(a) $\begin{bmatrix} 5 & 0 \\ 0 & 2 \end{bmatrix}$ (b) $\begin{bmatrix} 3 & 0 & 0 \\ 0 & -8 & 0 \\ 0 & 0 & \frac{1}{2} \end{bmatrix}$ (c) $\begin{bmatrix} 0 & 0 \\ 1 & 0 \end{bmatrix}$

(d) $\begin{bmatrix} 0 & 0 & 1 \\ 0 & 1 & 0 \\ 1 & 0 & 0 \end{bmatrix}$ (e) $\begin{bmatrix} 1 & 0 & 0 \\ 0 & 0 & 1 \\ 0 & 1 & 0 \end{bmatrix}$ (f) $\begin{bmatrix} 4 & 0 & 0 & 0 \\ 0 & \frac{3}{4} & 0 & 0 \\ 0 & 0 & -5 & 0 \\ 0 & 0 & 0 & 2 \end{bmatrix}$

(g) $\begin{bmatrix} 0 & 0 & 1 & 0 \\ 0 & 1 & 0 & 0 \\ 1 & 0 & 0 & 0 \\ 0 & 0 & 0 & 1 \end{bmatrix}$ (h) $\begin{bmatrix} 7 & 5 \\ 4 & 3 \end{bmatrix}$

6. Same as problem 6 in Exercises 1.5.A.

EXERCISES 1.5. C

1. Find four 2 by 2 matrices not mentioned in the text which are their own inverses.
2. The uniqueness of the identity is trickier to prove for a set of mathematical objects which is not commutative (such as matrices under multiplication). However, on page 37 we proved that the identity matrix I for 2 by 2 and 3 by 3 matrices commutes with all matrices of their own dimension. Suppose that there were another 2 by 2 matrix, E, such that $EA = A$ for all 2 by 2 matrices A. Prove that $E = I$.

ANSWERS 1.5. A

1. (a) $\begin{bmatrix} 0 & 0 \\ 0 & 0 \end{bmatrix}$ (b) $\begin{bmatrix} 0 & 0 & 0 \\ 0 & 0 & 0 \end{bmatrix}$ (c) $\begin{bmatrix} 0 & 0 & 0 & 0 & 0 \\ 0 & 0 & 0 & 0 & 0 \\ 0 & 0 & 0 & 0 & 0 \\ 0 & 0 & 0 & 0 & 0 \\ 0 & 0 & 0 & 0 & 0 \end{bmatrix}$

2. (a) $\begin{bmatrix} -2 & -3 \\ -107 & 6 \end{bmatrix}$ (b) $\begin{bmatrix} -6 & -7 & 9 \\ 0 & 2 & -1 \end{bmatrix}$

3. $\begin{bmatrix} 1 & 0 \\ 0 & 1 \end{bmatrix}\begin{bmatrix} a & b & c \\ d & e & f \end{bmatrix} = \begin{bmatrix} a & b & c \\ d & e & f \end{bmatrix}$

$\begin{bmatrix} a & b & c \\ d & e & f \end{bmatrix}\begin{bmatrix} 1 & 0 & 0 \\ 0 & 1 & 0 \\ 0 & 0 & 1 \end{bmatrix} = \begin{bmatrix} a & b & c \\ d & e & f \end{bmatrix}$

4. $\begin{bmatrix} 1 & 0 & 0 & 0 \\ 0 & 1 & 0 & 0 \\ 0 & 0 & 1 & 0 \\ 0 & 0 & 0 & 1 \end{bmatrix}$ $\begin{bmatrix} 1 & 0 & 0 & 0 & 0 \\ 0 & 1 & 0 & 0 & 0 \\ 0 & 0 & 1 & 0 & 0 \\ 0 & 0 & 0 & 1 & 0 \\ 0 & 0 & 0 & 0 & 1 \end{bmatrix}$

5. (a) $\begin{bmatrix} \frac{1}{3} & 0 \\ 0 & \frac{1}{4} \end{bmatrix}$ (b) $\begin{bmatrix} -1 & 0 & 0 \\ 0 & \frac{1}{2} & 0 \\ 0 & 0 & \frac{1}{3} \end{bmatrix}$ (c) $\begin{bmatrix} 1 & 0 & 0 \\ 0 & 1 & 0 \\ 0 & 0 & 1 \end{bmatrix}$

(d) no inverse (e) its own inverse (f) its own inverse

(g) $\begin{bmatrix} \frac{1}{3} & 0 & 0 & 0 \\ 0 & -\frac{1}{4} & 0 & 0 \\ 0 & 0 & \frac{3}{2} & 0 \\ 0 & 0 & 0 & \frac{1}{8} \end{bmatrix}$ (h) $\begin{bmatrix} 8 & -5 \\ -3 & 2 \end{bmatrix}$

6. (a) $\begin{bmatrix} d & e & f \\ a & b & c \\ g & h & i \end{bmatrix}$ (b) $\begin{bmatrix} a & b & c \\ g & h & i \\ d & e & f \end{bmatrix}$ (c) $\begin{bmatrix} g & h & i \\ d & e & f \\ a & b & c \end{bmatrix}$

(d) $\begin{bmatrix} 2a & 2b & 2c \\ d & e & f \\ g & h & i \end{bmatrix}$ (e) $\begin{bmatrix} a & b & c \\ d & e & f \\ 4g & 4h & 4i \end{bmatrix}$

(f) $\begin{bmatrix} a & b & c \\ 3d & 3e & 3f \\ g & h & i \end{bmatrix}$ (g) $\begin{bmatrix} a + 2d & b + 2e & c + 2f \\ d & e & f \\ g & h & i \end{bmatrix}$

(h) $\begin{bmatrix} a & b & c \\ 2a + d & 2b + e & 2c + f \\ g & h & i \end{bmatrix}$

ANSWERS 1.5. C

1. $\begin{bmatrix} -1 & 0 \\ 0 & -1 \end{bmatrix}$ $\begin{bmatrix} 0 & -1 \\ -1 & 0 \end{bmatrix}$ $\begin{bmatrix} -1 & 0 \\ 0 & 1 \end{bmatrix}$ $\begin{bmatrix} 1 & 0 \\ 0 & -1 \end{bmatrix}$
2. $I = EI = E$; the first equality is because of the property of E and the second is because I is the identity.

*1.6 Using Inverses in Cryptography

The multiplicative inverse of a matrix can be used in various applications to "undo" whatever the matrix itself has done. This section shows how multiplication by a matrix can be used to intentionally confuse the reader, and how a reader who knows the inverse can break the code. Such techniques are widely used in intelligence work; you can easily see how the coded message would be baffling to someone who does not know the inverse.

The basis of this technique in cryptography is the familiar code

a	b	c	d	e	f	g	h	i	j	k	l	m	n	o
1	2	3	4	5	6	7	8	9	10	11	12	13	14	15

p	q	r	s	t	u	v	w	x	y	z
16	17	18	19	20	21	22	23	24	25	26

Each letter of our message corresponds to a number according to this pattern, and then the numbers of our message are separated into vectors that are the same length as the dimension of the matrix we will use for the coding.

To see how this will be used, let V be any vector. (In cryptography examples, V might be two letters of the message written in numerical form using the above pattern.) Let A be any matrix that has an inverse and can be multiplied by V. Then the coded message will be AV, and the recipient can find the original message by multipliying by A^{-1} because

$$A^{-1}(AV) = (A^{-1}A)V = IV = V$$

1.6.1 Example

Encode the message WE ALL ERR by using the matrix $\begin{bmatrix} 5 & 7 \\ 2 & 3 \end{bmatrix}$.

Solution:

We first separate the letters into pairs (adding an extra Z at the end if needed to make an even number of letters), and then we convert the letters to numbers using the chart above:

W E	A L	L E	R R
23 5	1 12	12 5	18 18

We form each pair into a vector, and multiply it by the code matrix:

$$\begin{bmatrix} 5 & 7 \\ 2 & 3 \end{bmatrix}\begin{bmatrix} 23 \\ 5 \end{bmatrix} = \begin{bmatrix} 150 \\ 61 \end{bmatrix} \qquad \begin{bmatrix} 5 & 7 \\ 2 & 3 \end{bmatrix}\begin{bmatrix} 1 \\ 12 \end{bmatrix} = \begin{bmatrix} 89 \\ 38 \end{bmatrix}$$

$$\begin{bmatrix} 5 & 7 \\ 2 & 3 \end{bmatrix}\begin{bmatrix} 12 \\ 5 \end{bmatrix} = \begin{bmatrix} 95 \\ 39 \end{bmatrix} \qquad \begin{bmatrix} 5 & 7 \\ 2 & 3 \end{bmatrix}\begin{bmatrix} 18 \\ 18 \end{bmatrix} = \begin{bmatrix} 216 \\ 90 \end{bmatrix}$$

1.6.2 Example

Show how to break the coded message in Example 1.6.1.

Solution:

It is essential that we know the inverse of the code matrix. It is $\begin{bmatrix} 3 & -7 \\ -2 & 5 \end{bmatrix}$, as you can easily check. You might be able to compute it by setting up simultaneous equations, as shown in the previous section, but in this section you will always be given the inverse matrices. (Computing them will be postponed until Chapter 2.) To find the original numbers, we merely multiply the inverse of the original matrix by the pairs in the coded message:

$$\begin{bmatrix} 3 & -7 \\ -2 & 5 \end{bmatrix}\begin{bmatrix} 150 \\ 61 \end{bmatrix} = \begin{bmatrix} 23 \\ 5 \end{bmatrix} \qquad \begin{bmatrix} 3 & -7 \\ -2 & 5 \end{bmatrix}\begin{bmatrix} 89 \\ 38 \end{bmatrix} = \begin{bmatrix} 1 \\ 12 \end{bmatrix}$$

$$\begin{bmatrix} 3 & -7 \\ -2 & 5 \end{bmatrix}\begin{bmatrix} 95 \\ 39 \end{bmatrix} = \begin{bmatrix} 12 \\ 5 \end{bmatrix} \qquad \begin{bmatrix} 3 & -7 \\ -2 & 5 \end{bmatrix}\begin{bmatrix} 216 \\ 90 \end{bmatrix} = \begin{bmatrix} 18 \\ 18 \end{bmatrix}$$

SUMMARY

It is possible to make up difficult codes easily by multiplying pairs (or triples, quadruples, etc.) of numbers by a known matrix. The code can then be broken easily by someone who knows the inverse of the matrix, but only with great difficulty otherwise. This "undoing" of what has been done is fundamental to the application of inverses in many other fields, too.

EXERCISES 1.6. A

In each of the following problems, you are given the coded message and the matrix inverse needed to decode it. Decode!

1. $\begin{bmatrix} 3 & -1 \\ -5 & 2 \end{bmatrix}$ 41, 103, 27, 70, 7, 18, 41, 105

2. $\begin{bmatrix} 0 & 2 & 1 \\ 1 & 0 & 0 \\ 0 & 5 & 3 \end{bmatrix}$ 15, 14, −16, 3, 0, 5, 20, 38, −57

3. $\begin{bmatrix} 1 & 2 & 0 & 0 \\ 0 & 1 & 0 & 3 \\ 0 & 0 & 1 & 0 \\ 0 & 0 & 0 & 1 \end{bmatrix}$ 31, −11, 12, 5, 35, −14, 13, 5

EXERCISES 1.6. B

Decipher each of the following messages using the given matrix inverse.

1. $\begin{bmatrix} 4 & 7 \\ 2 & 4 \end{bmatrix}$ $-13.5, 8, -31, 20, -46, 27, -61, 35, -4, 3$

2. $\begin{bmatrix} -3 & 2 & 9 \\ 0 & 1 & 4 \\ 1 & 0 & 0 \end{bmatrix}$ $13, -155, 41, 12, -110, 29, 19, -219, 56$

3. $\begin{bmatrix} 1 & 0 & 0 & 0 \\ 2 & 1 & 0 & 0 \\ 3 & 1 & 1 & 0 \\ 6 & 3 & 0 & 1 \end{bmatrix}$ $16, -27, -20, -12, 5, 4, -4, -19$

ANSWERS 1.6. A

1. Take care 2. Love costs 3. Idle game

VOCABULARY

Learning the vocabulary of any subject is an important aspect of becoming familiar with it. To help you review, we list here some new words introduced in the chapter in the order in which they were defined: matrix, linear programming, row, column, 2 by 3 matrix, dimension, input–output matrix, matrix addition, scalar multiplication, assembly graph, quantity matrix, technological order, major diagonal, diagonal matrix, lower triangular matrix, symmetric matrix, upper triangular matrix, commutative, vector, inner product, model, primary inputs, final demands, Leontief matrices, technology matrix, technological equation, coefficient matrix, solution vector, constant vector, right-hand vector, square matrix, associative, additive identity, additive inverse, distributive, multiplicative identity, multiplicative inverse, negative, inverse.

SAMPLE TEST

Chapter 1

1. (12 pts) Let $A = \begin{bmatrix} 1 & 3 \\ 2 & 4 \end{bmatrix}$ and $B = \begin{bmatrix} -1 & 5 \\ 3 & 7 \end{bmatrix}$. Find:

   (a) a_{12} (b) $A + B$ (c) AB (d) $3A$
2. (6 pts) Write the additive identity and the multiplicative identity for 3 by 3 matrices.
3. (12 pts) Give an example of:

   (a) a symmetric matrix (b) a column vector (c) a lower triangular matrix (d) a 3 by 2 matrix (Do not give a matrix with all zero elements.)

4. (8 pts) Find the additive inverse and multiplicative inverse of $\begin{bmatrix} \frac{1}{4} & 0 \\ 0 & 3 \end{bmatrix}$ and $\begin{bmatrix} 0 & 1 \\ 1 & 0 \end{bmatrix}$.
5. (6 pts) Write the following systems of equations in matrix form.
   (a) $x_1 + 3x_2 = 7$, $4x_1 - 6x_2 = 8$ (b) $x_1 - x_3 = 7$, $2x_2 + 3x_3 = 4$
6. (6 pts) If $a_{ij} = 0$ for $i > j$, the matrix is a ________ ________ matrix. If $a_{ij} = a_{ji}$ for all i and j, the matrix is a ________ matrix.
7. (10 pts)
   (a) Give an example of a nonzero matrix that does not have a multiplicative inverse.
   (b) Give an example of two nonzero matrices whose product is the additive identity.
8. (10 pts) For the following table of a hypothetical economy, write the corresponding technology matrix and technological equation.

	Agriculture	Manufacturing	Final demands	Gross outputs
Agriculture	3	6	1	10
Manufacturing	5	3	4	12
Primary inputs	2	3		

9. (10 pts) To make a shadow box (S) we use 8 bolts (B) to fasten together 4 rectangles (R) of wood and 1 floor piece (F). The floor piece is itself made of a rectangle (R), 4 legs (L), each of which is fastened on by a bolt

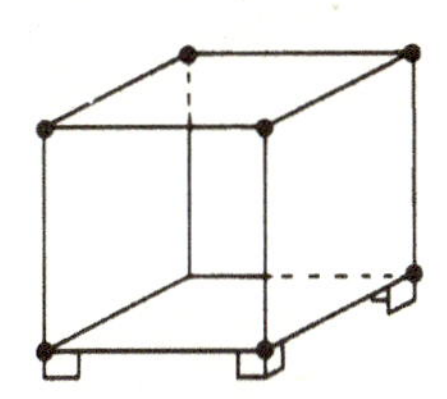

(B) to the floor piece. (a) Make an assembly graph and a quantity matrix that describe the construction of this shadow box. (b) Use the quantity matrix to calculate what parts are necessary to make 3 floor pieces and 5 shadow boxes. Show your work!
10. (10 pts) Suppose that apples sell for 7 cents each, oranges for 10 cents each, and pears for 15 cents each. Write a row vector describing this situation.

    If there is an order for 3 oranges, 2 pears, and 8 apples, write a *consistent* column vector for this order. Use an inner product to compute the total cost of this order. Show your work.

11. (10 pts) Decode 9, −8, 12, −3, 9, 9, −13, −104, using

$$\begin{bmatrix} 1 & 0 & 0 & 0 \\ 1 & 1 & 0 & 0 \\ 1 & 1 & 1 & 0 \\ 7 & 5 & 0 & 1 \end{bmatrix}$$

The answers are at the back of the book.

Chapter 2

GAUSS–JORDAN ROW OPERATIONS

2.1 Linear Equations with a Unique Solution

We interrupt the study of matrix inverses briefly to examine systems of linear equations. A linear equation is an equation of the form $a_1x_1 + a_2x_2 + \cdots + a_nx_n = b$, where not all $a_i = 0$; each term consists of either a number or a number times a variable. There are no products of variables and no powers (other than one or zero) or roots. The word "linear" refers to the fact that the graph of a linear equation in two unknowns is a straight line; you probably remember graphing them in elementary algebra. We postpone the discussion of graphs in higher dimensions until Section 3.4. Examples of linear equations include $3x - 2y = 7$ and $5x + 4y - z = 17$. Examples of equations that are not linear would be $xy = 4$, $x^2 + x = 4$, and $\sqrt{x} + y = 14$.

Systems of equations are closely related to matrices, and to find the inverse of a matrix we shall use techniques that can be most quickly learned by solving systems of simultaneous equations. In the following examples we assume that the reader has had some (perhaps long ago) experience with solving two equations in two unknowns, but we do not presuppose any prior knowledge of three equations in three unknowns.

In the first example the techniques you have used before to solve two equations in two unknowns are used to solve a system of three equations in three unknowns. Then we show how such systems can be solved faster by using matrices.

2.1.1 Example

Suppose that a store sells nuts in three types of containers: regular cans, holiday cans, and party cans. Each regular can contains 1 lb of cashews and 1 lb of peanuts; each holiday can contains 6 lb of cashews, 6 lb of peanuts, and 3 lb of walnuts; and each party can contains 1 lb of cashews and 2 lb of peanuts. If an order is received for 31 lb of cashews, 36 lb of peanuts, and 12 lb of walnuts, how can it be filled?

Solution:

To put such an applied problem into mathematical language, the first step is to ask ourselves, "What are the unknowns?" or "What, precisely, is the *quantity*(s) we seek?" The question "How can the order be filled?" means, in more mathematical terms, "What quantity of each container is needed to provide precisely 31 lb of cashews, 36 lb of peanuts, and 12 lb of walnuts?" It is the quantity of each container that is sought; we give these quantities symbolic names so we can refer to them and manipulate them easily. We shall let

x = number of regular cans to be used

y = number of holiday cans to be used

z = number of party cans to be used

Then we use these symbols to make equations, one for each part of the order. The "cashew equation" is made by realizing that the total number of pounds of cashews will be equal to the number of regular cans, plus 6 times the number of holiday cans (because there are 6 lb of cashews in each), plus the number of party cans:

$$x + 6y + z = 31$$

The "peanut equation" is obtained by adding the number of regular cans, 6 times the number of holiday cans, and 2 times the number of party cans:

$$x + 6y + 2z = 36$$

Since walnuts come only in the holiday can, and there are 3 lb of walnuts in each, the "walnut equation" is

$$3y = 12$$

Summarizing these, we get

$$\begin{aligned} x + 6y + z &= 31 \quad \text{(cashews)} \\ x + 6y + 2z &= 36 \quad \text{(peanuts)} \\ 3y &= 12 \quad \text{(walnuts)} \end{aligned} \tag{1}$$

The goal is to modify this system of equations until we get an equivalent system (that is, one with the same solution) of the form

$$\begin{aligned} x &= a \\ y &= b \\ z &= c \end{aligned}$$

where a, b, and c are easily visible numbers.

If we subtract the first equation in (1) from the second, the first will be the only equation with an x in it:

$$\begin{aligned} x + 6y + z &= 31 \\ z &= 5 \\ 3y &= 12 \end{aligned} \tag{2}$$

To place the y and z where they belong in the pattern given above, we interchange the second and third equations:

$$\begin{aligned} x + 6y + z &= 31 \\ 3y &= 12 \\ z &= 5 \end{aligned} \tag{3}$$

To solve for y, we now divide the second equation by 3:

$$\begin{aligned} x + 6y + z &= 31 \\ y &= 4 \\ z &= 5 \end{aligned} \tag{4}$$

To eliminate y from the first equation, we subtract 6 times the second equation from the first:

$$\begin{aligned} x + z &= 31 - 24 = 7 \\ y &= 4 \\ z &= 5 \end{aligned} \tag{5}$$

And, finally, to eliminate z from the first equation, we subtract the third equation from the first:

$$\begin{aligned} x &= 2 \\ y &= 4 \\ z &= 5 \end{aligned} \tag{6}$$

Thus the answer to the problem is to take 2 regular cans, 4 holiday cans, and 5 party cans. Substituting these quantities into the original equations, we get $2 + 6 \cdot 4 + 5 = 31$, $2 + 6 \cdot 4 + 2 \cdot 5 = 36$, and $3 \cdot 4 = 12$. Such problems are easier to check than to solve.

Now we translate the previous computations into matrix operations. The operations are called Gauss–Jordan row operations. System (1) above can be written in the matrix form

$$\begin{bmatrix} 1 & 6 & 1 \\ 1 & 6 & 2 \\ 0 & 3 & 0 \end{bmatrix} \begin{bmatrix} x \\ y \\ z \end{bmatrix} = \begin{bmatrix} 31 \\ 36 \\ 12 \end{bmatrix}$$

To simplify the writing, it is customary to omit the variables (they are understood) and to attach the right-hand vector of constants onto the right of the coefficient matrix, obtaining what is called the <u>augmented matrix</u> of system (1):

$$\begin{bmatrix} 1 & 6 & 1 & 31 \\ 1 & 6 & 2 & 36 \\ 0 & 3 & 0 & 12 \end{bmatrix}$$

Then we perform steps called Gauss–Jordan row operations on the augmented matrix which mimic the steps we used to solve the equations before. The three types of Gauss–Jordan row operations are:

1. Any row can be multiplied or divided by any number different from 0.
2. Any multiple of one row can be added to (or subtracted from) another row.
3. Any two rows can be interchanged.

In the following series of matrices, the numbers to the left of each matrix are there purely for your convenience in comparing the Gauss–Jordan operations with the algebra above; such numbers will not be included in future problems. We also include natural symbols which describe what is happening at each step. R_1 designates the first row, R_2 the second row, and R_3 the third.

(1) $$\begin{bmatrix} 1 & 6 & 1 & 31 \\ 1 & 6 & 2 & 36 \\ 0 & 3 & 0 & 12 \end{bmatrix}$$ $R_2 - R_1 \longrightarrow R_2$ replace R_2 by $R_2 - R_1$

(2) $$\begin{bmatrix} 1 & 6 & 1 & 31 \\ 0 & 0 & 1 & 5 \\ 0 & 3 & 0 & 12 \end{bmatrix}$$ $R_2 \longleftrightarrow R_3$ interchange R_2 and R_3

(3) $$\begin{bmatrix} 1 & 6 & 1 & 31 \\ 0 & 3 & 0 & 12 \\ 0 & 0 & 1 & 5 \end{bmatrix}$$ $\frac{1}{3}R_2 \longrightarrow R_2$ multiply R_2 by $\frac{1}{3}$

(4) $$\begin{bmatrix} 1 & 6 & 1 & 31 \\ 0 & 1 & 0 & 4 \\ 0 & 0 & 1 & 5 \end{bmatrix}$$ $R_1 - 6R_2 \longrightarrow R_1$ replace R_1 with $R_1 - 6R_2$

(5) $$\begin{bmatrix} 1 & 0 & 1 & 7 \\ 0 & 1 & 0 & 4 \\ 0 & 0 & 1 & 5 \end{bmatrix}$$ $R_1 - R_3 \longrightarrow R_1$ replace R_1 with $R_1 - R_3$

(6) $$\begin{bmatrix} 1 & 0 & 0 & 2 \\ 0 & 1 & 0 & 4 \\ 0 & 0 & 1 & 5 \end{bmatrix}$$

We can now put this back into the original matrix form and multiply out to obtain the answers to the original equation. With a little practice you can learn to do these final steps quickly and accurately in your head.

$$\begin{bmatrix} 1 & 0 & 0 \\ 0 & 1 & 0 \\ 0 & 0 & 1 \end{bmatrix}\begin{bmatrix} x \\ y \\ z \end{bmatrix} = \begin{bmatrix} 2 \\ 4 \\ 5 \end{bmatrix}$$

$$x = 2$$
$$y = 4$$
$$z = 5$$

In this example you can check that the third step was accomplished by the first type of Gauss–Jordan row operation listed on page 52; the first, fourth, and fifth steps were done by using the second type of Gauss–Jordan row operation; and the second step used the third type. Notice that we operate only on *rows*, never columns.

The next three examples (two in this section and one in the next) demonstrate further the technique of solving a system of simultaneous linear equations by applying Gauss–Jordan row operations to the augmented matrix of the system.

2.1.2 Example

Suppose that you have access to the same cans of nuts as in Example 2.1.1 and you receive an order for 34 lb of cashews, 37 lb of peanuts, and 15 lb of walnuts. How can you fill this order?

Solution:

Again we specify the variables by

$$x = \text{number of regular cans}$$
$$y = \text{number of holiday cans}$$
$$z = \text{number of party cans}$$

When we symbolize the need to fill the three parts of the order, we get three equations:

$$\begin{aligned} x + 6y + z &= 34 \\ x + 6y + 2z &= 37 \\ 3y &= 15 \end{aligned}$$

Then we write the augmented equation and proceed with Gauss–Jordan row operations as before:

$$\begin{bmatrix} 1 & 6 & 1 & 34 \\ 1 & 6 & 2 & 37 \\ 0 & 3 & 0 & 15 \end{bmatrix} \quad R_2 - R_1 \longrightarrow R_2$$

$$\begin{bmatrix} 1 & 6 & 1 & 34 \\ 0 & 0 & 1 & 3 \\ 0 & 3 & 0 & 15 \end{bmatrix} \quad R_2 \longleftrightarrow R_3$$

$$\begin{bmatrix} 1 & 6 & 1 & 34 \\ 0 & 3 & 0 & 15 \\ 0 & 0 & 1 & 3 \end{bmatrix} \quad \tfrac{1}{3}R_2 \longrightarrow R_2$$

$$\begin{bmatrix} 1 & 6 & 1 & 34 \\ 0 & 1 & 0 & 5 \\ 0 & 0 & 1 & 3 \end{bmatrix} \quad R_1 - 6R_2 \longrightarrow R_1$$

$$\begin{bmatrix} 1 & 0 & 1 & 4 \\ 0 & 1 & 0 & 5 \\ 0 & 0 & 1 & 3 \end{bmatrix} \qquad R_1 - R_3 \longrightarrow R_1$$

$$\begin{bmatrix} 1 & 0 & 0 & 1 \\ 0 & 1 & 0 & 5 \\ 0 & 0 & 1 & 3 \end{bmatrix}$$

So this time the order is filled by 1 regular can, 5 holiday cans, and 3 party cans. It is easy to substitute back in the original equations and check that $1 + 6 \cdot 5 + 3 = 34$, $1 + 6 \cdot 5 + 2 \cdot 3 = 37$, and $3 \cdot 5 = 15$.

This example may have seemed familiar to you, because it merely repeated the same steps as Example 2.1.1. This is because the variables were related in the same way—only the quantities ordered were different. You could probably learn to do such a problem for one fixed store very quickly; a computer could, too—even more quickly! The same techniques can also be applied to different, and harder, problems.

The solution to each such problem can be divided into two parts: (1) writing the problem in mathematical symbols, and (2) manipulating the symbols to get the answer. Both parts are important.

For the first part (translating English into mathematical language), the following steps are suggested:

1. Ask yourself, "What quantities am I looking for?"
2. Write "x = ________," "y = ________," and so on in English for each quantity that you want to find.
3. Write a set of equations, each one of which describes something that needs to be true. Often it helps to label each equation with one word telling how you obtained that equation (for example, "peanuts").

There are many correct ways to use the Gauss–Jordan row operations to change the augmented matrix to reduced form (as in the last matrix above), but the following provides one possible procedure.

1. Try to manipulate the matrix until there is a 1 in the upper left.
2. Add appropriate multiples of the first row to each other row so that zeros appear in the rest of the first column.
3. Without disturbing the first row, manipulate the matrix so that $a_{22} = 1$: use this to make the rest of the second column all zeros.
4. Without disturbing the rows above, proceed down the matrix from the upper left to lower right, making each $a_{ii} = 1$ in turn and use each 1 to clear that column of other nonzero numbers.
5. When the left part of the matrix looks like the multiplicative identity, merely read off the answers.

(*Important Hint*: You can save yourself and others grief if you *make it a habit to check every problem*. This is especially prudent when checking the answer to a problem is so much easier than solving it in the first place.)

In Chapter 3 we shall examine sets of linear equations that present surprises not covered by these rules, but in this chapter the examples and exercises have been chosen so that the outline above will suffice.

The following type of example can be generalized to larger, more realistic ecosystems comprised of dozens or hundreds of species.

2.1.3 Example in Biology

Three species of bacteria are to be kept in one test tube, where they will feed on three resources. Each day the test tube will be supplied with 14,000 units of the first resource, 6000 units of the second, and 8000 units of the third. Suppose that each of the first species of bacteria consumes 2 units a day of the first resource and 1 each of the second and third; each of the second species consumes 2 units each of the first and second resource and 3 of the third; and each member of the third species consumes 4 units of the first resource and 1 unit of the third. What are the populations of each species that can coexist in equilibrium in this test tube and consume all of the supplied resources?

Solution:

First we label our variables carefully:

x = number of the first species of bacteria

y = number of the second species of bacteria

z = number of the third species of bacteria

We use the three resources as the basis for three equations:

$$\begin{aligned} 2x + 2y + 4z &= 14{,}000 \quad &\text{(first resource)} \\ x + 2y &= 6{,}000 \quad &\text{(second resource)} \\ x + 3y + z &= 8{,}000 \quad &\text{(third resource)} \end{aligned}$$

Then we convert this set of equations into an augmented matrix and begin using Gauss–Jordan row operations:

$$\begin{bmatrix} 2 & 2 & 4 & 14{,}000 \\ 1 & 2 & 0 & 6{,}000 \\ 1 & 3 & 1 & 8{,}000 \end{bmatrix} \qquad R_1 - R_2 \longrightarrow R_1 \qquad \text{step 1 on page 54}$$

$$\begin{bmatrix} 1 & 0 & 4 & 8{,}000 \\ 1 & 2 & 0 & 6{,}000 \\ 1 & 3 & 1 & 8{,}000 \end{bmatrix} \qquad R_2 - R_1 \longrightarrow R_2 \qquad \text{step 2}$$

$$\begin{bmatrix} 1 & 0 & 4 & 8{,}000 \\ 0 & 2 & -4 & -2{,}000 \\ 1 & 3 & 1 & 8{,}000 \end{bmatrix} \qquad R_3 - R_1 \longrightarrow R_3 \qquad \text{step 2 (cont.)}$$

$$\begin{bmatrix} 1 & 0 & 4 & 8{,}000 \\ 0 & 2 & -4 & -2{,}000 \\ 0 & 3 & -3 & 0 \end{bmatrix} \quad \tfrac{1}{2}R_2 \longrightarrow R_2 \qquad \text{step 3}$$

$$\begin{bmatrix} 1 & 0 & 4 & 8{,}000 \\ 0 & 1 & -2 & -1{,}000 \\ 0 & 3 & -3 & 0 \end{bmatrix} \quad R_3 - 3R_2 \longrightarrow R_3 \qquad \text{step 3 (cont.)}$$

$$\begin{bmatrix} 1 & 0 & 4 & 8{,}000 \\ 0 & 1 & -2 & -1{,}000 \\ 0 & 0 & 3 & 3{,}000 \end{bmatrix} \quad \tfrac{1}{3}R_3 \longrightarrow R_3 \qquad \text{step 4}$$

$$\begin{bmatrix} 1 & 0 & 4 & 8{,}000 \\ 0 & 1 & -2 & -1{,}000 \\ 0 & 0 & 1 & 1{,}000 \end{bmatrix} \quad R_1 - 4R_3 \longrightarrow R_1 \qquad \text{step 4 (cont.)}$$

$$\begin{bmatrix} 1 & 0 & 0 & 4{,}000 \\ 0 & 1 & -2 & -1{,}000 \\ 0 & 0 & 1 & 1{,}000 \end{bmatrix} \quad R_2 + 2R_3 \longrightarrow R_2 \qquad \text{step 4 (cont.)}$$

$$\begin{bmatrix} 1 & 0 & 0 & 4{,}000 \\ 0 & 1 & 0 & 1{,}000 \\ 0 & 0 & 1 & 1{,}000 \end{bmatrix} \qquad \text{step 5}$$

Thus there must be 4000 of the first species of bacteria and 1000 of each of the other two species in that test tube if the given conditions are to all be true. No other combination will work.

SUMMARY

This section discussed both (1) techniques for setting up linear equations given a certain type of word problem, and (2) a method for solving a system of linear equations by using an augmented matrix and Gauss–Jordan row operations. These are two quite different skills, but both are essential for solving real-world problems of this type. The exercises of this section, in deference to human frailty, are fairly elementary; the next section gives slightly harder problems of the same type. These exercises may be challenging to you, but they are worth considerable effort, for in just trying them, you will learn something, and when you get the knack of analyzing and solving them, you will have absorbed some skills that may be applicable to more than just math problems.

Before the answers are given to the following exercise sets, there are "Hints" for each set, which consist of the equations that arise from each word problem. You may want to check your mathematical statement of each problem with the "Hints" before proceeding to find the solution using Gauss–Jordan row operations. If you have trouble setting up the word problems in mathematical language, you may want to do the alternative set of problems only up to the "Hints." On the other hand, if you find solving the systems of equations more difficult than setting them up, you may want to do some extra problems starting with the "Hint."

EXERCISES 2.1. A

1. Suppose that the store described in Example 2.1.1 receives an order for 18 lb of cashews, 20 lb of peanuts, and 6 lb of walnuts. How can this order be filled? (If you have trouble with this, compare your work to that shown in Example 2.1.1.)
2. A recreation center wants to purchase albums to be used in the center. There is no requirement as to the artists. The only requirement is that they purchase 40 rock albums, 32 Western albums, and 14 blues albums. There are three different shipping packages offered by the record company. They are an assorted carton containing 2 rock albums, 4 Western albums, and 1 blues album; a mixed case containing 4 rock and 2 Western albums; and a single shipment which contains 2 blues albums. What combination of these packages is needed to fill the center's order?
3. Suppose that a store has three sizes of cans of nuts. The large size contains 2 lb of peanuts and 1 lb of cashews. The mammoth size contains 1 lb of walnuts, 6 lb of peanuts, and 2 lb of cashews. The giant size contains 1 lb of walnuts, 4 lb of peanuts, and 2 lb of cashews. Suppose that the store receives an order for 5 lb of walnuts, 26 lb of peanuts, and 12 lb of cashews. How can it fill this order with the given cans?
4. There are three popular shapes of wooden blocks for children—rectangles, squares, and triangles. Suppose that a nursery school wants to be equipped with 160 rectangles, 34 squares, and 64 triangles, and they come in three types of boxes. A starter set contains 2 triangles, 2 squares, and 8 rectangles. A family set contains 10 triangles, 5 squares, and 20 rectangles. And rectangles alone can be bought with a dozen per box. How many of each type of box should the nursery school order?

EXERCISES 2.1. B

1. Suppose that the store described in Example 2.1.1 receives an order for 22 lb of cashews, 23 lb of peanuts, and 9 lb of walnuts. How can this order be filled? (If you have trouble with this, compare your work to that shown in Example 2.1.1.)
2. Suppose that the wrench department of a tool manufacturer produces three sets of wrench assortments: the Backyard Mechanic assortment, consisting of 2 combination wrenches and 1 pipe wrench; the Grease Monkey assortment, consisting of 1 crescent wrench, 6 combination wrenches, and 2 pipe wrenches; and the Professional assortment, consisting of 1 crescent wrench, 4 combination wrenches, and 2 pipe wrenches. They receive an order from a service station for 5 crescent wrenches, 26 combination wrenches, and 12 pipe wrenches. How many of each of its ready-packaged assortments should it send?

3. Suppose that a store has three sizes of cans of nuts. The large size contains 2 lb of peanuts and 1 lb of cashews. The mammoth size contains 1 lb of walnuts, 6 lb of peanuts, and 2 lb of cashews. And the giant size contains 1 lb of walnuts, 4 lb of peanuts, and 2 lb of cashews. Suppose that the store receives an order for 6 lb of walnuts, 34 lb of peanuts, and 15 lb of cashews. How can this order be filled?
4. Suppose that a dressmaking factory knows it will need 60 size 12 needles, 240 size 14 needles, and 70 size 16 needles. If the sample packs contain 2 of each size; the regular packs contain 2 each of size 12 and 16 and 6 of size 14; and the Sewer's Special contains 20 of size 12, 100 of size 14, and 25 of size 16, how many of each type of pack should the factory order?

EXERCISES 2.1. C

1. Three species of bacteria will be kept in one test tube and will feed on three resources. Each member of the first species consumes 1 unit of the first resource and 2 units of the second resource each day. Each member of the second species consumes 3 units of the first resource, 4 units of the second, and 3 units of the third each day. And each member of the third species consumes 2 units of the first resource, 3 units of the second resource, and 1 unit of the third resource each day. If the test tube is supplied daily with 31,000 units of the first resource, 46,000 units of the second, and 20,000 units of the third resource, how many of each species may coexist in equilibrium in the test tube and consume all of the supplied resources?
2. The ideas of this chapter apply easily to problems with more than three variables. The following set of equations is a bit more complicated to solve than the previous problems, but no new skills are needed. Solve simultaneously:

$$\begin{aligned} x_1 + x_3 + x_4 &= 1 \\ 2x_1 + 3x_2 + 4x_3 + 2x_4 &= 3 \\ x_2 + 3x_3 + 4x_4 &= -2 \\ x_1 + 2x_2 + x_3 &= 3 \end{aligned}$$

HINTS 2.1. A

1. x = number of regular cans; y = number of holiday cans; z = number of party cans

$$\begin{aligned} x + 6y + z &= 18 \quad \text{(cashews)} \\ x + 6y + 2z &= 20 \quad \text{(peanuts)} \\ 3y &= 6 \quad \text{(walnuts)} \end{aligned}$$

2. x = number of assorted cartons; y = number of mixed cases; z = number of single shipments

$$\begin{aligned} 2x + 4y &= 40 \quad \text{(rock)} \\ 4x + 2y &= 32 \quad \text{(Western)} \\ x + 2z &= 14 \quad \text{(blues)} \end{aligned}$$

3. x = number of large cans; y = number of mammoth cans; z = number of giant cans

$$\begin{aligned} y + z &= 5 \quad \text{(walnuts)} \\ 2x + 6y + 4z &= 26 \quad \text{(peanuts)} \\ x + 2y + 2z &= 12 \quad \text{(cashews)} \end{aligned}$$

4. x = number of starter sets; y = number of family sets; z = number of rectangle boxes

$$\begin{aligned} 2x + 10y &= 64 \quad \text{(triangles)} \\ 2x + 5y &= 34 \quad \text{(squares)} \\ 8x + 20y + 12z &= 160 \quad \text{(rectangles)} \end{aligned}$$

HINTS 2.1. B

1. x = number of regular cans; y = number of holiday cans; z = number of party cans

$$\begin{aligned} x + 6y + z &= 22 \quad \text{(cashews)} \\ x + 6y + 2z &= 23 \quad \text{(peanuts)} \\ 3y &= 9 \quad \text{(walnuts)} \end{aligned}$$

2. x = number of Backyard Mechanic assortments; y = number of Grease Monkey assortments; z = number of Professional assortments

$$\begin{aligned} y + z &= 5 \quad \text{(crescent wrenches)} \\ 2x + 6y + 4z &= 26 \quad \text{(combination wrenches)} \\ x + 2y + 2z &= 12 \quad \text{(pipe wrenches)} \end{aligned}$$

3. x = number of large cans; y = number of mammoth cans; z = number of giant cans

$$\begin{aligned} y + z &= 6 \quad \text{(walnuts)} \\ 2x + 6y + 4z &= 34 \quad \text{(peanuts)} \\ x + 2y + 2z &= 15 \quad \text{(cashews)} \end{aligned}$$

4. x = number of sample packs; y = number of regular packs; z = number of Sewer's Specials

$$\begin{aligned} 2x + 2y + 20z &= 60 \quad \text{(size 12)} \\ 2x + 6y + 100z &= 240 \quad \text{(size 14)} \\ 2x + 2y + 25z &= 70 \quad \text{(size 16)} \end{aligned}$$

HINTS 2.1. C

1. x = number of first species; y = number of second species; z = number of third species

$$\begin{aligned} x + 3y + 2z &= 31{,}000 \quad \text{(1st resource)} \\ 2x + 4y + 3z &= 46{,}000 \quad \text{(2nd resource)} \\ 3y + z &= 20{,}000 \quad \text{(3rd resource)} \end{aligned}$$

ANSWERS 2.1. A

1. 4 regular cans, 2 holiday cans, 2 party cans ($x = 4$, $y = 2$, $z = 2$)
2. 4 assorted cartons, 8 mixed cases, 5 single shipments ($x = 4$, $y = 8$, $z = 5$)
3. 2 large cans, 1 mammoth can, 4 giant cans ($x = 2$, $y = 1$, $z = 4$)
4. 2 starter sets, 6 family sets, 2 boxes of rectangles ($x = 2$, $y = 6$, $z = 2$)

ANSWERS 2.1. C

1. 3000 of first species, 4000 of second species, and 8000 of third species ($x = 3000$, $y = 4000$, $z = 8000$)
2. $x_1 = 2$, $x_2 = 1$, $x_3 = -1$, $x_4 = 0$

2.2 Linear Equations with a Unique Solution (continued)

In Section 2.1 we introduced two related but distinct skills—transforming word problems into matrix notation, and using Gauss–Jordan row operations to solve such problems. You may have learned both skills all in one gulp, but many students need a day of review and digestion. This section is designed to give you time to digest; it presents one example of the same type as those in the previous section but computationally more difficult.

When symbolizing the second type of the Gauss–Jordan row operations, some errors can be eliminated if *the row on the right side of the arrow is also the row appearing first on the left* side of the arrow; this is the row being changed.

2.2.1 Example

Suppose that a shipping company has three types of ships: an oil tanker, which carries 3 units of oil, 1 unit of general cargo products, and 10 passengers; a multipurpose transport, which carries 1 unit of oil, 2 units of general cargo, and 100 passengers; and a luxury liner, which takes 1 unit of general cargo and 1500 passengers. If during some period of time the company wants to transport 11 units of oil, 13 units of general cargo, and 2020 passengers, how many of each type of ship should be scheduled?

Solution:

We define

x = number of oil tankers
y = number of multipurpose transports
z = number of luxury liners

Then we have

$$\begin{aligned} 3x + y &= 11 \quad \text{(oil)} \\ x + 2y + z &= 13 \quad \text{(general cargo)} \\ 10x + 100y + 1500z &= 2020 \quad \text{(passengers)} \end{aligned}$$

We now convert the linear equations to an augmented matrix and begin the Gauss–Jordan row operations.

$$\begin{bmatrix} 3 & 1 & 0 & 11 \\ 1 & 2 & 1 & 13 \\ 10 & 100 & 1500 & 2020 \end{bmatrix} \quad \begin{matrix} R_1 \longleftrightarrow R_2 \\ \frac{1}{10}R_3 \longrightarrow R_3 \end{matrix}$$

In general, it is not permissible to do two row operations at once because we may perform something that is not really a row operation and therefore get a wrong answer. But if the two operations do not affect each other in any way, it is "safe" to do the two at once.

$$\begin{bmatrix} 1 & 2 & 1 & 13 \\ 3 & 1 & 0 & 11 \\ 1 & 10 & 150 & 202 \end{bmatrix} \quad \begin{matrix} R_2 - 3R_1 \longrightarrow R_2 \\ R_3 - R_1 \longrightarrow R_3 \end{matrix}$$

$$\begin{bmatrix} 1 & 2 & 1 & 13 \\ 0 & -5 & -3 & -28 \\ 0 & 8 & 149 & 189 \end{bmatrix} \quad \begin{matrix} 8R_2 \longrightarrow R_2 \\ 5R_3 \longrightarrow R_3 \end{matrix}$$

This step does not follow the suggestion on page 54 that says we should next make $a_{22} = 1$. Because this is a textbook problem, we can avoid fractions (which most people dislike) by first making $a_{32} = 0$ and then returning to divide R_2 by a_{22}. Such observations can save much time when doing "hand computations" (as opposed to computer programming) using Gauss–Jordan row operations; we continue to use insight instead of rules as we complete this example.

$$\begin{bmatrix} 1 & 2 & 1 & 13 \\ 0 & -40 & -24 & -224 \\ 0 & 40 & 745 & 945 \end{bmatrix} \qquad R_3 + R_2 \longrightarrow R_3$$

$$\begin{bmatrix} 1 & 2 & 1 & 13 \\ 0 & -40 & -24 & -224 \\ 0 & 0 & 721 & 721 \end{bmatrix} \qquad \begin{matrix} -\frac{1}{8}R_2 \longrightarrow R_2 \\ \frac{1}{721}R_3 \longrightarrow R_3 \end{matrix}$$

$$\begin{bmatrix} 1 & 2 & 1 & 13 \\ 0 & 5 & 3 & 28 \\ 0 & 0 & 1 & 1 \end{bmatrix} \qquad \begin{matrix} R_1 - R_3 \longrightarrow R_1 \\ R_2 - 3R_3 \longrightarrow R_2 \end{matrix}$$

$$\begin{bmatrix} 1 & 2 & 0 & 12 \\ 0 & 5 & 0 & 25 \\ 0 & 0 & 1 & 1 \end{bmatrix} \qquad \tfrac{1}{5}R_2 \longrightarrow R_2$$

$$\begin{bmatrix} 1 & 2 & 0 & 12 \\ 0 & 1 & 0 & 5 \\ 0 & 0 & 1 & 1 \end{bmatrix} \qquad R_1 - 2R_2 \longrightarrow R_1$$

$$\begin{bmatrix} 1 & 0 & 0 & 2 \\ 0 & 1 & 0 & 5 \\ 0 & 0 & 1 & 1 \end{bmatrix}$$

Thus the company needs 2 oil tankers, 5 general transports, and 1 luxury liner.

SUMMARY

This section provides an opportunity to review the ideas presented in Section 2.1. The exercises also use the same skills as those in the previous section, but the numbers are more realistic and thus more difficult. If you need to continue practicing with easy numbers, you can do the alternative set of Section 2.1.

Notice that the techniques of this section can be used to solve large systems of equations—200 equations in 200 unknowns, for example. All you would need would be lots of patience or a computer. But there are advantages in beginning with smaller systems.

EXERCISES 2.2. A

1. A brokerage house offers three stock portfolios. Portfolio *A* consists of 1 block of common stock only. Portfolio *B* offers 1 secured city bond, 4 blocks of common stock, and 3 blocks of preferred stock. Portfolio *C*

contains 1 city bond, 3 blocks of common stock, and 2 blocks of preferred stock. If the broker receives a request from a client wishing to purchase 2 bonds, 7 blocks of common stock, and 4 blocks of preferred stock, how can he meet the client's request?

2. Sally's Girl Scout troop is selling cookies for the Christmas season. There are 3 different kinds of cookies in 3 different containers: bags which hold 1 dozen chocolate chip and 1 dozen oatmeal; gift boxes which hold 2 dozen chocolate chip, 1 dozen mints, and 1 dozen oatmeal; and cookie tins which hold 3 dozen mints and 2 dozen chocolate chip. Sally's mother is having a Christmas party and wants 6 dozen oatmeal cookies, 10 dozen mints, and 14 dozen chocolate chip cookies. How can Sally fill her mother's order?
3. The new art teacher finds that colored paper can be bought in three different packages. The first package has 20 sheets of white paper, 15 sheets of blue paper, and 1 sheet of red paper. The second package has 3 sheets of blue paper and 1 sheet of red paper. The last package has 40 sheets of white paper and 30 sheets of blue paper. Suppose that he needs 200 sheets of white paper, 180 sheets of blue paper, and 12 sheets of red paper. How many of each type of package should he order?
4. Three types of marble sets are available on the store shelf. Type 1 contains 3 red marbles and 1 green marble; type 2 contains 1 red and 2 blue marbles; and type 3 contains 4 blue and 2 green marbles. How many of each set would it take to total 13 red marbles, 28 blue, and 13 green marbles?
5. A soap manufacturer decides to spend $6 million on radio, magazine, and TV advertising. If it spends as much on TV advertising as on magazines and radio together, and the amount spent on magazines and TV combined equals five times that spent on radio, what is the amount to be spent on each type of advertising?

EXERCISES 2.2. B

1. A brokerage house offers three stock portfolios. Portfolio *A* consists of 5 blocks of secured city bonds. Portfolio *B* offers 2 blocks of common stock and 10 blocks of preferred stock. Portfolio *C* contains 4 blocks of common stock and 10 blocks of preferred stock. Suppose that a broker receives a request for 10 blocks of city bonds, 12 blocks of common stock, and 40 blocks of preferred stock. How can the broker fill the order?
2. The doctor has recommended that the patient would feel better by eating each week 1 lb 15 oz of apricots, 9 oz of dates, and 1 lb 5 oz of prunes. The local health food store sells dried-fruit mixtures containing 3 oz of dates, 5 oz of prunes, and 7 oz of apricots. They also sell special boxes containing 3 oz of prunes and fancy bags containing 2 oz of apricots. How many of each should the patient buy?
3. An art teacher has decided to do a project which requires 88 red sheets of paper, 51 green sheets, and 69 blue sheets. The jumbo packs contain 24 red sheets, 12 blue sheets, and 12 green sheets. The assorted packs have 15 blue sheets and 9 green sheets. The specialized packs have 10 red sheets. How many of each type of pack should be ordered?

4. An interior decorator has ordered 12 cans of sunset paint, 35 cans of brown, and 18 cans of fuchsia. The paint store has special pair packs, containing 1 can each of sunset and fuchsia; darkening packs, containing 2 cans of sunset, 5 cans of brown, and 2 cans of fuchsia; and economy packs, containing 3 cans of sunset, 15 cans of brown, and 6 cans of fuchsia. How many of each type of pack should the paint store send to the interior decorator?
5. A clothing manufacturer decides to spend $6 million on radio, magazine, and TV advertising. If it spends as much on TV advertising as on magazines and radio together, and the amount spent on magazines and TV combined equals three times that spent on radio, what is the amount to be spent on each type of advertising?

EXERCISES 2.2. C

1. Suppose that the shipping company described in Example 2.2.1 wants to transport 6 units of oil, 8 units of general cargo, and 1810 passengers. How many of each type of ship should be scheduled?
2. Suppose that the shipping company described in Example 2.2.1 wants to transport 13 units of oil, 12 units of general cargo, and 1930 passengers. How many of each type of ship should be scheduled?
3. A glass of skim milk contains about 0.07 mg of vitamin B_1, 0.1 mg of iron, and 8.5 g of protein. A slice of whole wheat bread contains 0.18 mg of vitamin B_1, 1.1 mg of iron, and 3 g of protein. One quarter pound of beef provides 0.14 mg of vitamin B_1, 3.4 mg of iron, and 22 g of protein. If you want to consume 1.14 mg of vitamin B_1, 11.4 mg of iron, and 73 g of protein from these foods, how much of each should you eat?
4. A glass of skim milk contains about 0.07 mg of vitamin B_1, 0.1 mg of iron, and 8.5 g of protein. A slice of whole wheat bread contains 0.18 mg of vitamin B_1, 1.1 mg of iron, and 3 g of protein. One quarter pound of beef provides 0.14 mg of vitamin B_1, 3.4 mg of iron, and 22 g of protein. If you want to consume 0.71 mg of vitamin B_1, 9.1 mg of iron, and 58.5 g of protein from these foods, how much of each should you eat?

HINTS 2.2. A

1. x = number of portfolio A
   y = number of portfolio B
   z = number of portfolio C

$$\begin{aligned} y + z &= 2 \quad \text{(bonds)} \\ x + 4y + 3z &= 7 \quad \text{(common stock)} \\ 3y + 2z &= 4 \quad \text{(preferred stock)} \end{aligned}$$

2. x = number of bags
   y = number of gift boxes
   z = number of tins

$$\begin{aligned} x + y &= 6 \quad \text{(oatmeal)} \\ y + 3z &= 10 \quad \text{(mints)} \\ x + 2y + 2z &= 14 \quad \text{(chocolate chip)} \end{aligned}$$

3. x = number of 1st package
   y = number of 2nd package
   z = number of 3rd package

$$\begin{aligned} 20x + 40z &= 200 \quad \text{(white)} \\ 15x + 3y + 30z &= 180 \quad \text{(blue)} \\ x + y &= 12 \quad \text{(red)} \end{aligned}$$

4. x = number of type 1
   y = number of type 2
   z = number of type 3

$$\begin{aligned} 3x + y &= 13 \quad \text{(red)} \\ 2y + 4z &= 28 \quad \text{(blue)} \\ x + 2z &= 13 \quad \text{(green)} \end{aligned}$$

5. x = amount spent on radio; y = amount spent on magazines; z = amount spent on TV

$$\begin{aligned} x + y + z &= 6{,}000{,}000 \\ z &= x + y \\ y + z &= 5x \end{aligned}$$

HINTS 2.2. B

1. x = number of portfolio A; y = number of portfolio B; z = number of portfolio C

$$\begin{aligned} 5x \qquad\qquad &= 10 \quad \text{(bonds)} \\ 2y + 4z &= 12 \quad \text{(common stock)} \\ 10y + 10z &= 40 \quad \text{(preferred stock)} \end{aligned}$$

2. x = number of mixtures; y = number of boxes; z = number of bags

$$\begin{aligned} 7x + 2z &= 31 \quad \text{(apricots)} \\ 3x &= 9 \quad \text{(dates)} \\ 5x + 3y &= 21 \quad \text{(prunes)} \end{aligned}$$

3. x = number of jumbo packs; y = number of assorted packs; z = number of specialized packs

$$\begin{aligned} 24x + 10z &= 88 \quad \text{(red)} \\ 12x + 9y &= 51 \quad \text{(green)} \\ 12x + 15y &= 69 \quad \text{(blue)} \end{aligned}$$

4. x = number of pair packs; y = number of darkening packs; z = number of economy packs

$$\begin{aligned} x + 2y + 3z &= 12 \quad \text{(sunset)} \\ 5y + 15z &= 35 \quad \text{(brown)} \\ x + 2y + 6z &= 18 \quad \text{(fuchsia)} \end{aligned}$$

5. x = amount spent on radio; y = amount spent on magazines; z = amount spent on TV

$$\begin{aligned} x + y + z &= 6{,}000{,}000 \\ z &= x + y \\ y + z &= 3x \end{aligned}$$

HINTS 2.2. C

1.
$$\begin{aligned} 3x + y \qquad\quad &= 6 \\ x + 2y + z &= 8 \\ 10x + 100y + 1500z &= 1810 \end{aligned}$$

2.
$$\begin{aligned} 3x + y \qquad\quad &= 13 \\ x + 2y + z &= 12 \\ 10x + 100y + 1500z &= 1930 \end{aligned}$$

3. x = number of glasses of milk; y = number of slices of bread; z = number of quarter pounds of beef

$$\begin{aligned} 0.07x + 0.18y + 0.14z &= 1.14 \quad (B_1) \\ 0.1x + 1.1y + 3.4z &= 11.4 \quad \text{(iron)} \\ 8.5x + 3y + 22z &= 73 \quad \text{(protein)} \end{aligned}$$

4. Same variables as in previous problem

$$\begin{aligned} 0.07x + 0.18y + 0.14z &= 0.71 \quad (B_1) \\ 0.1x + 1.1y + 3.4z &= 9.1 \quad \text{(iron)} \\ 8.5x + 3y + 22z &= 58.5 \quad \text{(protein)} \end{aligned}$$

ANSWERS 2.2. A

1. $x = 1,\ y = 0,\ z = 2$
2. $x = 2,\ y = 4,\ z = 2$
3. $x = 2,\ y = 10,\ z = 4$
4. $x = 3,\ y = 4,\ z = 5$
5. $x = 1{,}000{,}000,\ y = 2{,}000{,}000,\ z = 3{,}000{,}000$

ANSWERS 2.2. C

1. 1 oil tanker, 3 multipurpose transports, and 1 luxury liner
2. 3 oil tankers, 4 multipurpose transports, and 1 luxury liner
3. 2 glasses of milk, 4 slices of bread, 2 servings of meat

Solution:

$$\begin{bmatrix} 0.07 & 0.18 & 0.14 & 1.14 \\ 0.1 & 1.1 & 3.4 & 11.4 \\ 8.5 & 3 & 22 & 73 \end{bmatrix}$$

$$\begin{matrix} 100R_1 \longrightarrow R_1 \\ 10R_2 \longrightarrow R_2 \\ 2R_3 \longrightarrow R_3 \end{matrix} \begin{bmatrix} 7 & 18 & 14 & 114 \\ 1 & 11 & 34 & 114 \\ 17 & 6 & 44 & 146 \end{bmatrix}$$

$$R_1 \longleftrightarrow R_2 \begin{bmatrix} 1 & 11 & 34 & 114 \\ 7 & 18 & 14 & 114 \\ 17 & 6 & 44 & 146 \end{bmatrix}$$

$$R_3 - R_2 \longrightarrow R_3 \begin{bmatrix} 1 & 11 & 34 & 114 \\ 7 & 18 & 14 & 114 \\ 10 & -12 & 30 & 32 \end{bmatrix}$$

$$\tfrac{1}{2}R_3 \longrightarrow R_3 \begin{bmatrix} 1 & 11 & 34 & 114 \\ 7 & 18 & 14 & 114 \\ 5 & -6 & 15 & 16 \end{bmatrix}$$

$$R_2 - R_3 \longrightarrow R_2 \begin{bmatrix} 1 & 11 & 34 & 114 \\ 2 & 24 & -1 & 98 \\ 5 & -6 & 15 & 16 \end{bmatrix}$$

$$R_2 - 2R_1 \longrightarrow R_2 \begin{bmatrix} 1 & 11 & 34 & 114 \\ 0 & 2 & -69 & -130 \\ 5 & -6 & 15 & 16 \end{bmatrix}$$

$$R_3 - 5R_1 \longrightarrow R_3 \begin{bmatrix} 1 & 11 & 34 & 114 \\ 0 & 2 & -69 & -130 \\ 0 & -61 & -155 & -554 \end{bmatrix}$$

$$\tfrac{1}{2}R_2 \longrightarrow R_2 \begin{bmatrix} 1 & 11 & 34 & 114 \\ 0 & 1 & -34.5 & -65 \\ 0 & -61 & -155 & -554 \end{bmatrix}$$

$$R_3 + 61R_2 \longrightarrow R_3 \begin{bmatrix} 1 & 11 & 34 & 114 \\ 0 & 1 & -34.5 & -65 \\ 0 & 0 & -2259.5 & -4519 \end{bmatrix}$$

$$-\tfrac{1}{2259.5}R_3 \longrightarrow R_3 \begin{bmatrix} 1 & 11 & 34 & 114 \\ 0 & 1 & -34.5 & -65 \\ 0 & 0 & 1 & 2 \end{bmatrix}$$

$$R_2 + 34.5R_3 \longrightarrow R_2 \begin{bmatrix} 1 & 11 & 34 & 114 \\ 0 & 1 & 0 & 4 \\ 0 & 0 & 1 & 2 \end{bmatrix}$$

$$R_1 - 11R_2 \longrightarrow R_1 \quad \begin{bmatrix} 1 & 0 & 34 & 70 \\ 0 & 1 & 0 & 4 \\ 0 & 0 & 1 & 2 \end{bmatrix}$$

$$R_1 - 34R_3 \longrightarrow R_1 \quad \begin{bmatrix} 1 & 0 & 0 & 2 \\ 0 & 1 & 0 & 4 \\ 0 & 0 & 1 & 2 \end{bmatrix}$$

4. 1 glass of milk, 2 slices of bread, 2 servings of meat

2.3 Elementary Matrices

Gauss–Jordan row operations appear in many aspects of mathematics. This section will examine some special types of matrices (called *elementary matrices*) that are closely related to Gauss–Jordan operations and the inverses of such matrices. The next section will show how to use these ideas to calculate the inverse of any matrix that has an inverse.

We list again here, for your convenience, the Gauss–Jordan row operations:

1. Multiplying any row by any number (possibly a fraction) other than 0.
2. Adding (or subtracting) a multiple of one row to another.
3. Interchanging any two rows.

2.3.1 Definition

When we take the product AB, we say that A is premultiplying B and that B is postmultiplying A.

It happens that any of the Gauss–Jordan row operations can be performed on a matrix by premultiplying that matrix by a judiciously chosen "elementary matrix." We shall not present a formal proof that this is so, but we urge you to study the following examples in order to see the patterns. In the exercises of this section you will have an opportunity to try your own hand at finding the matrix that performs a given row operation.

2.3.2 Example

Find a 3 by 3 matrix that interchanges the first two rows of any other matrix when it premultiplies the other matrix.

Solution:

$$\begin{bmatrix} 0 & 1 & 0 \\ 1 & 0 & 0 \\ 0 & 0 & 1 \end{bmatrix} \begin{bmatrix} a & b & c \\ d & e & f \\ g & h & i \end{bmatrix} = \begin{bmatrix} d & e & f \\ a & b & c \\ g & h & i \end{bmatrix}$$

Notice how the 1's in the matrix pick out just the row we want and put it where we want it to go. Can you guess the matrix that interchanges the last two rows? (The answer appears among the answers to this section's exercises.) It may take you a bit longer to fiddle around until you get the matrix that interchanges the first and third rows, but you will gain insight during such "fiddling" that no book or teacher can give you.

2.3.3 Example

Find the matrix that adds the first row to the second and puts the result in the second row.

Solution:

$$\begin{bmatrix} 1 & 0 & 0 \\ 1 & 1 & 0 \\ 0 & 0 & 1 \end{bmatrix} \begin{bmatrix} a & b & c \\ d & e & f \\ g & h & i \end{bmatrix} = \begin{bmatrix} a & b & c \\ a + d & b + e & c + f \\ g & h & i \end{bmatrix}$$

2.3.4 Example

Find the matrix that adds twice the first row to the third and puts the result in the third row.

Solution:

$$\begin{bmatrix} 1 & 0 & 0 \\ 0 & 1 & 0 \\ 2 & 0 & 1 \end{bmatrix} \begin{bmatrix} a & b & c \\ d & e & f \\ g & h & i \end{bmatrix} = \begin{bmatrix} a & b & c \\ d & e & f \\ 2a + g & 2b + h & 2c + i \end{bmatrix}$$

2.3.5 Example

Find the matrix that adds three times the second row to the first.

Solution:

$$\begin{bmatrix} 1 & 3 & 0 \\ 0 & 1 & 0 \\ 0 & 0 & 1 \end{bmatrix} \begin{bmatrix} a & b & c \\ d & e & f \\ g & h & i \end{bmatrix} = \begin{bmatrix} a + 3d & b + 3e & c + 3f \\ d & e & f \\ g & h & i \end{bmatrix}$$

It is worth your time to pose yourself similar questions and see if you can answer them. The exercises of this section provide for such experimentation.

2.3.6 Example

Find the matrix that multiplies the first row by 2.

Solution:

$$\begin{bmatrix} 2 & 0 & 0 \\ 0 & 1 & 0 \\ 0 & 0 & 1 \end{bmatrix} \begin{bmatrix} a & b & c \\ d & e & f \\ g & h & i \end{bmatrix} = \begin{bmatrix} 2a & 2b & 2c \\ d & e & f \\ g & h & i \end{bmatrix}$$

2.3.7 Example

Find the matrix that divides the third row by 2.

Solution:

Since dividing by 2 is the same as multiplying by $\frac{1}{2}$, the following matrix does the trick:

$$\begin{bmatrix} 1 & 0 & 0 \\ 0 & 1 & 0 \\ 0 & 0 & \frac{1}{2} \end{bmatrix} \begin{bmatrix} a & b & c \\ d & e & f \\ g & h & i \end{bmatrix} = \begin{bmatrix} a & b & c \\ d & e & f \\ g/2 & h/2 & i/2 \end{bmatrix}$$

Matrices such as those shown in Examples 2.3.2 to 2.3.7 are so important to matrix theory that they are given a special name.

2.3.8 Definition

A square matrix E that performs a Gauss–Jordan row operation on any other matrix A when it premultiplies A (that is, when EA is computed) is called an elementary matrix.

One very useful fact about elementary matrices is that every elementary matrix has an inverse. This fact is useful for finding the inverses of other matrices, as we shall see in the next section.

To find the inverse of an elementary matrix, it is necessary to find that matrix that reverses whatever the given elementary matrix does. The inverse of any elementary matrix that interchanges two rows (as in Example 2.3.2) is itself. This is because "itself" will interchange the same two rows back! For example, using the matrix of Example 2.3.2, we see that

$$\begin{bmatrix} 0 & 1 & 0 \\ 1 & 0 & 0 \\ 0 & 0 & 1 \end{bmatrix} \begin{bmatrix} 0 & 1 & 0 \\ 1 & 0 & 0 \\ 0 & 0 & 1 \end{bmatrix} = \begin{bmatrix} 1 & 0 & 0 \\ 0 & 1 & 0 \\ 0 & 0 & 1 \end{bmatrix}$$

There are only three 3 by 3 elementary matrices of this type: the one that interchanges the first and second rows, the one that interchanges the first and third rows, and the one that interchanges the second and third rows. But there are many elementary 3 by 3 matrices of the first and second types listed on page 66.

It is easy to find the inverse of matrices of the first type (the type that multiplies one row of a matrix by a nonzero scalar) because they are all diagonal matrices. (Remember that Definition 1.2.3 stated that a diagonal matrix is one whose elements are all zero except on the major diagonal.) If a diagonal matrix has any zero on the major diagonal, it does not have an inverse. But if all the diagonal elements are nonzero, we can easily find the inverse. Convince yourself that

$$\begin{bmatrix} a & 0 & 0 \\ 0 & b & 0 \\ 0 & 0 & c \end{bmatrix} \begin{bmatrix} 1/a & 0 & 0 \\ 0 & 1/b & 0 \\ 0 & 0 & 1/c \end{bmatrix} = \begin{bmatrix} 1 & 0 & 0 \\ 0 & 1 & 0 \\ 0 & 0 & 1 \end{bmatrix}$$

2.3.9 Rule

The inverse of any diagonal matrix with no zeros on the major diagonal is the diagonal matrix that has reciprocals of the original numbers in the corresponding positions.

It is not immediately clear how to find the inverse of an elementary matrix of the second type (the type that adds a multiple of one row to another row), but with a bit of ingenuity it can always be done. Here we show the inverse of the matrices given in Examples 2.3.3, 2.3.4, and 2.3.5. Can you find the rule? (See Exercises 2.3.C.)

2.3.10 Example

$$\begin{bmatrix} 1 & 0 & 0 \\ 1 & 1 & 0 \\ 0 & 0 & 1 \end{bmatrix} \begin{bmatrix} 1 & 0 & 0 \\ -1 & 1 & 0 \\ 0 & 0 & 1 \end{bmatrix} = \begin{bmatrix} 1 & 0 & 0 \\ 0 & 1 & 0 \\ 0 & 0 & 1 \end{bmatrix}$$

$$\begin{bmatrix} 1 & 0 & 0 \\ 0 & 1 & 0 \\ 2 & 0 & 1 \end{bmatrix} \begin{bmatrix} 1 & 0 & 0 \\ 0 & 1 & 0 \\ -2 & 0 & 1 \end{bmatrix} = \begin{bmatrix} 1 & 0 & 0 \\ 0 & 1 & 0 \\ 0 & 0 & 1 \end{bmatrix}$$

$$\begin{bmatrix} 1 & 3 & 0 \\ 0 & 1 & 0 \\ 0 & 0 & 1 \end{bmatrix} \begin{bmatrix} 1 & -3 & 0 \\ 0 & 1 & 0 \\ 0 & 0 & 1 \end{bmatrix} = \begin{bmatrix} 1 & 0 & 0 \\ 0 & 1 & 0 \\ 0 & 0 & 1 \end{bmatrix}$$

In this section we have always premultiplied general matrices by elementary matrices. If we reverse the order, putting the elementary matrix on the right of the other matrix, the result is to perform a "column operation" on the other matrix. We shall have no need for column operations in this text, so we merely mention briefly that they exist. If you have a flair for tracing patterns, you can verify that most of the facts about row operations described in this book are symmetrically true for column operations.

SUMMARY

In this section elementary matrices were shown to perform Gauss–Jordan row operations when premultiplying another matrix and the inverses of elementary matrices were computed. Now it is your turn to premultiply by elementary matrices.

EXERCISES 2.3. A

Find the elementary 3 by 3 matrix that, when it premultiplies another 3 by 3 matrix, performs on it the operation described below.

1. Triples the third row.
2. Divides the third row by 3.

3. Multiplies the first row by π.
4. Interchanges the first and third rows.
5. Interchanges the second and third rows.
6. Adds the third row to the second.
7. Adds the third row to the first.
8. Adds the second row to the third.
9. Adds twice the second row to the third.
10. Adds three times the second row to the third.
11. Adds one half the second row to the third.
12. Adds four times the second row to the first.
13. Adds four times the first row to the third.

14–26. Write the inverse of each matrix you found in problems 1–13.

Consider the set of 4 by 4 matrices. Find the elementary 4 by 4 matrix that, when it premultiplies another 4 by 4 matrix, performs on it the operations described below.

27. Interchanges the middle two rows.
28. Interchanges the first and fourth rows.
29. Interchanges both the middle two rows and also the first and fourth rows.
30. Multiplies the second row by 5.
31. Adds the third row to the fourth.

32–36. Find the inverse (another 4 by 4 matrix) of each answer to problems 27–31.

EXERCISES 2.3. B

Find the elementary 3 by 3 matrix that, when it premultiplies another 3 by 3 matrix, performs on it the operation described below.

1. Doubles the second row.
2. Halves the second row.
3. Multiplies the third row by π.
4. Interchanges the first and third rows.
5. Interchanges the second and third rows.
6. Adds the first row to the third. (*Hint:* See both Examples 2.3.3 and 2.3.4.)
7. Adds the second row to the first.
8. Adds the third row to the first.
9. Adds twice the third row to the first.
10. Adds three times the third row to the first.
11. Adds one half the third row to the first.
12. Adds five times the second row to the third.
13. Adds five times the third row to the second.

14–26. Write the inverse of each matrix you found in problems 1–13.

Consider the set of 4 by 4 matrices. Find the elementary 4 by 4 matrix that, when it premultiplies another 4 by 4 matrix, performs on it the operations described below.

27. Interchanges the first two rows.
28. Interchanges the last two rows.

29. Interchanges both the first two rows and the last two rows.
30. Multiplies the first row by 4.
31. Adds the first row to the second.
32–36. Find the inverse (another 4 by 4 matrix) to each answer to problems 27–31.

EXERCISES 2.3. C

1. Tell what elementary matrix adds b times the ith row to the jth row in the matrix that it premultiplies.
2. What is the multiplicative inverse of the matrix that answers problem 1?

ANSWERS 2.3. A

1. $\begin{bmatrix} 1 & 0 & 0 \\ 0 & 1 & 0 \\ 0 & 0 & 3 \end{bmatrix}$ 2. $\begin{bmatrix} 1 & 0 & 0 \\ 0 & 1 & 0 \\ 0 & 0 & \frac{1}{3} \end{bmatrix}$ 3. $\begin{bmatrix} \pi & 0 & 0 \\ 0 & 1 & 0 \\ 0 & 0 & 1 \end{bmatrix}$

4. $\begin{bmatrix} 0 & 0 & 1 \\ 0 & 1 & 0 \\ 1 & 0 & 0 \end{bmatrix}$ 5. $\begin{bmatrix} 1 & 0 & 0 \\ 0 & 0 & 1 \\ 0 & 1 & 0 \end{bmatrix}$ 6. $\begin{bmatrix} 1 & 0 & 0 \\ 0 & 1 & 1 \\ 0 & 0 & 1 \end{bmatrix}$

7. $\begin{bmatrix} 1 & 0 & 1 \\ 0 & 1 & 0 \\ 0 & 0 & 1 \end{bmatrix}$ 8. $\begin{bmatrix} 1 & 0 & 0 \\ 0 & 1 & 0 \\ 0 & 1 & 1 \end{bmatrix}$ 9. $\begin{bmatrix} 1 & 0 & 0 \\ 0 & 1 & 0 \\ 0 & 2 & 1 \end{bmatrix}$

10. $\begin{bmatrix} 1 & 0 & 0 \\ 0 & 1 & 0 \\ 0 & 3 & 1 \end{bmatrix}$ 11. $\begin{bmatrix} 1 & 0 & 0 \\ 0 & 1 & 0 \\ 0 & \frac{1}{2} & 1 \end{bmatrix}$ 12. $\begin{bmatrix} 1 & 4 & 0 \\ 0 & 1 & 0 \\ 0 & 0 & 1 \end{bmatrix}$

13. $\begin{bmatrix} 1 & 0 & 0 \\ 0 & 1 & 0 \\ 4 & 0 & 1 \end{bmatrix}$ 14. $\begin{bmatrix} 1 & 0 & 0 \\ 0 & 1 & 0 \\ 0 & 0 & \frac{1}{3} \end{bmatrix}$ 15. $\begin{bmatrix} 1 & 0 & 0 \\ 0 & 1 & 0 \\ 0 & 0 & 3 \end{bmatrix}$

16. $\begin{bmatrix} 1/\pi & 0 & 0 \\ 0 & 1 & 0 \\ 0 & 0 & 1 \end{bmatrix}$ 17. $\begin{bmatrix} 0 & 0 & 1 \\ 0 & 1 & 0 \\ 1 & 0 & 0 \end{bmatrix}$ 18. $\begin{bmatrix} 1 & 0 & 0 \\ 0 & 0 & 1 \\ 0 & 1 & 0 \end{bmatrix}$

19. $\begin{bmatrix} 1 & 0 & 0 \\ 0 & 1 & -1 \\ 0 & 0 & 1 \end{bmatrix}$ 20. $\begin{bmatrix} 1 & 0 & -1 \\ 0 & 1 & 0 \\ 0 & 0 & 1 \end{bmatrix}$ 21. $\begin{bmatrix} 1 & 0 & 0 \\ 0 & 1 & 0 \\ 0 & -1 & 1 \end{bmatrix}$

22. $\begin{bmatrix} 1 & 0 & 0 \\ 0 & 1 & 0 \\ 0 & -2 & 1 \end{bmatrix}$ 23. $\begin{bmatrix} 1 & 0 & 0 \\ 0 & 1 & 0 \\ 0 & -3 & 1 \end{bmatrix}$ 24. $\begin{bmatrix} 1 & 0 & 0 \\ 0 & 1 & 0 \\ 0 & -\frac{1}{2} & 1 \end{bmatrix}$

25. $\begin{bmatrix} 1 & -4 & 0 \\ 0 & 1 & 0 \\ 0 & 0 & 1 \end{bmatrix}$ 26. $\begin{bmatrix} 1 & 0 & 0 \\ 0 & 1 & 0 \\ -4 & 0 & 1 \end{bmatrix}$ 27. $\begin{bmatrix} 1 & 0 & 0 & 0 \\ 0 & 0 & 1 & 0 \\ 0 & 1 & 0 & 0 \\ 0 & 0 & 0 & 1 \end{bmatrix}$

28. $\begin{bmatrix} 0 & 0 & 0 & 1 \\ 0 & 1 & 0 & 0 \\ 0 & 0 & 1 & 0 \\ 1 & 0 & 0 & 0 \end{bmatrix}$ 29. $\begin{bmatrix} 0 & 0 & 0 & 1 \\ 0 & 0 & 1 & 0 \\ 0 & 1 & 0 & 0 \\ 1 & 0 & 0 & 0 \end{bmatrix}$

30. $\begin{bmatrix} 1 & 0 & 0 & 0 \\ 0 & 5 & 0 & 0 \\ 0 & 0 & 1 & 0 \\ 0 & 0 & 0 & 1 \end{bmatrix}$ 31. $\begin{bmatrix} 1 & 0 & 0 & 0 \\ 0 & 1 & 0 & 0 \\ 0 & 0 & 1 & 0 \\ 0 & 0 & 1 & 1 \end{bmatrix}$

32. $\begin{bmatrix} 1 & 0 & 0 & 0 \\ 0 & 0 & 1 & 0 \\ 0 & 1 & 0 & 0 \\ 0 & 0 & 0 & 1 \end{bmatrix}$ 33. $\begin{bmatrix} 0 & 0 & 0 & 1 \\ 0 & 1 & 0 & 0 \\ 0 & 0 & 1 & 0 \\ 1 & 0 & 0 & 0 \end{bmatrix}$

34. $\begin{bmatrix} 0 & 0 & 0 & 1 \\ 0 & 0 & 1 & 0 \\ 0 & 1 & 0 & 0 \\ 1 & 0 & 0 & 0 \end{bmatrix}$ 35. $\begin{bmatrix} 1 & 0 & 0 & 0 \\ 0 & \frac{1}{5} & 0 & 0 \\ 0 & 0 & 1 & 0 \\ 0 & 0 & 0 & 1 \end{bmatrix}$

36. $\begin{bmatrix} 1 & 0 & 0 & 0 \\ 0 & 1 & 0 & 0 \\ 0 & 0 & 1 & 0 \\ 0 & 0 & -1 & 1 \end{bmatrix}$

ANSWERS 2.3. C

1. The identity matrix plus the matrix with $a_{ji} = b$ and zeros elsewhere.
2. The identity matrix plus the matrix with $a_{ji} = -b$ and zeros elsewhere.

2.4 Finding the Multiplicative Inverse of a Matrix

At last we are ready to compute the inverse of a matrix! We start with examples showing how it is done; later we shall see why the given method works.

2.4.1 Rule

To find the inverse of a square matrix, write the matrix to the left of the multiplicative identity of the same dimension. Perform appropriate row operations on the given matrix to obtain the identity; by doing the same row operations simultaneously on the identity, you will obtain the inverse of the original matrix if it has one.

2.4.2 Example

Find the inverse of the matrix $\begin{bmatrix} 1 & 0 & 3 \\ 0 & -2 & 1 \\ 5 & 1 & 15 \end{bmatrix}$.

Solution:

Following Rule 2.4.1, we first write

$$\left[\begin{array}{ccc|ccc} 1 & 0 & 3 & 1 & 0 & 0 \\ 0 & -2 & 1 & 0 & 1 & 0 \\ 5 & 1 & 15 & 0 & 0 & 1 \end{array}\right] \qquad R_3 - 5R_1 \longrightarrow R_3$$

We then proceed to change the matrix on the left to the identity as in the previous sections, but each time we apply a Gauss–Jordan operation to the left matrix, we apply the same operation to the right matrix, too.

$$\left[\begin{array}{rrr|rrr} 1 & 0 & 3 & 1 & 0 & 0 \\ 0 & -2 & 1 & 0 & 1 & 0 \\ 0 & 1 & 0 & -5 & 0 & 1 \end{array}\right] \qquad R_2 \longleftrightarrow R_3$$

$$\left[\begin{array}{rrr|rrr} 1 & 0 & 3 & 1 & 0 & 0 \\ 0 & 1 & 0 & -5 & 0 & 1 \\ 0 & -2 & 1 & 0 & 1 & 0 \end{array}\right] \qquad R_3 + 2R_2 \longrightarrow R_3$$

$$\left[\begin{array}{rrr|rrr} 1 & 0 & 3 & 1 & 0 & 0 \\ 0 & 1 & 0 & -5 & 0 & 1 \\ 0 & 0 & 1 & -10 & 1 & 2 \end{array}\right] \qquad R_1 - 3R_3 \longrightarrow R_1$$

$$\left[\begin{array}{rrr|rrr} 1 & 0 & 0 & 31 & -3 & -6 \\ 0 & 1 & 0 & -5 & 0 & 1 \\ 0 & 0 & 1 & -10 & 1 & 2 \end{array}\right]$$

Rule 2.4.1 tells us that the matrix on the right will be the inverse of the original matrix. If you doubt that such a strange method works, just multiply and see:

$$\begin{bmatrix} 1 & 0 & 3 \\ 0 & -2 & 1 \\ 5 & 1 & 15 \end{bmatrix} \begin{bmatrix} 31 & -3 & -6 \\ -5 & 0 & 1 \\ -10 & 1 & 2 \end{bmatrix} = \begin{bmatrix} 1 & 0 & 0 \\ 0 & 1 & 0 \\ 0 & 0 & 1 \end{bmatrix}$$

It checks! It is probably safe to say that you would not have guessed this answer. We shall do another example so you have a firmer grasp of the method, and then we shall see why it works.

Sometimes when computing a matrix inverse, fractions appear. People tend to be afraid of fractions, but sometimes they are unavoidable. They behave much like other numbers and you do know how to handle them.

2.4.3 Example

Find the multiplicative inverse of $\begin{bmatrix} 4 & 0 & 0 \\ 1 & 3 & 1 \\ 0 & 2 & 2 \end{bmatrix}$.

Solution:

$$\left[\begin{array}{rrr|rrr} 4 & 0 & 0 & 1 & 0 & 0 \\ 1 & 3 & 1 & 0 & 1 & 0 \\ 0 & 2 & 2 & 0 & 0 & 1 \end{array}\right] \qquad \tfrac{1}{4}R_1 \longrightarrow R_1$$

$$\left[\begin{array}{rrr|rrr} 1 & 0 & 0 & \frac{1}{4} & 0 & 0 \\ 1 & 3 & 1 & 0 & 1 & 0 \\ 0 & 2 & 2 & 0 & 0 & 1 \end{array}\right] \qquad R_2 - R_1 \longrightarrow R_2$$

$$\left[\begin{array}{ccc|ccc} 1 & 0 & 0 & \frac{1}{4} & 0 & 0 \\ 0 & 3 & 1 & -\frac{1}{4} & 1 & 0 \\ 0 & 2 & 2 & 0 & 0 & 1 \end{array}\right] \quad R_2 - R_3 \longrightarrow R_2$$

$$\left[\begin{array}{ccc|ccc} 1 & 0 & 0 & \frac{1}{4} & 0 & 0 \\ 0 & 1 & -1 & -\frac{1}{4} & 1 & -1 \\ 0 & 2 & 2 & 0 & 0 & 1 \end{array}\right] \quad R_3 - 2R_2 \longrightarrow R_3$$

$$\left[\begin{array}{ccc|ccc} 1 & 0 & 0 & \frac{1}{4} & 0 & 0 \\ 0 & 1 & -1 & -\frac{1}{4} & 1 & -1 \\ 0 & 0 & 4 & \frac{1}{2} & -2 & 3 \end{array}\right] \quad \tfrac{1}{4}R_3 \longrightarrow R_3$$

$$\left[\begin{array}{ccc|ccc} 1 & 0 & 0 & \frac{1}{4} & 0 & 0 \\ 0 & 1 & -1 & -\frac{1}{4} & 1 & -1 \\ 0 & 0 & 1 & \frac{1}{8} & -\frac{1}{2} & \frac{3}{4} \end{array}\right] \quad R_2 + R_3 \longrightarrow R_2$$

$$\left[\begin{array}{ccc|ccc} 1 & 0 & 0 & \frac{1}{4} & 0 & 0 \\ 0 & 1 & 0 & -\frac{1}{8} & \frac{1}{2} & -\frac{1}{4} \\ 0 & 0 & 1 & \frac{1}{8} & -\frac{1}{2} & \frac{3}{4} \end{array}\right] \quad \text{inverse of the original matrix}$$

(Check this by multiplying it by the original matrix.)

Let us now see why this method of finding inverses works. Suppose we have a matrix A that has an inverse. If we perform a row operation on A, the result is the same as premultiplying it by a properly chosen elementary matrix E_1 (as shown in Section 2.3). In either case, we obtain E_1A as the result. If we do another row operation, we are premultiplying by another elementary matrix E_2 and we obtain E_2E_1A. As you know from the previous two examples and Rule 2.4.1, we keep doing this until we obtain the identity. In mathematical language this means that

$$E_n \cdots E_3 E_2 E_1 A = I$$

Postmultiplying both sides by A^{-1},

$$E_n \cdots E_3 E_2 E_1 A A^{-1} = I A^{-1}$$

which is

$$E_n \cdots E_3 E_2 E_1 I = A^{-1}$$

Since the left side of this equation describes what we have obtained to the right of the vertical line in the problems above, we have computed A^{-1} there.

2.4.4 Rule

If the matrices A and B both have an inverse, their product AB has an inverse and it is $B^{-1}A^{-1}$.

It is easy to check that this is so:

$$\begin{aligned}(AB)(B^{-1}A^{-1}) &= A(BB^{-1})A^{-1} && \text{since multiplication is associative}\\ &= A(I)A^{-1} && \text{since } B^{-1} \text{ is the inverse of } B\\ &= AA^{-1} && \text{since } I \text{ is the identity}\\ &= I && \text{since } A^{-1} \text{ is the inverse of } A\end{aligned}$$

It follows that $(AB)(B^{-1}A^{-1}) = I$. You can check similarly that $(B^{-1}A^{-1})(AB) = I$.

2.4.5 Rule

Every matrix that has an inverse commutes with its inverse. That is, $AA^{-1} = A^{-1}A = I$.

Perhaps you can convince yourself that this is true for elementary matrices. If so, note that by the argument just preceding Rule 2.4.4 and the answer to problem 6 in Exercises 2.4.A, every matrix that has an inverse is the product of elementary matrices. Then Rule 2.4.4 can be used to prove Rule 2.4.5.

The proof of statements such as Rule 2.4.5 forms the theme of formal books on linear algebra. This text, designed to emphasize ideas and applications, does not do justice to the beauty of the rigorous patterns. If you feel cheated, you should study a course in modern algebra or abstract linear algebra.

The following example will show how to use the inverse of a matrix to solve several systems of linear equations, each of the form $AX = B$, where the constant vector B varies from system to system but the matrix A remains the same.

2.4.6 Example

Solve the following sets of equations.

(a)
$$\begin{aligned} x + \quad\quad 3z &= 4\\ -2y + \quad z &= 10\\ 5x + y + 15z &= 6\end{aligned}$$

(b)
$$\begin{aligned} x + \quad\quad 3z &= 7\\ -2y + \quad z &= -3\\ 5x + y + 15z &= 0\end{aligned}$$

(c)
$$\begin{aligned} x + \quad\quad 3z &= 1\\ -2y + \quad z &= 3\\ 5x + y + 15z &= 4\end{aligned}$$

(d)
$$\begin{aligned} x + \quad\quad 3z &= 2\\ -2y + \quad z &= 0\\ 5x + y + 15z &= -1\end{aligned}$$

Solution:

It does not take unusual powers of observation to notice that these four examples follow a pattern. They are all of the form $AX = B$, where

$$A = \begin{bmatrix} 1 & 0 & 3\\ 0 & -2 & 1\\ 5 & 1 & 15\end{bmatrix} \qquad X = \begin{bmatrix} x\\ y\\ z\end{bmatrix}$$

and B varies with the examples. Premultiplying both sides of the equation $AX = B$ by A^{-1}, we get $A^{-1}AX = A^{-1}B$, which implies that $X = A^{-1}B$. Thus we can find the solution vector X in each set of equations by premultiplying the right-hand vector by A^{-1}. If you are unusually observant, you may remember that this is the matrix whose inverse we found in Example 2.4.2. Looking back, we see that

$$A^{-1} = \begin{bmatrix} 31 & -3 & -6 \\ -5 & 0 & 1 \\ -10 & 1 & 2 \end{bmatrix}$$

Thus the answers are:

(a)
$$X = \begin{bmatrix} 31 & -3 & -6 \\ -5 & 0 & 1 \\ -10 & 1 & 2 \end{bmatrix} \begin{bmatrix} 4 \\ 10 \\ 6 \end{bmatrix} = \begin{bmatrix} 58 \\ -14 \\ -18 \end{bmatrix}$$

(b)
$$X = \begin{bmatrix} 31 & -3 & -6 \\ -5 & 0 & 1 \\ -10 & 1 & 2 \end{bmatrix} \begin{bmatrix} 7 \\ -3 \\ 0 \end{bmatrix} = \begin{bmatrix} 226 \\ -35 \\ -73 \end{bmatrix}$$

(c)
$$X = \begin{bmatrix} 31 & -3 & -6 \\ -5 & 0 & 1 \\ -10 & 1 & 2 \end{bmatrix} \begin{bmatrix} 1 \\ 3 \\ 4 \end{bmatrix} = \begin{bmatrix} -2 \\ -1 \\ 1 \end{bmatrix}$$

(d)
$$X = \begin{bmatrix} 31 & -3 & -6 \\ -5 & 0 & 1 \\ -10 & 1 & 2 \end{bmatrix} \begin{bmatrix} 2 \\ 0 \\ -1 \end{bmatrix} = \begin{bmatrix} 68 \\ -11 \\ -22 \end{bmatrix}$$

SUMMARY

Finding the inverse of most matrices is more difficult than finding the inverses of the elementary matrices given in Section 2.3, because each of the more general matrices is a product of elementary matrices and must be broken into its component elementary matrices in order to calculate the inverse. A systematic method for doing this is presented in this section, as well as some other important facts about multiplicative inverses. In the next two sections we shall see how these skills can be applied to real-world problems.

EXERCISES 2.4. A

Find the multiplicative inverse of the following matrices.

1. $\begin{bmatrix} -3 & 2 & 9 \\ 0 & 1 & 4 \\ 1 & 0 & 0 \end{bmatrix}$ 2. $\begin{bmatrix} 2 & 2 & 0 \\ 1 & 0 & 0 \\ 0 & -10 & 1 \end{bmatrix}$ 3. $\begin{bmatrix} 1 & 2 & 0 & 0 \\ 0 & 1 & 0 & 3 \\ 0 & 0 & 1 & 0 \\ 0 & 0 & 0 & 1 \end{bmatrix}$

4. Find $(I - A)^{-1}$ if $A = \begin{bmatrix} 0 & 0 & 0 & 0 \\ 1 & 0 & 0 & 0 \\ 0 & 1 & 0 & 0 \\ 2 & 5 & 0 & 0 \end{bmatrix}$.

5. What is the multiplicative inverse of A^{-1}?

6. If $A^{-1} = E_3 E_2 E_1$, write A itself in terms of the E's.
7. Prove that $(B^{-1}A^{-1})(AB) = I$.
8. If A, B, and C are matrices that have inverses, $(ABC)^{-1} =$
9. Find the multiplicative inverse of $\begin{bmatrix} 2 & 3 & 1 \\ 1 & 2 & -1 \\ 3 & -4 & 2 \end{bmatrix}$.
10. Write the following sets of equations in the form $AX = B$ and solve, using the result of the previous exercise and the fact that if $AX = B$, then $X = A^{-1}B$.

(a) $\begin{aligned} 2x + 3y + z &= 3 \\ x + 2y - z &= 0 \\ 3x - 4y + 2z &= -4 \end{aligned}$ (b) $\begin{aligned} 2x + 3y + z &= -1 \\ x + 2y - z &= 4 \\ 3x - 4y + 2z &= 0 \end{aligned}$

11. Find $(AB)^{-1}$, where A is the matrix given in problem 1 and B is the matrix given in problem 2.

EXERCISES 2.4. B

Find the multiplicative inverse of the following matrices.

1. $\begin{bmatrix} 1 & 2 & -3 \\ 0 & 1 & 2 \\ 1 & 3 & 0 \end{bmatrix}$ 2. $\begin{bmatrix} 4 & 0 & 0 \\ 2 & 1 & 1 \\ 3 & 0 & 1 \end{bmatrix}$ 3. $\begin{bmatrix} 1 & 0 & 0 & 0 \\ 3 & 1 & 0 & 0 \\ 5 & 6 & 1 & 0 \\ 2 & 8 & 1 & 1 \end{bmatrix}$

4. Find $(I - A)^{-1}$ if $A = \begin{bmatrix} 0 & 0 & 0 & 0 \\ 1 & 0 & 0 & 0 \\ 0 & 1 & 0 & 0 \\ 3 & 4 & 0 & 0 \end{bmatrix}$.
5. What is the multiplicative inverse of A^{-1}?
6. If $A^{-1} = E_3 E_2 E_1$, write A itself in terms of the E's.
7. Prove that $(B^{-1}A^{-1})(AB) = I$.
8. If R, S, and T are matrices that have inverses, $(RST)^{-1} =$
9. Find the multiplicative inverse of $\begin{bmatrix} 2 & 0 & 4 \\ 3 & 1 & 2 \\ 0 & -2 & 2 \end{bmatrix}$.
10. Write the following sets of equations in the form $AX = B$ and solve, using the result of the previous exercise and the fact that if $AX = B$, then $X = A^{-1}B$.

(a) $\begin{aligned} 2x + \phantom{y + {}} 4z &= 4 \\ 3x + y + 2z &= -2 \\ -2y + 2z &= 0 \end{aligned}$ (b) $\begin{aligned} 2x + \phantom{y + {}} 4z &= -1 \\ 3x + y + 2z &= 6 \\ -2y + 2z &= 1 \end{aligned}$

11. Find $(AB)^{-1}$, where A is the matrix given in problem 1 and B is the matrix given in problem 2.

ANSWERS 2.4. A

To help you learn how to do these problems, we include here not only the answers, but also suggestions as to how they may be obtained. Remember! There are many other correct ways of getting these answers.

1. $\begin{bmatrix} 0 & 0 & 1 \\ -4 & 9 & -12 \\ 1 & -2 & 3 \end{bmatrix}$ $R_1 \leftrightarrow R_3$;
$R_3 + 3R_1 \rightarrow R_3$; $R_3 - 2R_2 \rightarrow R_3$;
$R_2 - 4R_3 \rightarrow R_2$

2. $\begin{bmatrix} 0 & 1 & 0 \\ \frac{1}{2} & -1 & 0 \\ 5 & -10 & 1 \end{bmatrix}$ $R_1 \longleftrightarrow R_2$; $R_2 - 2R_1 \longrightarrow R_2$; $\frac{1}{2}R_2 \longrightarrow R_2$; $R_3 + 10R_2 \longrightarrow R_3$

3. $\begin{bmatrix} 1 & -2 & 0 & 6 \\ 0 & 1 & 0 & -3 \\ 0 & 0 & 1 & 0 \\ 0 & 0 & 0 & 1 \end{bmatrix}$ $R_1 - 2R_2 \longrightarrow R_1$; $R_1 + 6R_4 \longrightarrow R_1$; $R_2 - 3R_4 \longrightarrow R_2$

4. $\begin{bmatrix} 1 & 0 & 0 & 0 \\ 1 & 1 & 0 & 0 \\ 1 & 1 & 1 & 0 \\ 7 & 5 & 0 & 1 \end{bmatrix}$

5. A

6. $E_1^{-1}E_2^{-1}E_3^{-1}$

7. $(B^{-1}A^{-1})(AB) = B^{-1}(A^{-1}A)B = B^{-1}IB = B^{-1}B = I$

8. $C^{-1}B^{-1}A^{-1}$

9. $\begin{bmatrix} 0 & 0.4 & 0.2 \\ 0.2 & -0.04 & -0.12 \\ 0.4 & -0.68 & -0.04 \end{bmatrix}$

10. (a) $\begin{bmatrix} 2 & 3 & 1 \\ 1 & 2 & -1 \\ 3 & -4 & 2 \end{bmatrix}\begin{bmatrix} x \\ y \\ z \end{bmatrix} = \begin{bmatrix} 3 \\ 0 \\ -4 \end{bmatrix}$ $x = -0.8$, $y = 1.08$, $z = 1.36$

(b) $\begin{bmatrix} 2 & 3 & 1 \\ 1 & 2 & -1 \\ 3 & -4 & 2 \end{bmatrix}\begin{bmatrix} x \\ y \\ z \end{bmatrix} = \begin{bmatrix} -1 \\ 4 \\ 0 \end{bmatrix}$ $x = 1.6$, $y = -0.36$, $z = -3.12$

11. $\begin{bmatrix} 0 & 1 & 0 \\ \frac{1}{2} & -1 & 0 \\ 5 & -10 & 1 \end{bmatrix}\begin{bmatrix} 0 & 0 & 1 \\ -4 & 9 & -12 \\ 1 & -2 & 3 \end{bmatrix} = \begin{bmatrix} -4 & 9 & -12 \\ 4 & -9 & \frac{25}{2} \\ 41 & -92 & 128 \end{bmatrix}$

2.5 Using Inverses in Leontief Models

In Section 1.4 we examined hypothetical examples of the open Leontief model of an economy. (The "open" refers to the fact that a "final demands" vector was included in the technological equation.) The technological equation for an open Leontief model can be expressed in English as

$$\begin{bmatrix} \text{technological} \\ \text{matrix} \end{bmatrix}\begin{bmatrix} \text{gross} \\ \text{outputs} \end{bmatrix} + \begin{bmatrix} \text{final} \\ \text{demands} \end{bmatrix} = \begin{bmatrix} \text{gross} \\ \text{outputs} \end{bmatrix}$$

and in mathematical notation as $AX + D = X$, where A denotes the technology matrix, D denotes the final demands vector, and X denotes the gross outputs vector.

Suppose that we have found the technology matrix for a particular economic system (Section 1.4 considered how to make a technology matrix from the raw economic data) and that we know what we want the demand vector to be (that is, what we want to be "left over" for households, government, and export). To find what gross outputs are necessary, we must find what vector X makes the technological equation true. Thus we solve the equation for X.

Because of the similarity of matrices to numbers, ordinary algebraic procedures can be used for the first few steps:

$$\begin{aligned} AX + D &= X \\ D &= X - AX && \text{(subtracting)} \\ &= IX - AX && \text{(because } I \text{ is the identity)} \\ &= (I - A)X && \text{(distributive law)} \end{aligned}$$

If the letters in this last expression signified numbers, we could merely multiply both sides of the equation

$$D = (I - A)X$$

by $(I - A)^{-1}$ and have an expression for X. But since we are working with matrices, we must be more careful. We can say that *if* $(I - A)$ has an inverse $(I - A)^{-1}$, then we multiply both sides of the equation on the left by $(I - A)^{-1}$ and obtain

$$(I - A)^{-1}D = X$$

This is the expression that we use to solve for the gross outputs, given the technology matrix and the final demands.

It is common to give the technology matrix a new name when we use it in mathematics.

2.5.1 Definition

Suppose that the elements of a matrix A are all positive or zero and that the sum of the elements in each column does not exceed 1. Then we say that A is a Leontief matrix.

2.5.2 Definition

If A is a Leontief matrix, and if $(I - A)^{-1}$ exists and contains only nonnegative elements, then $(I - A)^{-1}$ is called the Leontief inverse of A.

2.5.3 Example

Find the Leontief inverse of $\begin{bmatrix} 0.3 & 0.4 \\ 0.6 & 0.2 \end{bmatrix}$.

Solution:

First we compute $I - A$:

$$\begin{bmatrix} 1 & 0 \\ 0 & 1 \end{bmatrix} - \begin{bmatrix} 0.3 & 0.4 \\ 0.6 & 0.2 \end{bmatrix} = \begin{bmatrix} 0.7 & -0.4 \\ -0.6 & 0.8 \end{bmatrix}$$

Then to compute $(I - A)^{-1}$ we set up the double matrix and use Gauss–Jordan row operations:

$$\left[\begin{array}{rr|rr} 0.7 & -0.4 & 1 & 0 \\ -0.6 & 0.8 & 0 & 1 \end{array}\right] \quad \begin{array}{l} 10R_1 \longrightarrow R_1 \\ 10R_2 \longrightarrow R_2 \end{array}$$

$$\left[\begin{array}{rr|rr} 7 & -4 & 10 & 0 \\ -6 & 8 & 0 & 10 \end{array}\right] \quad R_1 + R_2 \longrightarrow R_1$$

$$\left[\begin{array}{rr|rr} 1 & 4 & 10 & 10 \\ -6 & 8 & 0 & 10 \end{array}\right] \quad R_2 + 6R_1 \longrightarrow R_2$$

$$\left[\begin{array}{rr|rr} 1 & 4 & 10 & 10 \\ 0 & 32 & 60 & 70 \end{array}\right] \quad \tfrac{1}{32}R_2 \longrightarrow R_2$$

$$\left[\begin{array}{rr|rr} 1 & 4 & 10 & 10 \\ 0 & 1 & \frac{15}{8} & \frac{35}{16} \end{array}\right] \quad R_1 - 4R_2 \longrightarrow R_1$$

$$\left[\begin{array}{rr|rr} 1 & 0 & \frac{5}{2} & \frac{5}{4} \\ 0 & 1 & \frac{15}{8} & \frac{35}{16} \end{array}\right]$$

At this point it is wise to check and be sure our computations are correct:

$$\begin{bmatrix} \frac{5}{2} & \frac{5}{4} \\ \frac{15}{8} & \frac{35}{16} \end{bmatrix} \begin{bmatrix} \frac{7}{10} & -\frac{4}{10} \\ -\frac{6}{10} & \frac{8}{10} \end{bmatrix} = \begin{bmatrix} 1 & 0 \\ 0 & 1 \end{bmatrix}$$

2.5.4 Example

Let us look again at the model economy given in Example 1.4.1.

	Agriculture	Manufacturing	Final demands	Gross outputs
Agriculture	30	48	22	100
Manufacturing	60	24	36	120
Primary inputs	10	48		

Suppose that we want to increase the agricultural sector of the demand vector from 22 to 24 while keeping the manufacturing sector of the final demands constant at 36. What must the gross outputs be if the technology matrix remains the same?

Solution:

We write the technological equation as in Section 1.4:

$$\begin{bmatrix} 0.3 & 0.4 \\ 0.6 & 0.2 \end{bmatrix} \begin{bmatrix} 100 \\ 120 \end{bmatrix} + \begin{bmatrix} 22 \\ 36 \end{bmatrix} = \begin{bmatrix} 100 \\ 120 \end{bmatrix}$$

From this we can see that the technology matrix is precisely the Leontief matrix for which we found the Leontief inverse in the previous example. (Surprise!) Thus we can find the new gross outputs by solving the equation $AX + D = X$ for X by using the new demand vector in the equation

$$X = (I - A)^{-1}D = \begin{bmatrix} \frac{5}{2} & \frac{5}{4} \\ \frac{15}{8} & \frac{35}{16} \end{bmatrix} \begin{bmatrix} 24 \\ 36 \end{bmatrix} = \begin{bmatrix} 105 \\ 123.75 \end{bmatrix}$$

This work can be checked by setting up the new technological equation and verifying that it is true. Since

$$\begin{bmatrix} 0.3 & 0.4 \\ 0.6 & 0.2 \end{bmatrix} \begin{bmatrix} 105 \\ 123.75 \end{bmatrix} = \begin{bmatrix} 31.5 + 49.5 \\ 63 \quad + 24.75 \end{bmatrix} = \begin{bmatrix} 81 \\ 87.75 \end{bmatrix}$$

it is true that

$$\begin{bmatrix} 0.3 & 0.4 \\ 0.6 & 0.2 \end{bmatrix} \begin{bmatrix} 105 \\ 123.75 \end{bmatrix} + \begin{bmatrix} 24 \\ 36 \end{bmatrix} = \begin{bmatrix} 105 \\ 123.75 \end{bmatrix}$$

Thus the new table is

	Agriculture	Manufacturing	Final demands	Gross outputs
Agriculture	31.5	49.5	24	105
Manufacturing	63	24.75	36	123.75
Primary inputs	10.5	49.50		

If you compute the technology matrix for this table, you will discover (of course, since it was constructed that way) that it is the same as that for the table on the previous page.

Warning! Mathematical models are approximations at best, and if the economy alters very much, the errors in these models become significant. The computations here tend to be most nearly valid for forecasting or planning *minor* changes in an economy, but they do point the way "the wind is blowing."

SUMMARY

When an open Leontief model ($AX + D = X$) is solved for X, one obtains $X = (I - A)^{-1}D$. This section referred to the open Leontief model given in Section 1.4, found the Leontief inverse of its technology (Leontief) matrix, and used the inverse to calculate the model of a slightly expanded economy.

The following set of exercises is the first of several which refers back to problems earlier in the book. Using two bookmarks, one at the earlier exercises and one here near the answers, will enable you to flip back and forth quickly.

EXERCISES 2.5. A

For each of the model economies given in problems 3 and 4, Exercises 1.4.A (pages 33 and 34), find the new gross outputs needed to produce the new final demands given below. Write the new technological equation and the table for the new model economy similar to that on the previous page.
(3a) 15 units of agriculture and 18 units of manufacturing
(3b) 15 units of agriculture and 28 units of manufacturing
(4) 105 units of primary metals and 160 units of other metal goods

EXERCISES 2.5. B

For each model economy given in problems 3 and 4, Exercises 1.4.B (pages 34 and 35), find the new gross outputs needed to produce the new final demands given below. Write the new technological equation and the table for the new model economy similar to that on the previous page.
(3a) 12 units of agriculture and 6 units of manufacturing
(3b) 12 units of agriculture and 15 units of manufacturing
(4) 30 units at plant I and 26 units at plant II

EXERCISES 2.5. C

Tell what gross outputs are needed for the agriculture segment to produce 25 units, the manufacturing sector to produce 26 units, and the service sector to produce 21 units if the technological matrix is the same as that in problem 1, Exercises 1.4.C (page 35). Write the new technological equation and the table for the new model economy.

ANSWERS 2.5. A

(3a) New gross outputs:

$$\begin{bmatrix} 4 & 3 \\ 2 & 4 \end{bmatrix} \begin{bmatrix} 15 \\ 18 \end{bmatrix} = \begin{bmatrix} 114 \\ 102 \end{bmatrix}$$

New technological equation:

$$\begin{bmatrix} 0.6 & 0.3 \\ 0.2 & 0.6 \end{bmatrix} \begin{bmatrix} 114 \\ 102 \end{bmatrix} + \begin{bmatrix} 15 \\ 18 \end{bmatrix} = \begin{bmatrix} 114 \\ 102 \end{bmatrix}$$

New model:

	Agriculture	Manufacturing	Final demands	Gross outputs
Agriculture	68.4	30.6	15	114
Manufacturing	22.8	61.2	18	102
Primary inputs	22.8	10.2		

(3b) New gross outputs:

$$\begin{bmatrix} \frac{40}{13} & \frac{30}{13} \\ \frac{25}{13} & \frac{35}{13} \end{bmatrix} \begin{bmatrix} 15 \\ 28 \end{bmatrix} = \begin{bmatrix} \frac{1440}{13} \\ \frac{1355}{13} \end{bmatrix} \approx \begin{bmatrix} 111 \\ 104 \end{bmatrix}$$

New technological equation:

$$\begin{bmatrix} 0.3 & 0.6 \\ 0.5 & 0.2 \end{bmatrix} \begin{bmatrix} 111 \\ 104 \end{bmatrix} + \begin{bmatrix} 15 \\ 28 \end{bmatrix} = \begin{bmatrix} 111 \\ 104 \end{bmatrix}$$

New model:

	Agriculture	Manufacturing	Final demands	Gross outputs
Agriculture	33.3	62.4	15	111
Manufacturing	55.5	20.8	28	104
Primary inputs	22.2	20.8		

(4) New gross outputs:

$$\begin{bmatrix} \frac{18}{5} & \frac{12}{5} \\ \frac{14}{5} & \frac{16}{5} \end{bmatrix} \begin{bmatrix} 105 \\ 160 \end{bmatrix} = \begin{bmatrix} 762 \\ 806 \end{bmatrix}$$

New technological equation:

$$\begin{bmatrix} \frac{1}{3} & \frac{1}{2} \\ \frac{7}{12} & \frac{1}{4} \end{bmatrix} \begin{bmatrix} 762 \\ 806 \end{bmatrix} + \begin{bmatrix} 105 \\ 160 \end{bmatrix} = \begin{bmatrix} 762 \\ 806 \end{bmatrix}$$

New model:

	Primary metals	Other metal goods	Sales to nonmetal industries	Total budget of metal industry
Primary metals	254	403	105	762
Other metal goods	444.5	201.5	160	806
Nonmetal inputs	63.5	201.5		

ANSWERS 2.5. C

New gross outputs:

$$\begin{bmatrix} \frac{85}{41} & \frac{55}{41} & \frac{75}{82} \\ \frac{35}{41} & \frac{95}{41} & \frac{55}{82} \\ \frac{40}{41} & \frac{50}{41} & \frac{90}{41} \end{bmatrix} \begin{bmatrix} 25 \\ 26 \\ 21 \end{bmatrix} = \begin{bmatrix} \frac{8685}{82} \\ \frac{7845}{82} \\ \frac{4190}{41} \end{bmatrix}$$

New technological equation:

$$\begin{bmatrix} 0.3 & 0.3 & 0.2 \\ 0.2 & 0.4 & 0.1 \\ 0.2 & 0.2 & 0.4 \end{bmatrix} \begin{bmatrix} \frac{8685}{82} \\ \frac{7845}{82} \\ \frac{4190}{41} \end{bmatrix} + \begin{bmatrix} 25 \\ 26 \\ 21 \end{bmatrix} = \begin{bmatrix} \frac{8685}{82} \\ \frac{7845}{82} \\ \frac{4190}{41} \end{bmatrix}$$

New model:

	Agriculture	Manufacturing	Services	Final demands	Gross outputs
Agriculture	$\frac{5211}{164}$	$\frac{4707}{164}$	$\frac{838}{41}$	25	$\frac{8685}{82}$
Manufacturing	$\frac{1737}{82}$	$\frac{3138}{82}$	$\frac{419}{41}$	26	$\frac{7845}{82}$
Services	$\frac{1737}{82}$	$\frac{1569}{82}$	$\frac{1676}{41}$	21	$\frac{4190}{41}$
Primary inputs	$\frac{5211}{164}$	$\frac{1569}{164}$	$\frac{1257}{41}$		

*2.6 Parts-Listing Problem and Accounting Model

There are a variety of models of real-world situations that involve exactly the same mathematics as the open Leontief model. In this section we explore two such models.

An Accounting Model

In the following example we assume that the company has only two service departments, but the same reasoning applies to companies with many departments; the only difference is that there are more equations and larger matrices.

2.6.1 Example

Suppose that a company has two service departments, each of which charges a percentage of its costs to the other. Specifically, suppose the first charges 20 percent of its costs to the second and the second charges 10 percent of its costs to the first. Suppose further that the first department has direct monthly costs of $1000 and the second has direct costs of $1200. What are the total costs of each department?

Solution:

Let x_1 denote the total costs of the first department and x_2 denote the total costs of the second department. Then the total costs of the first department will be the sum of its direct cost plus the amount charged to it by the second department.

$$x_1 = 1000 + 0.1x_2$$

Similarly, the total costs of the second department will be its direct cost plus the amount charged to it by the first department:

$$x_2 = 1200 + 0.2x_1$$

We can summarize these two equations in matrix notation as

$$X = D + AX$$

if we set

$$X = \begin{bmatrix} x_1 \\ x_2 \end{bmatrix} \qquad D = \begin{bmatrix} 1000 \\ 1200 \end{bmatrix} \qquad A = \begin{bmatrix} 0 & 0.1 \\ 0.2 & 0 \end{bmatrix}$$

Notice that this equation is of the same form as the equation in the Leontief model with the sides of the equation reversed, so the solution method will be the same.

$$\begin{aligned} X &= D + AX \\ IX - AX &= D \\ (I - A)X &= D \\ X &= (I - A)^{-1}D \end{aligned}$$

To solve for X, therefore, compute $I - A = \begin{bmatrix} 1 & -0.1 \\ -0.2 & 1 \end{bmatrix}$ and find its inverse:

$$\left[\begin{array}{cc|cc} 1 & -0.1 & 1 & 0 \\ -0.2 & 1 & 0 & 1 \end{array}\right] \qquad 10R_2 \longrightarrow R_2$$

$$\left[\begin{array}{cc|cc} 1 & -0.1 & 1 & 0 \\ -2 & 10 & 0 & 10 \end{array}\right] \qquad R_2 + 2R_1 \longrightarrow R_2$$

$$\left[\begin{array}{cc|cc} 1 & -0.1 & 1 & 0 \\ 0 & 9.8 & 2 & 10 \end{array}\right] \qquad \frac{1}{9.8} R_2 \longrightarrow R_2$$

$$\left[\begin{array}{cc|cc} 1 & -0.1 & 1 & 0 \\ 0 & 1 & \frac{20}{98} & \frac{100}{98} \end{array}\right] \qquad R_1 + \frac{1}{10} R_2 \longrightarrow R_1$$

$$\left[\begin{array}{cc|cc} 1 & 0 & \frac{100}{98} & \frac{10}{98} \\ 0 & 1 & \frac{20}{98} & \frac{100}{98} \end{array}\right]$$

Therefore, to solve for X we set

$$X = (I - A)^{-1}D = \begin{bmatrix} \frac{100}{98} & \frac{10}{98} \\ \frac{20}{98} & \frac{100}{98} \end{bmatrix} \begin{bmatrix} 1000 \\ 1200 \end{bmatrix} = \begin{bmatrix} \frac{112{,}000}{98} \\ \frac{140{,}000}{98} \end{bmatrix} \approx \begin{bmatrix} 1143 \\ 1429 \end{bmatrix}$$

It turns out the total costs of the first department are about $1143 and the total costs of the second are about $1429.

Notice that if the direct monthly costs, D, change in the problem above but the relationship between the departments remains the same, we need not do all the computations again; we merely substitute the new D into the equation in the preceding paragraph.

Parts-Listing Problem

We can now extend our study of the parts-listing problem by introducing a production vector, P, which, it turns out, plays the same role in the parts-listing problem that the gross outputs vector does in the open Leontief model. In Example 1.2.1 (page 8) the production vector, P, that corresponds to a given order vector B signifies just enough tripods, legs, bolts, and rods so that if we send the request to each department in turn in chronological order (from bottom to top), each department will have precisely the number of items that appears in the order vector plus enough to produce all the items above it. That is, we want P to represent the number of items that are needed to produce P plus enough left over for the order vector B.

2.6.2 Example

Refer to Example 1.2.1. Suppose that an order has been received for 2 tripods, 3 legs, 4 bolts, and 5 rods. Find the production vector for this order.

Solution:
We need to find P such that $P = QP + B$. Solving for P, we have

$$IP - QP = B$$
$$(I - Q)P = B$$
$$P = (I - Q)^{-1}B$$

The letters are different, but the calculation should look familiar. In this example we have

$$Q = \begin{bmatrix} 0 & 0 & 0 & 0 \\ 3 & 0 & 0 & 0 \\ 1 & 2 & 0 & 0 \\ 0 & 3 & 0 & 0 \end{bmatrix} \quad \text{and} \quad B = \begin{bmatrix} 2 \\ 3 \\ 4 \\ 5 \end{bmatrix}$$

so we begin

$$I - Q = \begin{bmatrix} 1 & 0 & 0 & 0 \\ 0 & 1 & 0 & 0 \\ 0 & 0 & 1 & 0 \\ 0 & 0 & 0 & 1 \end{bmatrix} - \begin{bmatrix} 0 & 0 & 0 & 0 \\ 3 & 0 & 0 & 0 \\ 1 & 2 & 0 & 0 \\ 0 & 3 & 0 & 0 \end{bmatrix}$$
$$= \begin{bmatrix} 1 & 0 & 0 & 0 \\ -3 & 1 & 0 & 0 \\ -1 & -2 & 1 & 0 \\ 0 & -3 & 0 & 1 \end{bmatrix}$$

and then continue

$$\left[\begin{array}{rrrr|rrrr} 1 & 0 & 0 & 0 & 1 & 0 & 0 & 0 \\ -3 & 1 & 0 & 0 & 0 & 1 & 0 & 0 \\ -1 & -2 & 1 & 0 & 0 & 0 & 1 & 0 \\ 0 & -3 & 0 & 1 & 0 & 0 & 0 & 1 \end{array}\right] \quad \begin{array}{l} R_2 + 3R_1 \longrightarrow R_2 \\ R_3 + R_1 \longrightarrow R_3 \end{array}$$

$$\left[\begin{array}{rrrr|rrrr} 1 & 0 & 0 & 0 & 1 & 0 & 0 & 0 \\ 0 & 1 & 0 & 0 & 3 & 1 & 0 & 0 \\ 0 & -2 & 1 & 0 & 1 & 0 & 1 & 0 \\ 0 & -3 & 0 & 1 & 0 & 0 & 0 & 1 \end{array}\right] \quad \begin{array}{l} R_3 + 2R_2 \longrightarrow R_3 \\ R_4 + 3R_2 \longrightarrow R_4 \end{array}$$

$$\left[\begin{array}{rrrr|rrrr} 1 & 0 & 0 & 0 & 1 & 0 & 0 & 0 \\ 0 & 1 & 0 & 0 & 3 & 1 & 0 & 0 \\ 0 & 0 & 1 & 0 & 7 & 2 & 1 & 0 \\ 0 & 0 & 0 & 1 & 9 & 3 & 0 & 1 \end{array}\right]$$

We conclude that

$$P = (I - Q)^{-1}B = \begin{bmatrix} 1 & 0 & 0 & 0 \\ 3 & 1 & 0 & 0 \\ 7 & 2 & 1 & 0 \\ 9 & 3 & 0 & 1 \end{bmatrix} \begin{bmatrix} 2 \\ 3 \\ 4 \\ 5 \end{bmatrix} = \begin{bmatrix} 2 \\ 9 \\ 24 \\ 32 \end{bmatrix}$$

Thus the production vector is 2 tripods, 9 legs, 24 bolts, and 32 rods.

To check this, imagine the following scenario. The rod department receives the order and puts 32 rods into a waiting box, which goes to the bolt department. The bolt department donates 24 bolts and sends the box to the leg department. To make the 9 required legs, the leg department uses 18 bolts (leaving 6) and 27 rods (leaving 5). Thus the box contains 9 legs, 6 bolts, and 5 rods as it moves on to the tripod department. There 2 tripods are made, using up 6 legs and 2 bolts—and the box now contains the 2 tripods, 3 legs, 4 bolts, and 5 rods that were ordered!

EXERCISES 2.6. A

1. Suppose that a company has two service departments, each of which charges a percentage of its costs to the other. The first charges 20 percent of its cost to the second and the second charges 30 percent of its cost to the first. If the first has direct weekly costs of $4700 and the second has direct weekly costs of $9400, what are the total costs of each department?
2. In problem 8, Exercises 1.2.A (page 20), find the production vector for an order of 5 aprons, 3 frontpieces, no rectangles, 15 pompoms, and 6 ties.
3. In problem 2, find the production vector if the order was for 4 aprons, no frontpieces, 7 rectangles, 12 pompoms, and 8 ties.
4. In problem 9, Exercises 1.2.A (page 20), find the production vector for an order of 3 coat-hanger stands, 5 base plates, 10 rods, and 10 bolts (and no metal hooks, hook assemblies, or tiny top plates).

EXERCISES 2.6. B

1. Suppose that a company has two service departments, each of which charges a percentage of its costs to the other. The first charges 30 percent of its costs to the second and the second charges 10 percent of its costs to the first. If the first has direct weekly costs of $9700 and the second has direct weekly costs of $10,088, what are the total costs of each department?
2. In problem 8, Exercises 1.2.B (page 19), find the production vector for an order of 5 salt-and-pepper holders, 4 legs, no holder pieces, 3 circles, and 8 rods.
3. In problem 2, find the production vector for an order of 3 salt-and-pepper holders, no extra legs, 4 holder pieces, 2 circles, and 5 rods.
4. In problem 9, Exercises 1.2.B (page 19), find the production vector if the order is for 4 photo albums, 2 looseleaf folders, 10 picture pages, no blank pages, and 16 picture slots.

ANSWERS 2.6. A

1. First department: $8000
   Second department: $11,000
   $\left(Hint: (I - A)^{-1} = \begin{bmatrix} \frac{50}{47} & \frac{15}{47} \\ \frac{10}{47} & \frac{50}{47} \end{bmatrix}. \right)$
2. 5 aprons, 8 frontpieces, 8 rectangles, 65 pompoms, 16 ties
3. 4 aprons, 4 frontpieces, 11 rectangles, 40 pompoms, 16 ties
4. 3 coat-hanger stands, 3 hook assemblies, 13 rods, 8 base plates, 3 tiny top plates, 18 metal hooks, 40 bolts

VOCABULARY

unique solution, augmented matrix, Gauss–Jordan row operation, reduced form, premultiply, postmultiply, elementary matrix, open Leontief model, Leontief matrix, Leontief inverse

SAMPLE TEST Chapter 2

Do five of the following six problems. Each problem counts 20 points.

1. A store sells cheese in three different packages: the large assortment package, containing 4 lb of cheddar, 2 lb of swiss, and 3 lb of American; the Mix Variety package, containing 2 lb of cheddar, 1 lb of swiss, and 2 lb of American; and the small package, containing 2 lb of cheddar and 1 lb of American. How can the store fill an order for 54 lb of cheddar cheese, 20 lb of swiss cheese, and 42 lb of American cheese?
2. Find the 3 by 3 matrix that, when it premultiplies another 3 by 3 matrix, performs on it the operation indicated.
   (a) Triples the middle row.
   (b) Interchanges the second and third rows.
   (c) Adds 4 times the third row to the second row.
   (d), (e), (f) Write the multiplicative inverse of each matrix given in parts (a), (b), and (c), respectively.

3. (a) If A, B, and C are three matrices, each of which has an inverse, what is $(ABC)^{-1}$?
   (b) Each of the matrices in parts (a), (b), and (c) of problem 2 is called an ______________ matrix.
   (c) $(A^{-1})^{-1} =$
   (d) If $ABCDEF = I$, where each letter designates a matrix, then the multiplicative inverse of AB is ______________.

4. Find the multiplicative inverse of $\begin{bmatrix} 3 & 0 & 4 \\ 0 & 5 & 1 \\ 2 & 0 & 0 \end{bmatrix}$.

5. Suppose that the following technological equation has been found in a simple economy, where the final demands consist of 25 units in the agriculture sector and 35 units in the manufacturing sector. Write the new technological equation, including the new gross outputs, for the economy with the same Leontief matrix and a final demands vector with 27 units of agriculture and 36 units of manufacturing. (Do not convert fractions to decimals.)

$$\begin{bmatrix} 0.4 & 0.4 \\ 0.6 & 0.3 \end{bmatrix} \begin{bmatrix} 175 \\ 200 \end{bmatrix} + \begin{bmatrix} 25 \\ 35 \end{bmatrix} = \begin{bmatrix} 175 \\ 200 \end{bmatrix}$$

6. Suppose that $QX + B = X$, where

$$Q = \begin{bmatrix} 0 & 0 & 0 \\ 2 & 0 & 0 \\ 1 & 3 & 0 \end{bmatrix} \quad \text{and} \quad B = \begin{bmatrix} 3 \\ 4 \\ 5 \end{bmatrix}$$

Find X.

The answers are at the back of the book.

Chapter 3

SYSTEMS OF LINEAR EQUATIONS WITHOUT UNIQUE SOLUTIONS

3.1 Recognizing Nonunique Solutions

In Chapter 2 the systems of linear equations were carefully chosen so that each system of three equations in three unknowns had one unique "answer" for x, y, and z. We refer to this set of three answers as *the* "solution" to the system of linear equations.

3.1.1 Definition

A solution of a system of equations in n unknowns is an ordered set of n numbers that when substituted for the unknowns make all the equations of the system true. (A solution is often written in vector form.)

Not all systems of three linear equations in three unknowns have a unique solution. Some have more than one solution, and others have no solution at all. Actually, the situation is not so different from that for two equations in two unknowns. We pause to examine two types of systems that are passed over quickly in elementary algebra courses. You may even have forgotten about such systems!

3.1.2 Example

Find the solutions of the following systems of two linear equations in two unknowns.

(a) $\begin{aligned} x + y &= 1 \\ x + y &= 2 \end{aligned}$ (b) $\begin{aligned} x + y &= 1 \\ 2x + 2y &= 2 \end{aligned}$

Solution:

The left system has no solution at all; if the sum of two numbers is 1, it cannot be 2.

The right system has an infinite number of solutions; if we set $y = a$ (any number at all) and $x = 1 - a$, we will find that both equations are true. For example, if we set $a = 0$, we get the solution $\begin{bmatrix} 1 \\ 0 \end{bmatrix}$ and if we set $a = 5$, we get the solution $\begin{bmatrix} -4 \\ 5 \end{bmatrix}$.

3.1.3 Definition

A system of linear equations with no solution is called inconsistent.

3.1.4 Definition

If a system of linear equations has more than one solution, we say that it is redundant.

The terms "inconsistent" and "redundant" are common English words defined here in a precise way; we shall use them often in this chapter. You will note that in Example 3.1.2, part (a) shows an inconsistency and part (b) shows a redundancy.

Conceptually, the situation is no different for 3 linear equations in 3 unknowns or for 4 linear equations in 4 unknowns, or for 16 linear equations in 16 unknowns. If m and n are any two integers (possibly the same) and we have m linear equations in n unknowns, there are three possibilities:

1. The system has a unique solution.
2. The system is inconsistent.
3. The system is redundant.

But with more variables it is more challenging to recognize which of these three statements is true for a given system of equations—and considerably more difficult to find the solution(s) in the first and third situations. This chapter presents methods for recognizing which statement describes a given system of equations and for finding whatever solutions exist. We begin with a couple of easy problems.

3.1.5 Example

For each of the following systems of linear equations, find the solution(s), if any exist.

(a)
$$\begin{aligned} x + y + z &= 1 \\ x + y + z &= 2 \\ 2x + 2y + 3z &= 7 \end{aligned}$$

(b)
$$\begin{aligned} x + y + z &= 1 \\ 2x + 2y + 2z &= 2 \\ 2x + 2y + 3z &= 0 \end{aligned}$$

If you look at these for a moment, you may realize that they parrot the situations in Example 3.1.2.

The system in part (a) cannot have any solution because if the sum of three numbers is 1, it cannot be 2.

In part (b), if the sum of three numbers is 1, the sum of their doubles will always be 2; this implies that system (b) cannot have a unique solution. By subtracting the second equation from the third, we see that $z = -2$. Substituting this back in the first equation, we find that if $y = a$, $x = 3 - a$. Every solution to this system is of the form $\begin{bmatrix} 3 - a \\ a \\ -2 \end{bmatrix}$, where a can be any number at all.

When there is more than one solution, we often want to express the difference between giving *one* of the solutions and giving *all* the solutions. The following two definitions help us do this.

3.1.6 Definition

A vector containing particular numbers that satisfy a given system of linear equations is called a particular solution of that system.

3.1.7 Definition

A vector that symbolically represents *all* the solutions of a given system of linear equations is called a complete solution or a general solution of that system of equations.

In Example 3.1.5 we can obtain the particular solution $\begin{bmatrix} 3-0 \\ 0 \\ -2 \end{bmatrix} = \begin{bmatrix} 3 \\ 0 \\ -2 \end{bmatrix}$ by setting $a = 0$, and we can obtain the particular solution $\begin{bmatrix} 3-5 \\ 5 \\ -2 \end{bmatrix} = \begin{bmatrix} -2 \\ 5 \\ -2 \end{bmatrix}$ by setting $a = 5$. The complete solution is $\begin{bmatrix} 3-a \\ a \\ -2 \end{bmatrix}$ because every particular solution has this form for some a. Furthermore, no matter what value a takes, the following set of equations is true.

$$\begin{aligned} (3-a) + a - 2 &= 1 \\ 2(3-a) + 2a + 2(-2) &= 2 \\ 2(3-a) + 2a + 3(-2) &= 0 \end{aligned}$$

Example 3.1.5 was easy, but with a little manipulation, we can make the problems harder. (The goal here is not to be ornery, but to show that harder problems can be both posed and solved.) In part (a), suppose that we replace the first equation with the third minus the first; we know we can do this without affecting the solution set. Then we obtain

$$\begin{aligned} x + y + 2z &= 6 \\ x + y + z &= 2 \\ 2x + 2y + 3z &= 7 \end{aligned}$$

3.1.8 Example

If possible, solve the system of equations immediately above.

Solution:

This system must also be inconsistent, since it has the same solutions as Example 3.1.5(a). But if you had not been told how we obtained these equations, you might not have noticed that it is inconsistent. So you probably would have applied the Gauss–Jordan row operations as in earlier problems. We also proceed by first putting the problem into augmented matrix form, and then by applying row operations.

$$\begin{bmatrix} 1 & 1 & 2 & 6 \\ 1 & 1 & 1 & 2 \\ 2 & 2 & 3 & 7 \end{bmatrix} \qquad R_2 - R_1 \longrightarrow R_2$$

$$\begin{bmatrix} 1 & 1 & 2 & 6 \\ 0 & 0 & -1 & -4 \\ 2 & 2 & 3 & 7 \end{bmatrix} \qquad R_3 - 2R_1 \longrightarrow R_3$$

$$\begin{bmatrix} 1 & 1 & 2 & 6 \\ 0 & 0 & -1 & -4 \\ 0 & 0 & -1 & -5 \end{bmatrix} \qquad -R_2 \longrightarrow R_2$$

$$\begin{bmatrix} 1 & 1 & 2 & 6 \\ 0 & 0 & 1 & 4 \\ 0 & 0 & -1 & -5 \end{bmatrix} \qquad R_3 + R_2 \longrightarrow R_3$$

$$\begin{bmatrix} 1 & 1 & 2 & 6 \\ 0 & 0 & 1 & 4 \\ 0 & 0 & 0 & -1 \end{bmatrix}$$

If we arrive at a row that is all zeros except for a nonzero number on the right, we can conclude that the system is inconsistent. This is because (in this case) the last row says that

$$0x + 0y + 0z = -1$$

and no matter what numbers we choose for x, y, and z, the sum on the left will be zero, which is not equal to -1.

3.1.9 Example

Solve the following system of equations if possible:

$$\begin{aligned} x + y + 2z &= -1 \\ 2x + 2y + 2z &= 2 \\ 2x + 2y + 3z &= 0 \end{aligned}$$

Solution:

It is possible you noticed that this set was obtained from the set in Example 3.1.5(b) by replacing the first equation with the third minus the first. But it is also possible that you did not see this! If not, we can again set up an augmented matrix and apply Gauss–Jordan row operations.

$$\begin{bmatrix} 1 & 1 & 2 & -1 \\ 2 & 2 & 2 & 2 \\ 2 & 2 & 3 & 0 \end{bmatrix} \qquad R_2 - 2R_1 \longrightarrow R_2 \text{ and } R_3 - 2R_1 \longrightarrow R_3$$

$$\begin{bmatrix} 1 & 1 & 2 & -1 \\ 0 & 0 & -2 & 4 \\ 0 & 0 & -1 & 2 \end{bmatrix} \qquad -\tfrac{1}{2}R_2 \longrightarrow R_2$$

$$\begin{bmatrix} 1 & 1 & 2 & -1 \\ 0 & 0 & 1 & -2 \\ 0 & 0 & -1 & 2 \end{bmatrix} \qquad R_3 + R_2 \longrightarrow R_3$$

$$\begin{bmatrix} 1 & 1 & 2 & -1 \\ 0 & 0 & 1 & -2 \\ 0 & 0 & 0 & 0 \end{bmatrix}$$

The row of zeros at the bottom symbolizes the equation

$$0x + 0y + 0z = 0$$

which is clearly true for any x, y, or z that we choose. *This tells us that there is more than one solution to the original system of equations.* To find out what the solutions are, we put the last matrix back into equation form:

$$\begin{aligned} x + y + 2z &= -1 \\ z &= -2 \end{aligned}$$

Thus there is only one possibility for z: -2. We can pick anything for y; it is common to set $y = a$. Then we can solve for x in terms of y and z:

$$x + a + 2(-2) = -1 \qquad \text{or} \qquad x = -1 + 4 - a = 3 - a$$

The solution vector is $\begin{bmatrix} 3-a \\ a \\ -2 \end{bmatrix}$. It is easy to check these values in the original system:

$$\begin{aligned} (3-a) + a + 2(-2) &= -1 \\ 2(3-a) + 2a + 2(-2) &= 2 \\ 2(3-a) + 2a + 3(-2) &= 0 \end{aligned}$$

If we use Gauss–Jordan row operations on any 3 by 4 matrix derived from 3 equations in 3 unknowns as suggested in Section 2.1, we will eventually end up with one of the following three situations.

(a) $\begin{bmatrix} 1 & 0 & 0 & a \\ 0 & 1 & 0 & b \\ 0 & 0 & 1 & c \end{bmatrix}$ A unique solution: $x = a$, $y = b$, $z = c$

(b) There is a row $0 \quad 0 \quad 0 \quad a$ with $a \neq 0$. The original system is inconsistent.

(c) There is no row $0 \quad 0 \quad 0 \quad a$ with $a \neq 0$ and at least one row of all zeros. The original system is redundant.

The reasons behind this rule were illustrated in Examples 3.1.8 and 3.1.9 and will be developed further in the next section. The next section will also present a method for finding all the solutions in the third case. Section 3.4 interprets these results geometrically, and Section 3.5 explores relationships among the rows in the various types of matrices. Meanwhile, we conclude this section with an application.

3.1.10 Example in Biology

Three species of bacteria will be kept in one test tube and will feed on three resources. Each member of the first species consumes 1 unit each of the first and second resource and 2 units of the third. The second species similarly consumes 1 unit each of the first and second resource and 2 units of the third. The third species consumes 2 units each of the first and third resource only. If the test tube is supplied daily with 10,000 units of the first resource, 8000 units of the second, and 18,000 units of the third, how many of each species may coexist in equilibrium in the test tube so that all of the supplied resources are consumed?

Solution:

The example begins like those in Chapter 2. Let x = number of first species, y = number of second species, and z = number of third species:

$$\begin{aligned} x + y + 2z &= 10{,}000 \quad &\text{(first resource)} \\ x + y &= 8{,}000 \quad &\text{(second resource)} \\ 2x + 2y + 2z &= 18{,}000 \quad &\text{(third resource)} \end{aligned}$$

Setting this system up as an augmented matrix, we apply the Gauss–Jordan row operations:

$$\begin{bmatrix} 1 & 1 & 2 & 10{,}000 \\ 1 & 1 & 0 & 8{,}000 \\ 2 & 2 & 2 & 18{,}000 \end{bmatrix} \quad \begin{matrix} R_2 - R_1 \longrightarrow R_2 \\ R_3 - 2R_1 \longrightarrow R_3 \end{matrix}$$

$$\begin{bmatrix} 1 & 1 & 2 & 10{,}000 \\ 0 & 0 & -2 & -2{,}000 \\ 0 & 0 & -2 & -2{,}000 \end{bmatrix} \quad \begin{matrix} R_1 + R_2 \longrightarrow R_1 \\ R_3 - R_2 \longrightarrow R_3 \end{matrix}$$

$$\begin{bmatrix} 1 & 1 & 0 & 8{,}000 \\ 0 & 0 & -2 & -2{,}000 \\ 0 & 0 & 0 & 0 \end{bmatrix} \quad -\tfrac{1}{2}R_2 \longrightarrow R_2$$

$$\begin{bmatrix} 1 & 1 & 0 & 8{,}000 \\ 0 & 0 & 1 & 1{,}000 \\ 0 & 0 & 0 & 0 \end{bmatrix}$$

It is at this point that the example differs from the earlier ones. This augmented matrix tells us that

$$\begin{aligned} x + y &= 8000 \\ z &= 1000 \end{aligned}$$

so the answer is not unique. For equilibrium there must be 1000 of the third species, but there are many ways we can choose the number of the first and second species—just as long as their total number is 8000. For example, one particular solution would be to have 4000 of each of the first two species and 1000 of the third.

The ideas in the example can be applied to a much larger ecological system by increasing the number of variables and equations. Thus, within limitations, there may be some variation in the possible solutions that yield equilibrium in a game farm or a much larger system.

SUMMARY

A system of n linear equations in n unknowns need not have a unique solution. It may have no solutions (in which case it is called an inconsistency) or it may have many solutions (in which case it is called a redundancy). This section showed how to use Gauss–Jordan row operations to discover for a given system which of these situations occurs. The mathematics describing an economy or an ecosystem often has many solutions; such situations provide applications for the mathematics of this section and the next.

EXERCISES 3.1. A

Use Gauss–Jordan row operations to reduce each of the following matrices to one of the forms indicated on page 94. Tell whether it is of form (a), (b), or (c).

1. $\begin{bmatrix} 2 & 3 & 1 & 4 \\ 1 & 2 & 3 & 0 \\ 6 & 5 & 11 & -4 \end{bmatrix}$

2. $\begin{bmatrix} 3 & -1 & 0 & 2 \\ 2 & 1 & -3 & -1 \\ 1 & -2 & 3 & 1 \end{bmatrix}$

3. $\begin{bmatrix} 0 & 1 & -3 & 4 \\ 2 & -2 & 0 & 3 \\ -6 & 8 & -6 & -1 \end{bmatrix}$

4. $\begin{bmatrix} 4 & 3 & 0 & -1 \\ 3 & 2 & -2 & 0 \\ 1 & -2 & 0 & 3 \end{bmatrix}$

5. $\begin{bmatrix} 2 & -1 & 1 & 0 \\ 0 & 2 & 3 & -2 \\ 2 & -5 & -5 & 2 \end{bmatrix}$

6. $\begin{bmatrix} 3 & 0 & -2 & 5 \\ 2 & 1 & -2 & 0 \\ 2 & -2 & 0 & 5 \end{bmatrix}$

Tell whether each of the following sets of equations has a unique solution, no solutions, or many solutions. In other words, tell whether the solution set consists of one point, no points, or many points. In still other words, tell whether each system has a unique solution, is inconsistent, or is redundant.

Save your answers because you may want to refer to them while doing Exercises 3.2.A.

7. $\begin{aligned} 2x - y &= 7 \\ 2x \quad &= 5 \\ -y &= 4 \end{aligned}$

8. $\begin{aligned} 3x + \quad z &= 4 \\ x \quad &= 2 \\ 3x + 2y \quad &= 1 \end{aligned}$

9. $\begin{aligned} 2x \quad &= 4 \\ -2x + y - z &= -4 \\ 3x - y + z &= 7 \end{aligned}$

10. $\begin{aligned} x + \quad z &= 2 \\ x - 2y - 2z &= -2 \\ 2x - 2y - z &= 0 \end{aligned}$

11. $\begin{aligned} x + \quad z &= 2 \\ 2x + \quad 3z &= 3 \\ -2x + 3y \quad &= -4 \end{aligned}$

12. $\begin{aligned} x + 2y - z &= 4 \\ x + \quad 3z &= -1 \\ 3x + 2y + 5z &= 2 \end{aligned}$

13. $\begin{aligned} x + y + 2z &= 1 \\ 2x + 2y + 4z &= 2 \\ -3x - 3y - 6z &= -3 \end{aligned}$

14. $\begin{aligned} 5x + 2y + 3z &= 7 \\ 5x + 2y + 3z &= 1 \\ -x + 2y - 3z &= 6 \end{aligned}$

15. $\begin{aligned} x + 3y + 2z &= -4 \\ x + 2y - z &= -6 \\ 2x + 5y + z &= -10 \end{aligned}$

EXERCISES 3.1. B

Use Gauss–Jordan row operations to reduce each of the following matrices to one of the forms indicated on page 94. Tell whether it is of form (a), (b), or (c).

1. $\begin{bmatrix} 1 & 3 & 0 & -2 \\ 2 & 7 & -3 & 4 \\ 0 & 1 & -3 & 8 \end{bmatrix}$

2. $\begin{bmatrix} 2 & 0 & 1 & -3 \\ -3 & 4 & 0 & 2 \\ 0 & 2 & 1 & 1 \end{bmatrix}$

3. $\begin{bmatrix} 0 & 1 & -3 & 4 \\ 2 & 4 & 1 & 0 \\ 2 & 1 & 10 & 2 \end{bmatrix}$

4. $\begin{bmatrix} 2 & -1 & 3 & 0 \\ 0 & 3 & 4 & -2 \\ 6 & -9 & 1 & 4 \end{bmatrix}$

5. $\begin{bmatrix} -3 & 0 & 1 & 4 \\ 2 & -1 & -2 & 0 \\ 1 & -2 & -3 & 4 \end{bmatrix}$

6. $\begin{bmatrix} 2 & 0 & -1 & 2 \\ 3 & -2 & -1 & 0 \\ 2 & -2 & 0 & 1 \end{bmatrix}$

Tell whether each of the following sets of equations has a unique solution, no solutions, or many solutions. In other words, tell whether the solution set consists of one point, no points, or many points. In still other words, tell whether each system has a unique solution, is inconsistent, or is redundant. Save your answers because you may want to refer to them while doing Exercises 3.2.B.

7. $\begin{aligned} 2x + y \quad &= 3 \\ x + \quad z &= 2 \\ 3x + y + z &= 5 \end{aligned}$

8. $\begin{aligned} 3x - 2z &= 5 \\ 2x \quad &= 8 \\ z &= 3 \end{aligned}$

9. $\begin{aligned} 2x + y \quad &= 0 \\ 2x - \quad z &= 1 \\ 2y + 3z &= -1 \end{aligned}$

10. $\begin{aligned} 3x + 5y \quad &= 1 \\ x + 2y - 3z &= 1 \\ x + y + 6z &= -1 \end{aligned}$

11. $\begin{aligned} x + y - z &= 1 \\ 2x + 2y - 2z &= 2 \\ -3x - 3y + 3z &= -3 \end{aligned}$

12. $\begin{aligned} x + y + 4z &= -2 \\ x + 2y \quad &= 4 \\ x + y + 2z &= 0 \end{aligned}$

13.
$$\begin{aligned} -y + 9z &= -2 \\ x + 2y - 3z &= 1 \\ x + y + 6z &= 3 \end{aligned}$$

14.
$$\begin{aligned} 3x + 2y - z &= 1 \\ 3y + 2z &= 0 \\ 3x - 4y - 5z &= 3 \end{aligned}$$

15.
$$\begin{aligned} 2x + z &= 1 \\ x + y - z &= -1 \\ 3x - y + 3z &= 3 \end{aligned}$$

EXERCISES 3.1. C

Three species of bacteria will be kept in one test tube and will feed on three resources. Each member of the first species consumes 3 units of the first resource and 1 unit of the third. Each bacterium of the second type consumes 1 unit of the first resource and 2 units each of the second and third. Each bacterium of the third type consumes 2 units of the first resource and 4 each of the second and third. If the test tube is supplied daily with 12,000 units of the first resource, 12,000 units of the second, and 14,000 units of the third, how many of each species can coexist in equilibrium in the test tube so that all of the supplied resources are consumed?

ANSWERS 3.1. A

1. (a) 2. (b) 3. (c) 4. (a) 5. (b) 6. (b)
7. inconsistent 8. unique solution: $(2, -\frac{5}{2}, -2)$
9. inconsistent 10. redundant 11. unique solution: $(3, \frac{2}{3}, -1)$
12. redundant 13. redundant 14. inconsistent 15. redundant

ANSWERS 3.1. C

There must be 2000 bacteria of the first type. The sum of the number of the second type plus twice the number of the third type must equal 6000.

3.2 Finding Nonunique Solutions of a System of Linear Equations

In Section 3.1 we saw that some systems of linear equations have more than one solution, and we saw how to identify such systems. If we attempt to solve them, some row of the augmented matrix will be all zeros. This tells us that the system is redundant, but it does not tell what the various solutions are; in this section we discuss how to find the solutions of a redundant system of linear equations.

Using the Gauss–Jordan row operations defined in Section 2.1, we can change the augmented matrix corresponding to the system of equations to what we now call its reduced form. If the system has a unique solution, this reduced form is as in Section 2.1:

$$\begin{bmatrix} 1 & 0 & 0 & a \\ 0 & 1 & 0 & b \\ 0 & 0 & 1 & c \end{bmatrix}$$

If not, we compromise with the following form for a matrix:

1. The first nonzero number in each row is 1; we shall refer to these 1's as the leading 1's (read "leading ones").
2. Each column containing a leading 1 is all zeros except for the leading 1.

3.2.1 Definition

A matrix satisfying the above two conditions is called a *row-reduced matrix*.

Gauss–Jordan row operations can be used to change any augmented matrix to a row-reduced matrix. The steps are similar to those in Chapter 2; one works from the left column toward the right. But if the left column is all zeros, it is impossible to get a leading 1 in that column, so we inspect the second column. Otherwise, divide all the elements of some row, whose element in the left column is not zero, by the left element of that row and interchange that row with the first. Use that leading 1 to change the other elements in the first column to zeros, and then look at the second column. If it contains only zeros except in the first row, there is no way to get a leading 1 in that column, and we then inspect the third column. Whenever we can obtain a leading 1 we do so, and then use that leading 1 to change the other elements of the column to zero, working always from left to right.

It is relatively easy to write down the solution(s) of a row-reduced matrix. This was obvious when the solution was unique, as it always was in Chapter 2. We now discuss the technique for writing the solutions when they are not unique.

Consider any nonzero row of a row-reduced matrix. It will contain a leading 1, a number in the right-hand column (a constant), and, perhaps, some other nonzero numbers—but *only* in those columns which contain no leading 1. Thus it is easy to solve for the variable corresponding to the column of the leading 1 in terms of the variables in the columns which do not contain any leading 1; the other variables (those which have a leading 1 in the corresponding column) will be missing because their coefficients are zero.

It is customary to assign a parameter to each column that does not contain a leading 1. "Parameter" is a word that mathematicians use when they cannot decide if something is a constant or a variable. We use it here to denote that some of the variables have been assigned values, but we do not know (or care) what these values are; then the other variables can be found in terms of these parameters.

We summarize these ideas in the following two rules. The rest of the section will be devoted to showing how these rules can be applied to a variety of examples.

1. Each column (except the one on the right) that does *not* contain a leading 1 gives rise to a *parameter* in the solution; we can let the variable corresponding to such a column be any number at all.
2. Each row can be used to solve for the variable corresponding to its leading 1 in terms of the parameters and constants.

3.2.2 Example

If each of the following row-reduced matrices describe in the usual way a system of linear equations, find the complete solution to the system of equations.

(a) $\begin{bmatrix} 1 & 0 & 2 & 3 \\ 0 & 1 & 4 & 5 \\ 0 & 0 & 0 & 0 \end{bmatrix}$ (b) $\begin{bmatrix} 1 & 2 & 0 & 3 \\ 0 & 0 & 1 & 4 \\ 0 & 0 & 0 & 0 \end{bmatrix}$

(c) $\begin{bmatrix} 1 & 0 & 2 & 3 \\ 0 & 1 & 0 & 4 \\ 0 & 0 & 0 & 0 \end{bmatrix}$

Solution:

First we should check and make sure each matrix is in reduced form. Each row that is not all zeros begins with a 1, and each of these leading 1's is in a column that otherwise consists entirely of zeros. So they are indeed in reduced form.

(a) Since the third column has no leading 1, the third variable (which we shall call z) can be chosen to be any value. We set z to be equal to the parameter a. Then the top two rows symbolize

$$x + 0y + 2a = 3 \qquad \text{and} \qquad 0x + y + 4a = 5$$

so the complete solution is

$$x = 3 - 2a \qquad y = 5 - 4a \qquad z = a$$

Using vector notation, this is often written

$$X = \begin{bmatrix} 3 - 2a \\ 5 - 4a \\ a \end{bmatrix} \qquad \text{or} \qquad X = \begin{bmatrix} 3 \\ 5 \\ 0 \end{bmatrix} + a \begin{bmatrix} -2 \\ -4 \\ 1 \end{bmatrix}$$

(b) This time the second column has no leading 1, so we set the second variable (which we shall call y) equal to a. Then the two nonzero rows tell us that $x + 2a = 3$ and $z = 4$, so the complete solution is

$$x = 3 - 2a \qquad y = a \qquad z = 4$$

In vector form this can be written

$$X = \begin{bmatrix} 3 - 2a \\ a \\ 4 \end{bmatrix} \qquad \text{or} \qquad X = \begin{bmatrix} 3 \\ 0 \\ 4 \end{bmatrix} + a \begin{bmatrix} -2 \\ 1 \\ 0 \end{bmatrix}$$

(c) Again we see that the third column contains no leading 1, so we set $z = a$. Then we have $x + 2a = 3$ and $y = 4$, which gives

$$x = 3 - 2a \qquad y = 4 \qquad z = a$$

so the complete solution in vector form is

$$X = \begin{bmatrix} 3 - 2a \\ 4 \\ a \end{bmatrix} \quad \text{or} \quad X = \begin{bmatrix} 3 \\ 4 \\ 0 \end{bmatrix} + a\begin{bmatrix} -2 \\ 0 \\ 1 \end{bmatrix}$$

In this expression for the solution of part (c), the vector $\begin{bmatrix} 3 \\ 4 \\ 0 \end{bmatrix}$ is called a particular solution to the system of equations; if you put these numbers into the original system (or the one symbolized by the reduced matrix), they will make the equations true. However, they are not the only set of three numbers which do this; they are the numbers we get by setting $a = 0$. If we set $a = 1$, we find that

$$X = \begin{bmatrix} 3 \\ 4 \\ 0 \end{bmatrix} + \begin{bmatrix} -2 \\ 0 \\ 1 \end{bmatrix} = \begin{bmatrix} 1 \\ 4 \\ 1 \end{bmatrix}$$

is another particular solution to the system. If we take any other specific value for a, we will obtain another particular solution to this system. If you put any triple of numbers thus obtained into the equation, they will make it true.

On the other hand, the expression

$$X = \begin{bmatrix} 3 - 2a \\ 4 \\ a \end{bmatrix} = \begin{bmatrix} 3 \\ 4 \\ 0 \end{bmatrix} + a\begin{bmatrix} -2 \\ 0 \\ 1 \end{bmatrix}$$

is called a complete solution or general solution to the system, since every solution is of this form for some a. (See Definitions 3.1.6 and 3.1.7 on page 92.)

The reasoning in these sections applies to four linear equations in four unknowns, to five linear equations in five unknowns, and so on. We shall speak of "n linear equations in n unknowns" when we want to cover these cases—and also the case of 17 linear equations in 17 unknowns, 102 linear equations in 102 unknowns, and so on. We now turn to examples of larger systems.

In these systems it is common to label the first unknown x_1, the second x_2, and the nth x_n. In this way we never run out of symbols, no matter how large the system may be.

3.2.3 Example

Suppose that you have been given a system of four equations in four unknowns and by using the Gauss–Jordan reduction method you have changed the system's augmented matrix to the following reduced form. What is the solution(s)?

(a) $\begin{bmatrix} 1 & 0 & 0 & 2 & 3 \\ 0 & 1 & 0 & 4 & 5 \\ 0 & 0 & 1 & 6 & 7 \\ 0 & 0 & 0 & 0 & 0 \end{bmatrix}$ (b) $\begin{bmatrix} 1 & 0 & 2 & 0 & 3 \\ 0 & 1 & 4 & 0 & 5 \\ 0 & 0 & 0 & 1 & 6 \\ 0 & 0 & 0 & 0 & 0 \end{bmatrix}$

(c) $\begin{bmatrix} 0 & 1 & 0 & 2 & 3 \\ 0 & 0 & 1 & 4 & 5 \\ 0 & 0 & 0 & 0 & 0 \\ 0 & 0 & 0 & 0 & 0 \end{bmatrix}$

Solution:

(a) Since the fourth column has no leading 1, we set $x_4 = a$. The first three rows tell us that $x_1 + 2a = 3$, $x_2 + 4a = 5$, and $x_3 + 6a = 7$, so the solution in vector form is

$$X = \begin{bmatrix} 3 - 2a \\ 5 - 4a \\ 7 - 6a \\ a \end{bmatrix} = \underbrace{\underbrace{\begin{bmatrix} 3 \\ 5 \\ 7 \\ 0 \end{bmatrix}}_{\text{particular solution}} + a \begin{bmatrix} -2 \\ -4 \\ -6 \\ 1 \end{bmatrix}}_{\text{general solution}}$$

Notice that $\begin{bmatrix} 3 \\ 5 \\ 7 \\ 0 \end{bmatrix}$ is the particular solution obtained for this system by setting $a = 0$ in the general solution. If we choose to set $a = 1$, we get another particular solution, $\begin{bmatrix} 1 \\ 1 \\ 1 \\ 1 \end{bmatrix}$. By setting a equal to any other number in the general solution, we can obtain another particular solution.

(b) $x_3 = a$, $x_1 + 2a = 3$, $x_2 + 4a = 5$, and $x_4 = 6$, so the general solution is

$$X = \begin{bmatrix} 3 - 2a \\ 5 - 4a \\ a \\ 6 \end{bmatrix} = \begin{bmatrix} 3 \\ 5 \\ 0 \\ 6 \end{bmatrix} + a \begin{bmatrix} -2 \\ -4 \\ 1 \\ 0 \end{bmatrix}$$

(c) If there are two rows that are all zero, there will be two columns without leading 1's. In this case we can pick both x_1 and x_4 arbitrarily: $x_1 = a$, $x_4 = b$, $x_2 + 2b = 3$, and $x_3 + 4b = 5$ gives

$$X = \begin{bmatrix} a \\ 3 - 2b \\ 5 - 4b \\ b \end{bmatrix} = \begin{bmatrix} 0 \\ 3 \\ 5 \\ 0 \end{bmatrix} + a\begin{bmatrix} 1 \\ 0 \\ 0 \\ 0 \end{bmatrix} + b\begin{bmatrix} 0 \\ -2 \\ -4 \\ 1 \end{bmatrix}$$

The same approach can be used for 5 by 5 matrices, although the possibilities are more complex.

3.2.4 Example

Suppose that you have been given a system of five equations in five unknowns and by using the Gauss–Jordan reduction method, you have changed the system's augmented matrix to the following reduced form. What is the solution(s)?

(a) $\begin{bmatrix} 1 & 0 & 0 & 3 & 4 & 7 \\ 0 & 1 & 0 & 2 & 9 & 8 \\ 0 & 0 & 1 & -1 & 8 & 6 \\ 0 & 0 & 0 & 0 & 0 & 0 \\ 0 & 0 & 0 & 0 & 0 & 0 \end{bmatrix}$

(b) $\begin{bmatrix} 0 & 1 & 0 & 8 & 0 & 9 \\ 0 & 0 & 1 & -1 & 0 & 3 \\ 0 & 0 & 0 & 0 & 1 & -4 \\ 0 & 0 & 0 & 0 & 0 & 0 \\ 0 & 0 & 0 & 0 & 0 & 0 \end{bmatrix}$

(c) $\begin{bmatrix} 1 & 0 & 0 & 7 & -3 & 4 \\ 0 & 0 & 1 & 6 & 8 & 4 \\ 0 & 0 & 0 & 0 & 0 & 0 \\ 0 & 0 & 0 & 0 & 0 & 0 \\ 0 & 0 & 0 & 0 & 0 & 0 \end{bmatrix}$

Solution:

(a) $x_4 = a$, $x_5 = b$, $x_1 + 3a + 4b = 7$, $x_2 + 2a + 9b = 8$, $x_3 - a + 8b = 6$, so the general solution is

$$X = \begin{bmatrix} 7 - 3a - 4b \\ 8 - 2a - 9b \\ 6 + a - 8b \\ a \\ b \end{bmatrix} = \begin{bmatrix} 7 \\ 8 \\ 6 \\ 0 \\ 0 \end{bmatrix} + a\begin{bmatrix} -3 \\ -2 \\ 1 \\ 1 \\ 0 \end{bmatrix} + b\begin{bmatrix} -4 \\ -9 \\ -8 \\ 0 \\ 1 \end{bmatrix}$$

(b) $x_1 = a$, $x_4 = b$, $x_2 + 8b = 9$, $x_3 - b = 3$, $x_5 = -4$, so the general solution is

$$X = \begin{bmatrix} a \\ 9-8b \\ 3+b \\ b \\ -4 \end{bmatrix} = \begin{bmatrix} 0 \\ 9 \\ 3 \\ 0 \\ -4 \end{bmatrix} + a\begin{bmatrix} 1 \\ 0 \\ 0 \\ 0 \\ 0 \end{bmatrix} + b\begin{bmatrix} 0 \\ -8 \\ 1 \\ 1 \\ 0 \end{bmatrix}$$

(c) $x_2 = a$, $x_4 = b$, $x_5 = c$, $x_1 + 7b - 3c = 4$, $x_3 + 6b + 8c = 4$, so the general solution is

$$X = \begin{bmatrix} 4-7b+3c \\ a \\ 4-6b-8c \\ b \\ c \end{bmatrix} = \begin{bmatrix} 4 \\ 0 \\ 4 \\ 0 \\ 0 \end{bmatrix} + a\begin{bmatrix} 0 \\ 1 \\ 0 \\ 0 \\ 0 \end{bmatrix} + b\begin{bmatrix} -7 \\ 0 \\ -6 \\ 1 \\ 0 \end{bmatrix} + c\begin{bmatrix} 3 \\ 0 \\ -8 \\ 0 \\ 1 \end{bmatrix}$$

3.2.5 Example

In the previous chapters we considered "open" Leontief models, in which some inputs (called "primary inputs") and some outputs (called "final demands") were not included in the technological matrix of the model. Sometimes it is useful for economists to consider a closed Leontief model, which incorporates *all* inputs and outputs into the technological matrix. One step in doing this is to assume that the "labor" input is the same sector as the "households" output and to make a "labor" column in the matrix corresponding to the "households" row. In a closed Leontief model there is no "final demands" vector and no "primary input"; what goes into the economy is the same as what goes out. The technological equation is then

$$\begin{bmatrix} \text{technological} \\ \text{matrix} \end{bmatrix}\begin{bmatrix} \text{gross} \\ \text{outputs} \end{bmatrix} = \begin{bmatrix} \text{gross} \\ \text{outputs} \end{bmatrix}$$

or merely

$$AX = X$$

The following closed Leontief model might describe the economy of an entire country.

x = government's budget
y = value of industrial output
z = households' budget

$$\left\{\begin{matrix}\text{fraction of}\\\text{government budget}\\\text{that is}\\\text{consumed by}\\\text{government}\end{matrix}\right\}x + \left\{\begin{matrix}\text{fraction of}\\\text{industrial}\\\text{output that}\\\text{is paid in}\\\text{taxes}\end{matrix}\right\}y + \left\{\begin{matrix}\text{fraction of}\\\text{households'}\\\text{budget that}\\\text{is paid in}\\\text{taxes}\end{matrix}\right\}z = x$$

$$\left\{\begin{matrix}\text{fraction of}\\\text{government budget}\\\text{that is}\\\text{invested in}\\\text{industry}\end{matrix}\right\}x + \left\{\begin{matrix}\text{fraction of}\\\text{industrial}\\\text{output}\\\text{reinvested}\end{matrix}\right\}y + \left\{\begin{matrix}\text{fraction of}\\\text{households'}\\\text{budget that}\\\text{is invested}\\\text{(private savings)}\end{matrix}\right\}z = y$$

$$\left\{\begin{matrix}\text{fraction of}\\\text{government budget}\\\text{given to}\\\text{households}\end{matrix}\right\}x + \left\{\begin{matrix}\text{fraction of}\\\text{industrial}\\\text{output paid}\\\text{to households}\\\text{(wages and}\\\text{salaries)}\end{matrix}\right\}y + \left\{\begin{matrix}\text{fraction of}\\\text{households'}\\\text{budget that}\\\text{is consumed}\end{matrix}\right\}z = z$$

A little reflection reveals that *the sum of each column in the coefficient matrix is 1.*

Find all the gross output vectors that satisfy the closed Leontief model $AX = X$ when

$$A = \begin{bmatrix} 0.4 & 0.2 & 0.3 \\ 0.3 & 0.1 & 0.1 \\ 0.3 & 0.7 & 0.6 \end{bmatrix}$$

Solution:

The given equations are

$$\begin{aligned} 0.4x + 0.2y + 0.3z &= x \\ 0.3x + 0.1y + 0.1z &= y \\ 0.3x + 0.7y + 0.6z &= z \end{aligned} \qquad \text{or} \qquad \begin{aligned} 0 &= 0.6x - 0.2y - 0.3z \\ 0 &= -0.3x + 0.9y - 0.1z \\ 0 &= -0.3x - 0.7y + 0.4z \end{aligned}$$

Writing the equations on the right as an augmented matrix, we now proceed with the Gauss–Jordan reduction method.

$$\begin{bmatrix} 0.6 & -0.2 & -0.3 & 0 \\ -0.3 & 0.9 & -0.1 & 0 \\ -0.3 & -0.7 & 0.4 & 0 \end{bmatrix} \qquad \begin{matrix} \frac{10}{6}R_1 \longrightarrow R_1 \\ 10R_2 \longrightarrow R_2 \\ 10R_3 \longrightarrow R_3 \end{matrix}$$

$$\begin{bmatrix} 1 & -\frac{1}{3} & -\frac{1}{2} & 0 \\ -3 & 9 & -1 & 0 \\ -3 & -7 & 4 & 0 \end{bmatrix} \qquad \begin{matrix} R_2 + 3R_1 \longrightarrow R_2 \\ R_3 + 3R_1 \longrightarrow R_3 \end{matrix}$$

$$\begin{bmatrix} 1 & -\frac{1}{3} & -\frac{1}{2} & 0 \\ 0 & 8 & -\frac{5}{2} & 0 \\ 0 & -8 & \frac{5}{2} & 0 \end{bmatrix} \qquad R_3 + R_2 \longrightarrow R_3$$

$$\begin{bmatrix} 1 & -\frac{1}{3} & -\frac{1}{2} & 0 \\ 0 & 8 & -\frac{5}{2} & 0 \\ 0 & 0 & 0 & 0 \end{bmatrix} \qquad \tfrac{1}{8}R_2 \longrightarrow R_2$$

$$\begin{bmatrix} 1 & -\frac{1}{3} & -\frac{1}{2} & 0 \\ 0 & 1 & -\frac{5}{16} & 0 \\ 0 & 0 & 0 & 0 \end{bmatrix} \qquad R_1 + \tfrac{1}{3}R_2 \longrightarrow R_1$$

$$\begin{bmatrix} 1 & 0 & -\frac{29}{48} & 0 \\ 0 & 1 & -\frac{5}{16} & 0 \\ 0 & 0 & 0 & 0 \end{bmatrix}$$

From this last matrix we conclude that $z = a$, $x = 29a/48$, $y = 5a/16$.

The economic interpretation is that an economy will satisfy the given equation if the government budget is $\frac{29}{48}$ times the total households' budget and the industrial budget is $\frac{5}{16}$ times the households' budget.

SUMMARY

If a system of linear equations has more than one solution, it will have an infinite number of particular solutions. If we symbolically express all these particular solutions in one vector (which must, necessarily, contain at least one parameter), this expression is called a general solution to the system. To find the general solution of a system of linear equations, use Gauss–Jordan row operations to change the corresponding augmented matrix to its reduced form, and then pick off the solution using parameters. Several examples of how this is done were given in this section, including one application to a closed Leontief model.

EXERCISES 3.2. A

Save your answers and work from this exercise set because you may want to refer to them while working on Exercises 3.4.A.

1. Find the general solutions to problems 10, 12, 13, and 15 of Exercises 3.1.A.
2. Find the general solution to each of the following systems of equations.

(a)
$$\begin{aligned} x + 2y \phantom{{}-z} &= -1 \\ x + 2y - z &= -4 \\ 3x + 6y + z &= 0 \end{aligned}$$

(b)
$$\begin{aligned} 0x + 2y \phantom{{}+z} &= 4 \\ y + z &= 5 \end{aligned}$$

(c)
$$\begin{aligned} 2x_1 + \phantom{x_2 + {}} 4x_3 \phantom{{}+x_4} &= -2 \\ 3x_1 + x_2 + 4x_3 \phantom{{}+x_4} &= -1 \\ x_1 + \phantom{x_2 + {}} 2x_3 + x_4 &= 5 \\ 2x_2 - 4x_3 + x_4 &= 10 \end{aligned}$$

(d)
$$\begin{aligned} x_1 - 3x_2 + \phantom{x_3 + {}} 4x_4 &= 2 \\ x_1 - 3x_2 + x_3 + 2x_4 &= 9 \\ 2x_1 - 6x_2 - x_3 + 10x_4 &= -3 \\ 2x_3 - 4x_4 &= 14 \end{aligned}$$

3. Suppose that you have used the Gauss–Jordan reduction method on the augmented matrix corresponding to a given system of linear equations until you arrive at the reduced matrix given here. Give the general solution for the original system of equations.

(a) $\begin{bmatrix} 1 & 0 & 2 & -4 \\ 0 & 1 & -1 & 3 \\ 0 & 0 & 0 & 0 \end{bmatrix}$ (b) $\begin{bmatrix} 1 & 2 & 0 & -4 \\ 0 & 0 & 1 & 7 \\ 0 & 0 & 0 & 0 \end{bmatrix}$

(c) $\begin{bmatrix} 1 & 0 & 1 & 0 & 5 \\ 0 & 1 & 2 & 0 & -3 \\ 0 & 0 & 0 & 1 & 4 \\ 0 & 0 & 0 & 0 & 0 \end{bmatrix}$ (d) $\begin{bmatrix} 1 & 0 & -2 & 3 & 4 \\ 0 & 1 & 1 & -2 & 5 \\ 0 & 0 & 0 & 0 & 0 \\ 0 & 0 & 0 & 0 & 0 \end{bmatrix}$

(e) $\begin{bmatrix} 1 & 2 & 3 & 0 & -5 \\ 0 & 0 & 0 & 1 & 4 \\ 0 & 0 & 0 & 0 & 0 \\ 0 & 0 & 0 & 0 & 0 \end{bmatrix}$ (f) $\begin{bmatrix} 1 & 0 & 2 & 0 & 3 & 5 \\ 0 & 1 & 5 & 0 & 6 & 7 \\ 0 & 0 & 0 & 1 & 8 & 9 \\ 0 & 0 & 0 & 0 & 0 & 0 \\ 0 & 0 & 0 & 0 & 0 & 0 \end{bmatrix}$

EXERCISES 3.2. B

Save your answers and work from this exercise set because you will want to refer to them while working on Exercises 3.4.B.

1. Find the general solutions to problems 7, 10, 11, and 15 of Exercises 3.1.B.
2. Find the general solution to each of the following systems of equations.

(a) $\begin{aligned} x + y + z &= 7 \\ 2x + 4z &= 8 \\ 3x - 2y + 8z &= 6 \end{aligned}$ (b) $\begin{aligned} 2x - 4y &= 8 \\ 2x - 4y + z &= 6 \\ -3x + 6y &= -12 \end{aligned}$

(c) $\begin{aligned} x_1 + x_2 - x_3 &= 3 \\ 2x_2 + 2x_3 - x_4 &= 6 \\ x_1 - 2x_3 + 3x_4 &= -5 \\ 2x_1 - 4x_3 &= 2 \end{aligned}$ (d) $\begin{aligned} x_1 - 2x_2 + x_3 + 4x_4 &= -2 \\ 3x_1 - 6x_2 + 12x_4 &= -9 \\ 3x_1 - 6x_2 + 2x_3 + 12x_4 &= -7 \\ -2x_1 + 4x_2 + x_3 - 8x_4 &= 7 \end{aligned}$

3. Suppose that you have used the Gauss–Jordan reduction method on the augmented matrix corresponding to a given system of linear equations until you arrive at the matrix given below. Give the general solution for the original system of equations.

(a) $\begin{bmatrix} 1 & 0 & -3 & 2 \\ 0 & 1 & 2 & -1 \\ 0 & 0 & 0 & 0 \end{bmatrix}$ (b) $\begin{bmatrix} 1 & -4 & 0 & 2 \\ 0 & 0 & 1 & 5 \\ 0 & 0 & 0 & 0 \end{bmatrix}$

(c) $\begin{bmatrix} 1 & -2 & 0 & 0 & 7 \\ 0 & 0 & 1 & 0 & 5 \\ 0 & 0 & 0 & 1 & -2 \\ 0 & 0 & 0 & 0 & 0 \end{bmatrix}$

(d) $\begin{bmatrix} 1 & 10 & 0 & -3 & 6 \\ 0 & 0 & 1 & 4 & -5 \\ 0 & 0 & 0 & 0 & 0 \\ 0 & 0 & 0 & 0 & 0 \end{bmatrix}$

(e) $\begin{bmatrix} 1 & 0 & -4 & -1 & 8 \\ 0 & 1 & 2 & 3 & -7 \\ 0 & 0 & 0 & 0 & 0 \\ 0 & 0 & 0 & 0 & 0 \end{bmatrix}$

(f) $\begin{bmatrix} 1 & 1 & 0 & -2 & 0 & 6 \\ 0 & 0 & 1 & 3 & 0 & -2 \\ 0 & 0 & 0 & 0 & 1 & 4 \\ 0 & 0 & 0 & 0 & 0 & 0 \\ 0 & 0 & 0 & 0 & 0 & 0 \end{bmatrix}$

EXERCISES 3.2. C

1. Prove that any closed Leontief model ($AX = X$, where the sums of the columns of A are all 1 and the elements in A are all positive) has many solutions. That is, prove that any system of 3 equations in 3 unknowns such that the sum of the coefficients of each unknown is 1 is a redundancy. [*Hint:* Consider the set of equations

$$\begin{aligned} ax + by + cz &= x \\ dx + ey + fz &= y \\ (1 - a - d)x + (1 - b - e)y + (1 - c - f)z &= z \end{aligned}$$

and use Gauss–Jordan row operations to reduce the corresponding augmented matrix to one where there is a row of all zeros.]

ANSWERS 3.2. A

1. (10) $X = \begin{bmatrix} 2 \\ 2 \\ 0 \end{bmatrix} + a\begin{bmatrix} -1 \\ -\frac{3}{2} \\ 1 \end{bmatrix}$ (12) $X = \begin{bmatrix} -1 \\ \frac{5}{2} \\ 0 \end{bmatrix} + a\begin{bmatrix} -3 \\ 2 \\ 1 \end{bmatrix}$

(13) $X = \begin{bmatrix} 1 \\ 0 \\ 0 \end{bmatrix} + a\begin{bmatrix} -1 \\ 1 \\ 0 \end{bmatrix} + b\begin{bmatrix} -2 \\ 0 \\ 1 \end{bmatrix}$ (15) $X = \begin{bmatrix} -10 \\ 2 \\ 0 \end{bmatrix} + a\begin{bmatrix} 7 \\ -3 \\ 1 \end{bmatrix}$

2. (a) $X = \begin{bmatrix} -1 \\ 0 \\ 3 \end{bmatrix} + a\begin{bmatrix} -2 \\ 1 \\ 0 \end{bmatrix}$ (b) $X = \begin{bmatrix} 0 \\ 2 \\ 3 \end{bmatrix} + a\begin{bmatrix} 1 \\ 0 \\ 0 \end{bmatrix}$

(c) $X = \begin{bmatrix} -1 \\ 2 \\ 0 \\ 6 \end{bmatrix} + a\begin{bmatrix} -2 \\ 2 \\ 1 \\ 0 \end{bmatrix}$ (d) $X = \begin{bmatrix} 2 \\ 0 \\ 7 \\ 0 \end{bmatrix} + a\begin{bmatrix} 3 \\ 1 \\ 0 \\ 0 \end{bmatrix} + b\begin{bmatrix} -4 \\ 0 \\ 2 \\ 1 \end{bmatrix}$

3. (a) $X = \begin{bmatrix} -4 \\ 3 \\ 0 \end{bmatrix} + a\begin{bmatrix} -2 \\ 1 \\ 1 \end{bmatrix}$ (b) $X = \begin{bmatrix} -4 \\ 0 \\ 7 \end{bmatrix} + a\begin{bmatrix} -2 \\ 1 \\ 0 \end{bmatrix}$

(c) $X = \begin{bmatrix} 5 \\ -3 \\ 0 \\ 4 \end{bmatrix} + a\begin{bmatrix} -1 \\ -2 \\ 1 \\ 0 \end{bmatrix}$ (d) $X = \begin{bmatrix} 4 \\ 5 \\ 0 \\ 0 \end{bmatrix} + a\begin{bmatrix} 2 \\ -1 \\ 1 \\ 0 \end{bmatrix} + b\begin{bmatrix} -3 \\ 2 \\ 0 \\ 1 \end{bmatrix}$

(e)

$$X = \begin{bmatrix} -5 \\ 0 \\ 0 \\ 4 \end{bmatrix} + a\begin{bmatrix} -2 \\ 1 \\ 0 \\ 0 \end{bmatrix} + b\begin{bmatrix} -3 \\ 0 \\ 1 \\ 0 \end{bmatrix}$$

(f)

$$X = \begin{bmatrix} 5 \\ 7 \\ 0 \\ 9 \\ 0 \end{bmatrix} + a\begin{bmatrix} -2 \\ -5 \\ 1 \\ 0 \\ 0 \end{bmatrix} + b\begin{bmatrix} -3 \\ -6 \\ 0 \\ -8 \\ 1 \end{bmatrix}$$

ANSWERS 3.2. C

1. Bringing the variables to the left side, the system becomes

$$\begin{aligned} (a-1)x + by + cz &= 0 \\ dx + (e-1)y + fz &= 0 \\ (1-a-d)x + (1-b-e)y + (-c-f)z &= 0 \end{aligned}$$

The augmented matrix is

$$\begin{bmatrix} a-1 & b & c & 0 \\ d & e-1 & f & 0 \\ 1-a-d & 1-b-e & -c-f & 0 \end{bmatrix} \quad R_3 + R_1 \longrightarrow R_3$$

$$\begin{bmatrix} a-1 & b & c & 0 \\ d & e-1 & f & 0 \\ -d & 1-e & -f & 0 \end{bmatrix} \quad R_3 + R_2 \longrightarrow R_3$$

$$\begin{bmatrix} a-1 & b & c & 0 \\ d & e-1 & f & 0 \\ 0 & 0 & 0 & 0 \end{bmatrix}$$

Since the last row is all zeros, the original system was redundant and there are many solutions.

★3.3 Analysis of Traffic Flow Networks

Linear equations with many solutions can be used in a part of applied mathematics called network analysis. Network analysis had its origins in electrical engineering, but it is now often applied to other fields, including information theory and the study of transportation systems. In this section we explore an example in one such field that is relatively easy to understand.

3.3.1 Example

One part of a city's network of traffic is shown in Figure 3.3–1, with the number of cars that enter and leave during a peak evening hour as shown. (These numbers can be obtained by laying a cable across the street.) All the streets are one-way in the direction indicated. Notice that there are 700 cars entering the system each hour and 700 leaving, so there is hope of keeping them in equilibrium.

Figure 3.3–1

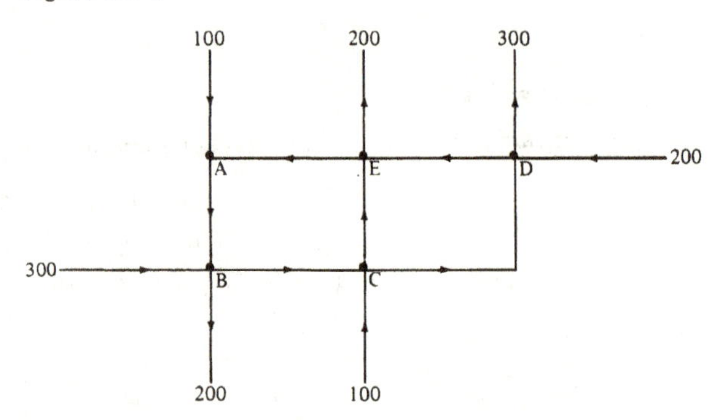

(a) Construct a mathematical model that describes this system using linear algebra.
(b) Suppose that your tax dollars are at work fixing road segment AE so that the stretch of road has been completely cut off. What will be the traffic flow along the other stretches?
(c) Suppose that, in consideration of those poor souls who want to travel from D to B, 100 cars are permitted to trickle each hour through path AE. How does this affect the other branches?

Figure 3.3–2

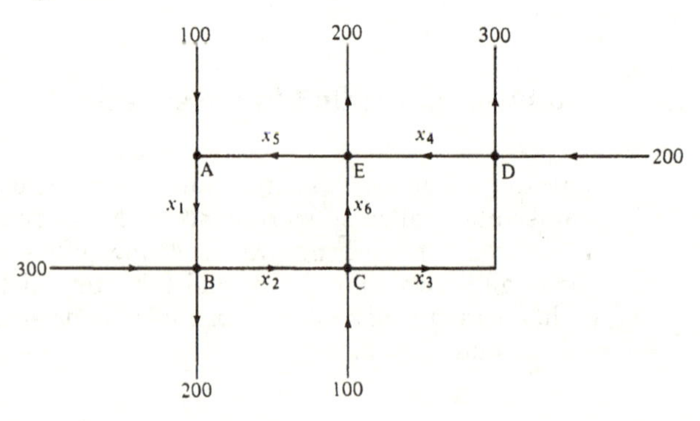

Solution:

(a) Since the variables that we will solve for are the number of cars that travel each branch of the network, we label the branches as indicated in Figure 3.3–2. The basis for setting up the equations in the analysis is the assumption that the number of cars that leave each intersection will be the same as the number that enter it. Thus $(100 + x_5)$ cars enter intersection A and x_1 cars leave it, and so it must be true that $x_1 = x_5 + 100$. We can write similar equations for each intersection as follows:

$$
\begin{aligned}
&A: && x_1 = x_5 + 100\\
&B: && x_1 + 300 = x_2 + 200\\
&C: && x_2 + 100 = x_3 + x_6\\
&D: && x_3 + 200 = x_4 + 300\\
&E: && x_4 + x_6 = x_5 + 200
\end{aligned}
$$

Preparing to write an augmented matrix, we can rearrange these equations:

$$
\begin{array}{rrrrrrcr}
x_1 & & & & - x_5 & & = & 100\\
x_1 & - x_2 & & & & & = & -100\\
& x_2 & - x_3 & & & - x_6 & = & -100\\
& & x_3 & - x_4 & & & = & 100\\
& & & x_4 & - x_5 & + x_6 & = & 200
\end{array}
$$

Writing this system as an augmented matrix, we then apply the Gauss–Jordan row operations.

$$
\begin{bmatrix}
1 & 0 & 0 & 0 & -1 & 0 & 100\\
1 & -1 & 0 & 0 & 0 & 0 & -100\\
0 & 1 & -1 & 0 & 0 & -1 & -100\\
0 & 0 & 1 & -1 & 0 & 0 & 100\\
0 & 0 & 0 & 1 & -1 & 1 & 200
\end{bmatrix}
\quad R_2 - R_1 \longrightarrow R_2
$$

$$
\begin{bmatrix}
1 & 0 & 0 & 0 & -1 & 0 & 100\\
0 & -1 & 0 & 0 & 1 & 0 & -200\\
0 & 1 & -1 & 0 & 0 & -1 & -100\\
0 & 0 & 1 & -1 & 0 & 0 & 100\\
0 & 0 & 0 & 1 & -1 & 1 & 200
\end{bmatrix}
\quad
\begin{array}{l}
-R_2 \longrightarrow R_2\\
-R_3 \longrightarrow R_3\\
-R_4 \longrightarrow R_4\\
-R_5 \longrightarrow R_5
\end{array}
$$

$$
\begin{bmatrix}
1 & 0 & 0 & 0 & -1 & 0 & 100\\
0 & 1 & 0 & 0 & -1 & 0 & 200\\
0 & -1 & 1 & 0 & 0 & 1 & 100\\
0 & 0 & -1 & 1 & 0 & 0 & -100\\
0 & 0 & 0 & -1 & 1 & -1 & -200
\end{bmatrix}
\quad R_3 + R_2 \longrightarrow R_3
$$

$$\begin{bmatrix} 1 & 0 & 0 & 0 & -1 & 0 & 100 \\ 0 & 1 & 0 & 0 & -1 & 0 & 200 \\ 0 & 0 & 1 & 0 & -1 & 1 & 300 \\ 0 & 0 & -1 & 1 & 0 & 0 & -100 \\ 0 & 0 & 0 & -1 & 1 & -1 & -200 \end{bmatrix} \quad R_4 + R_3 \longrightarrow R_4$$

$$\begin{bmatrix} 1 & 0 & 0 & 0 & -1 & 0 & 100 \\ 0 & 1 & 0 & 0 & -1 & 0 & 200 \\ 0 & 0 & 1 & 0 & -1 & 1 & 300 \\ 0 & 0 & 0 & 1 & -1 & 1 & 200 \\ 0 & 0 & 0 & -1 & 1 & -1 & -200 \end{bmatrix} \quad R_5 + R_4 \longrightarrow R_5$$

$$\begin{bmatrix} 1 & 0 & 0 & 0 & -1 & 0 & 100 \\ 0 & 1 & 0 & 0 & -1 & 0 & 200 \\ 0 & 0 & 1 & 0 & -1 & 1 & 300 \\ 0 & 0 & 0 & 1 & -1 & 1 & 200 \\ 0 & 0 & 0 & 0 & 0 & 0 & 0 \end{bmatrix}$$

Expressing this augmented matrix in terms of the original variables,

$$x_1 = x_5 + 100$$
$$x_2 = x_5 + 200$$
$$x_3 = x_5 - x_6 + 300$$
$$x_4 = x_5 - x_6 + 200$$

(b) If $x_5 = 0$, then $x_1 = 100$ cars per hour and $x_2 = 200$ cars per hour. Since x_4 must be at least zero (unless the signs are rearranged to reverse the traffic flow), we must have $x_6 \leq 200$. The driving public can cause x_6 to be any number between 0 and 200 without having a traffic jam. If it exceeds 200, there will be a pileup of cars wanting x_4 to be negative—unless someone quickly reverses the one-way signs on ED! Assuming that x_6 is indeed between 0 and 200, it follows that $x_3 = 300 - x_6$ and $x_4 = 200 - x_6$.

(c) If $x_5 = 100$, then $x_1 = 200$ and $x_2 = 300$. Also, $x_3 = 400 - x_6$ and $x_4 = 300 - x_6$. Thus x_6 must be some number between 0 and 300 and the others are as given.

Notice that it would be very strange if any of the variables exceeded 700, because that is the number entering the system in one hour. If there were more than 700 on any one path, that would imply that some cars had stayed around from the previous hour. Since most people are not that fond of driving, probably all variables will be less than 700.

(Example 3.3.1 was the first time this book showed how to reduce an m by n matrix where $m \neq n$, but no new techniques were needed. In general, you will find no more difficulty in the cases where the number of unknowns is different from the number of equations.)

SUMMARY

When the typical traffic flow in and out of some closed network has been measured, the flow within that network can be analyzed using

Gauss–Jordan row operations. The variables describe the number of cars traveling along each closed branch of the network and the equations are set up using the fact that the number of cars entering each intersection must equal the number leaving. There will generally be some parameters in the solution, reflecting the fact that there are many ways that traffic can flow through the network without causing congestion. There will, however, be bounds on the number of cars accommodated without tie-ups, and these bounds can be concluded under varying circumstances from the mathematical model.

EXERCISES 3.3. A

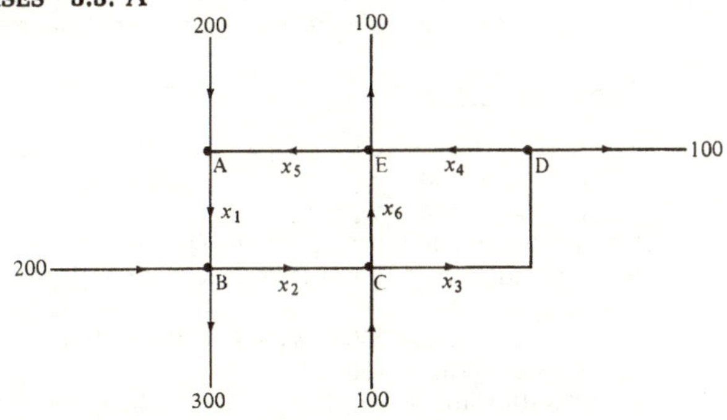

1. Write five equations describing the traffic flow shown in the diagram, one for each of the lettered intersections.
2. Use the Gauss–Jordan reduction process to write x_1, x_2, x_3, and x_4 in terms of the other two variables.
3. If the route AE has been flooded out, so that $x_5 = 0$, what is the maximum number of cars that x_6 can carry in order to avoid traffic congestion?
4. Under the conditions of problem 3, find the values of x_1, x_2, x_3, and x_4.
5. Suppose that a large carnival is planning to settle temporarily along road stretch AB. It will clearly slow down traffic. As a minimum, how large must x_1 be to avoid a terrible snarl?

EXERCISES 3.3. B

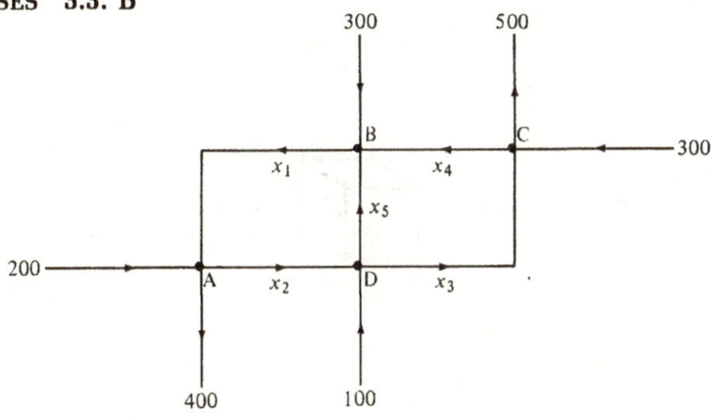

1. Write the four equations describing the traffic flow shown in the diagram, one for each of the lettered intersections.
2. Use the Gauss–Jordan reduction process to write x_1, x_2, and x_3 in terms of the other two variables.
3. Suppose the route AD is being repaired so that $x_2 \leq 200$. How large can x_4 be under this condition? If it is as large as possible, find the values of the other variables.
4. If AD is being repaired and $x_2 \leq 200$, how large could x_5 possibly be? If x_5 is as large as possible, find the values of the other variables.
5. Suppose that a large public function is being held at intersection C, after which at least 500 carloads of guests can be anticipated to drive out a one-way exit toward intersection B. What is the minimum number that x_1, x_2, and x_3 can be expected to be in that hour?

ANSWERS 3.3. A

1. A: $x_1 = x_5 + 200$ D: $x_3 = x_4 + 100$
   B: $x_1 + 200 = x_2 + 300$ E: $x_4 + x_6 = 100 + x_5$
   C: $x_2 + 100 = x_3 + x_6$
2. $x_1 = x_5 + 200$, $x_2 = x_5 + 100$, $x_3 = x_5 - x_6 + 200$, $x_4 = x_5 - x_6 + 100$
3. 100; since $x_4 \geq 0$ and $x_5 = 0$, $x_6 \leq 100$, so the maximum possible value is $x_6 = 100$.
4. $x_1 = 0 + 200 = 200$, $x_2 = 0 + 100 = 100$, $x_3 = 0 - 100 + 200 = 100$, $x_4 = 0 - 100 + 100 = 0$.
5. 200 (But notice that if $x_1 = 200$, nobody can plan on driving from E to A through this network!)

3.4 Geometric Interpretations of Linear Equations

During your study of two equations in two unknowns in previous algebra courses you probably drew graphs of the equations. Each linear equation in two unknowns corresponded to a straight line on the graph, and the point where the two straight lines intersected (if there was one) corresponded to the solution of the two equations. In this section we shall explore (similar?) geometric interpretations of systems of equations in more than two variables.

Figure 3.4–1

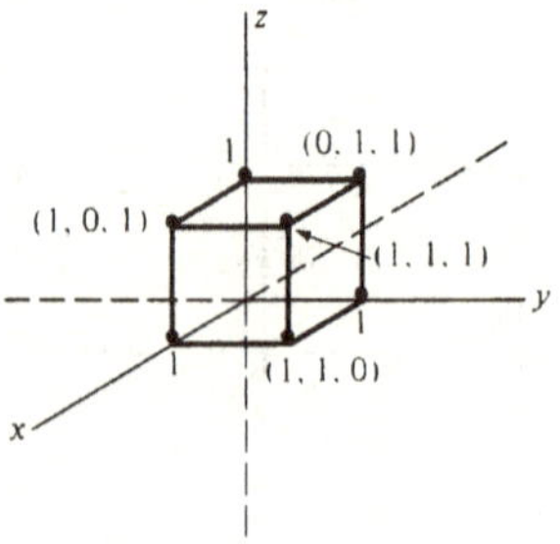

The situation quickly becomes complicated. If we have three variables, the graph must be in three dimensions. One way to picture a three-dimensional graph is to imagine the xy-plane down on a table and the z-axis rising up perpendicular to the table (Figure 3.4–1). To map a point (a, b, c) we travel a distance a in the direction of the x-axis, then a distance b in the direction of the y-axis, and then a distance c in the direction of the z-axis. It is tricky to draw pictures of three-dimensional objects on two-dimensional paper and expensive to print books in more than two dimensions. But Figures 3.4–1 and 3.4–2 should help you understand how we can indicate points in three-dimensional space on two-dimensional paper.

3.4.1 Example

Graph the points (2, 0, 0), (2, 3, 0), and (2, 3, 4).

Solution:

Figure 3.4–2

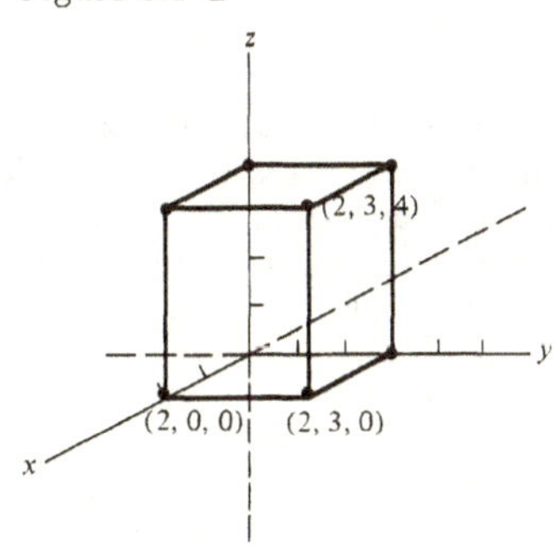

Next we hope to convince you that each linear equation in three unknowns corresponds to a specific plane in 3-space (that is, "three-dimensional space"). This may not be immediately obvious, but a little discussion should make it plausible.

Figure 3.4–3

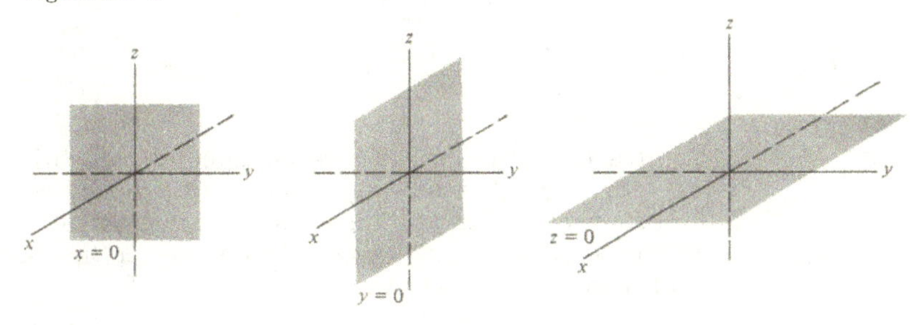

First, the equation $z = 0$ corresponds to the xy-plane—the plane "on the table" in Figure 3.4–3. The equation $x = 0$ corresponds to the yz-plane, and the equation $y = 0$ corresponds to the xz-plane (Figure 3.4–3).

As in two-dimensional theory, it is conventional to identify an equation with its graph and say that it "is" its graph. Thus the equation $z = 5$ "is" the plane 5 units above the xy-plane (Figure 3.4–4).

Figure 3.4–4

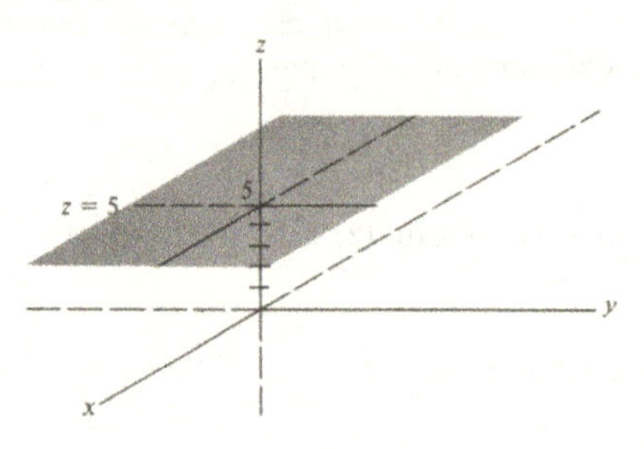

The equation $x = y$ goes through the z-axis and cuts the xy-plane perpendicularly at a 45° angle from both axes. Any other equation in x and y *only* will also be perpendicular to the xy-plane and will cut that plane in a line describing the two-dimensional graph of the given equation (Figure 3.4–5). Similarly, an equation in x and z only will cut the xz-plane at right angles.

Figure 3.4–5

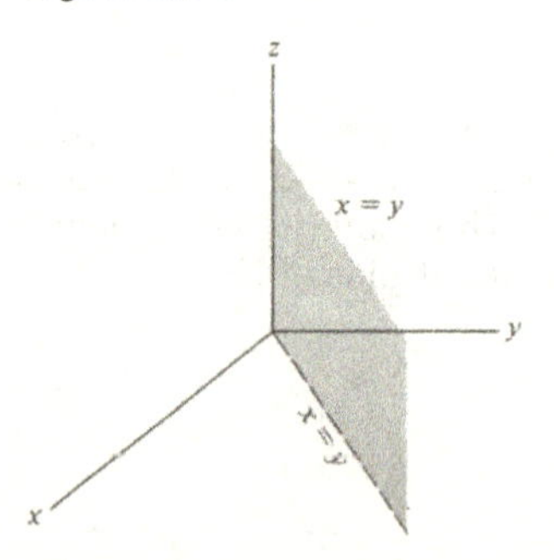

If a linear equation shows all three of the variables in a form $ax + by + cz = 0$, where a, b, and c are all *not* zero, then the corresponding plane cuts all three axes at a tilt. We can then set any two of the variables equal to zero and solve for the third variable to find where it cuts that axis.

3.4.2 Example

Graph the equation $x + y + z = 1$.

Solution:

We show where this plane cuts each of the axes. Setting $x = y = 0$, we find that $z = 1$, so the plane cuts the z-axis at $z = 1$. We make similar quick calculations in the other two cases and draw the triangle shown in Figure 3.4–6. If you imagine the triangle to be extended infinitely far in all directions, you are imagining a set of points (a plane) that corresponds to the equation $x + y + z = 1$.

Figure 3.4–6

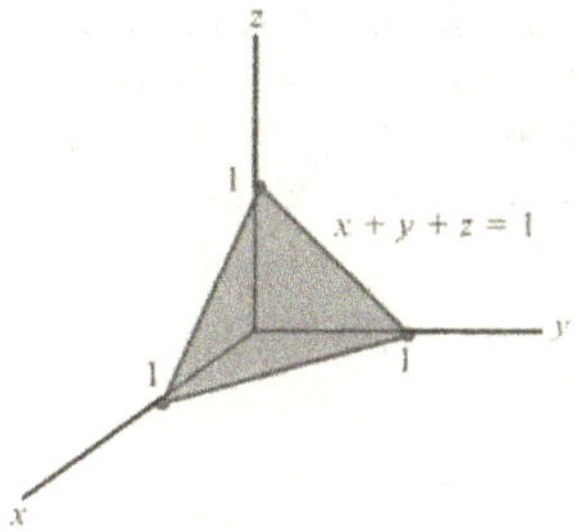

3.4.3 Example

Graph the equation $2x + y + 3z = 6$.

Solution:

First, we set $y = z = 0$ and solve for x. Since $2x = 6$ implies that $x = 3$, the point (3, 0, 0) is in the solution set: the plane cuts the x-axis where $x = 3$. Setting $x = z = 0$, we find that the plane cuts the y-axis where $y = 6$. And setting $x = y = 0$, we see that $3z = 6$, which implies that $z = 2$, so the plane cuts the z-axis where $z = 2$. Plotting these points, we get the picture in Figure 3.4–7.

Figure 3.4–7

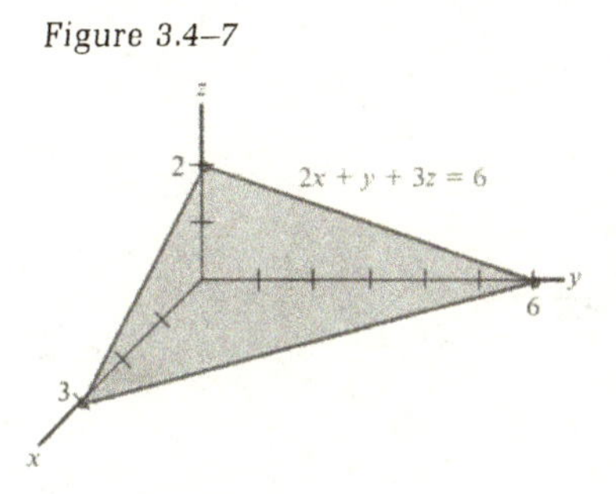

Since the coefficients of x, y, and z are all positive in the preceding examples, the triangles indicating their graphs lie in the first octant—that part of 3-space where all the variables are positive. If one or more of the coefficients are negative, the triangles are harder (but not impossibly difficult) to draw because they are not in the first octant. The exercises will provide you an opportunity to sketch the basic triangles yourself that indicate the location of other linear equations in 3-space.

Now we turn to *systems* of linear equations in three unknowns. A single linear equation, as we have indicated, has a plane as its solution set. What is the geometric figure corresponding to the solution set of 2 linear equations in 3 unknowns?

It depends! There are three possibilities.

1. If the planes are parallel, there are no common points (Figure 3.4–8). The solution set is the null set (that is, the empty set). In this case the set of equations is inconsistent.

Figure 3.4–8

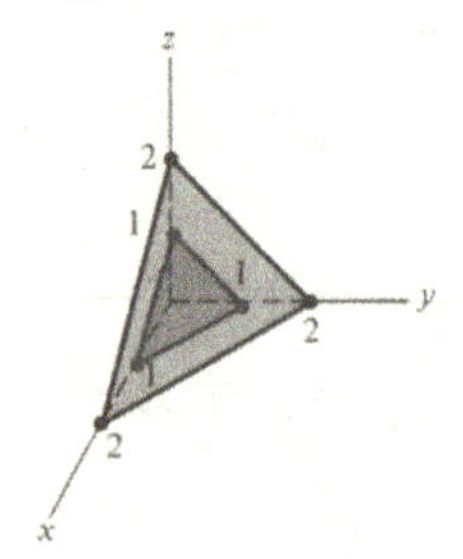

Example: $$\begin{aligned} x + y + z &= 1 \\ x + y + z &= 2 \end{aligned}$$

2. If the two equations describe the same plane, their solution set is that entire plane (Figure 3.4–9). In this case the set of equations is redundant.

Figure 3.4–9

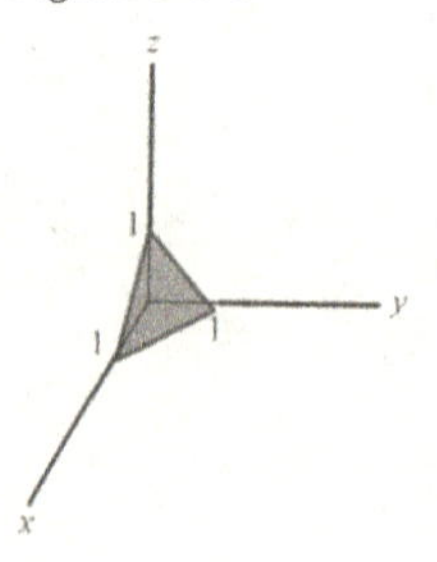

Example: $x + y + z = 1$
$2x + 2y + 2z = 2$

3. In the "typical" case the two planes will intersect in a straight line (Figure 3.4–10). This case is also a redundancy.

Figure 3.4–10

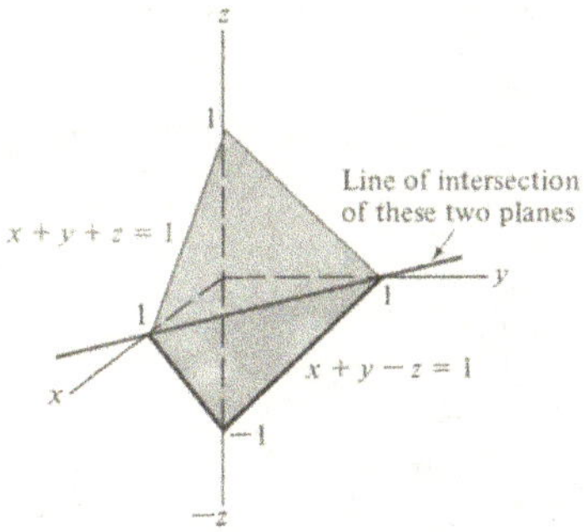

Example: $x + y + z = 1$
$x + y - z = 1$

Thus there are three possibilities for the solution set of 2 linear equations in 3 unknowns—the null set, a plane, and a straight line.

What is the geometric figure corresponding to the solution set of 3 linear equations in 3 unknowns?

Again, it depends. Since any two of these equations might describe the same plane and the third might have any of the relationships to that plane discussed above, all of the three possibilities above are still possibilities in this case. In addition, there is another kind of inconsistency and the "typical" case for 3 linear equations in 3 unknowns. This last, of course, is the case discussed in Chapter 2—where the three planes intersect at a point. We list here all five possibilities for the geometric interpretation of 3 linear equations in 3 unknowns.

1. Two planes (or perhaps all three) are parallel. The solution set is the null set.
2. All three equations describe the same plane. The solution set is that entire plane.
3. The three planes' intersection is a whole straight line. (If you open up your book, any three of its pages intersect in the binding.)
4. Each pair of the three planes intersects in a line, but the three lines thus formed are parallel in space. The solution set is then the null set. (Open up your book and put an extra sheet of paper over the opening. This paper and any two pages of the book suggest the situation in this case.)

5. The planes describing 3 linear equations in 3 unknowns may intersect at one point (Figure 3.4–11).

Example:
$$\begin{aligned} x + y + z &= 1 \\ x + y - z &= 1 \\ x &= y \end{aligned}$$

Figure 3.4–11

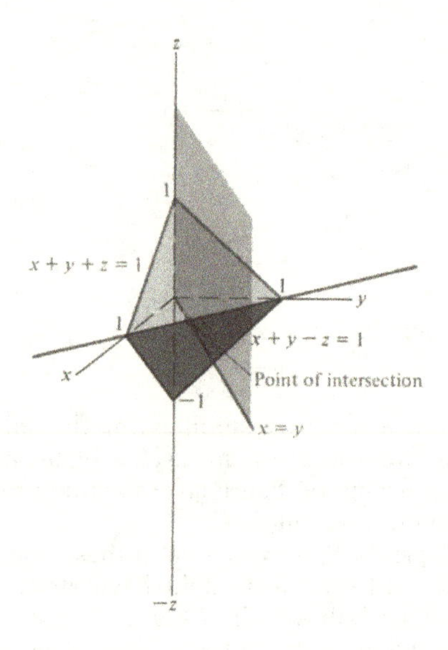

In this case we have a unique solution to the original three equations; this is the situation that was studied in Section 2.1.

If a set of three linear equations describes parallel planes, the Gauss–Jordan reduction method will yield an inconsistency. If all three describe the same plane, the reduced form of the augmented matrix will have two rows of zeros. The parameters of Section 3.2 (that is, the a, b, etc. multiplying the vectors in the solutions) tell the "degrees of freedom" in the solution set—that is, the number of parameters is the same as the number of dimensions of the solution set.

Thus a solution of the form $X = \begin{bmatrix} 3 \\ 5 \\ 0 \end{bmatrix} + a\begin{bmatrix} -2 \\ -4 \\ 1 \end{bmatrix}$, as on page 100, describes a line in 3-space: the point $\begin{bmatrix} 3 \\ 5 \\ 0 \end{bmatrix}$ is one point on the line and each value of a gives another point on the line. Similarly, the solution

$$X = \begin{bmatrix} 1 \\ 0 \\ 0 \end{bmatrix} + a \begin{bmatrix} -1 \\ 1 \\ 0 \end{bmatrix} + b \begin{bmatrix} -2 \\ 0 \\ 1 \end{bmatrix}$$

to problem 13 of Exercises 3.1.A on page 108 describes an entire plane. The two parameters a and b can be thought of as enabling the points to "move" in two dimensions around the plane.

In 4-space (that is, where there are 4 variables) it is impossible to draw pictures or to visualize in one's mind what is going on. You need not feel that the inability to do this is your personal lack—human beings cannot visualize more than three dimensions in space. But we *can* use the same algebra and the same language as we do in 3-space to describe 4-space (or, in fact, n-space).

For example, a vector such as $[2, 3, -1, 7]$ might be said to describe a point in 4-space. It can be the solution set of 4 linear equations in 4 unknowns if the system has a unique solution. A point can also be written as a column vector. The set $\left\{ a \begin{bmatrix} 3 \\ 3 \\ -2 \\ 5 \end{bmatrix} \right\}$, where a varies over all real numbers, is often called a line in 4-space. The solution set to Example 3.2.3(a) describes a line in 4-space. The solution to Example 3.2.3(c) describes a plane, since there are two dimensions of freedom.

The ideas in this section are sophisticated and are not needed for the rest of this course. We have introduced them quickly so you can talk to mathematicians who speak this language. If you want to learn the concepts thoroughly, you should take a full theoretical course in linear algebra.

SUMMARY

The graph of a linear equation in 3 unknowns is a *plane* in three-dimensional space. The graph of a system of 2 equations in 3 unknowns is often a *line*, and the graph of a system of 3 equations in 3 unknowns is often a point. But there are many degenerate cases, which are listed in this section. If there are 4 unknowns, we speak of a graph in 4-space; although such a situation cannot be actually visualized by human beings, it is a convenient mathematical tool.

EXERCISES 3.4. A

1. Sketch the following planes (that is, show the small triangles which, when extended, describe the planes).
   (a) $3x + 2y + z = 6$ (b) $2x + y + 4z = 4$
   (c) $x - y - z = 1$

2. Refer to the answers to Exercises 3.2.A (pages 108–109) and tell what geometric figure (that is, null set, point, line, plane) each solution set describes.

EXERCISES 3.4. B

1. Sketch the following planes (that is, show the small triangles which, when extended, describe the planes).
   (a) $2x + 3y + z = 6$ (b) $4x + y + 2z = 4$
   (c) $x - y + z = 1$
2. Refer to your answers to Exercises 3.2.B and tell what geometric figure (that is, null set, point, line, plane) each solution set describes.

EXERCISES 3.4. C

Sketch the following planes (that is, show the small triangles, which, when extended, describe the planes).
(a) $3x - 2y - z = 6$ (b) $-x + 4y - 2z = 4$

ANSWERS 3.4. A

1. (a)

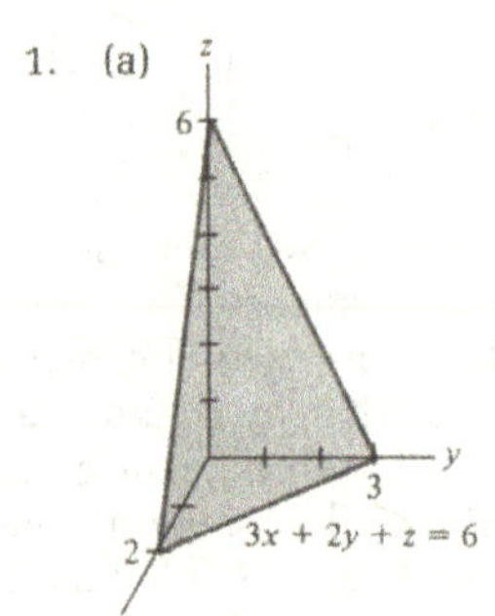

(b)

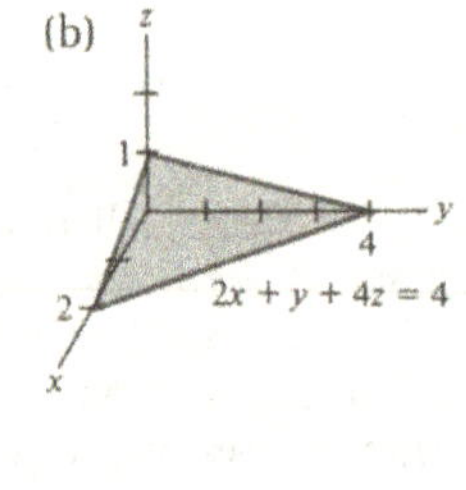

(c)

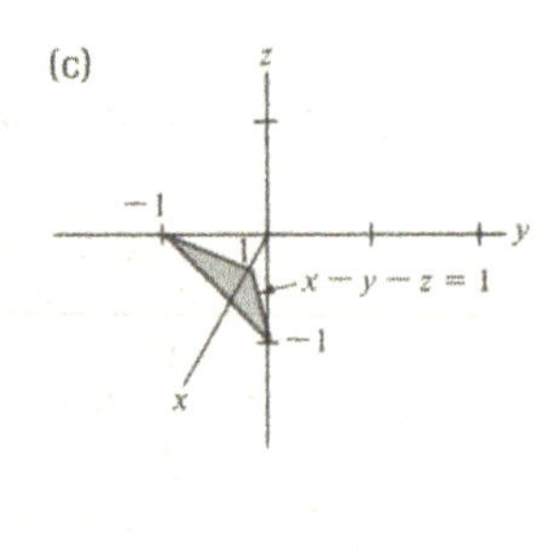

2. 1. (10) line (12) line (13) plane (15) line
   2. (a) line (b) line (c) line in 4-space (d) plane in 4-space
   3. (a) line (b) line (c) line in 4-space (d) plane in 4-space
      (e) plane in 4-space (f) plane in 5-space

ANSWERS 3.4. C

(a)

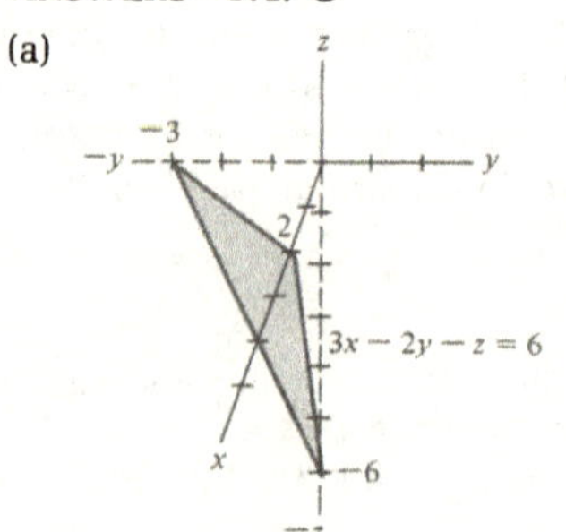

(b)

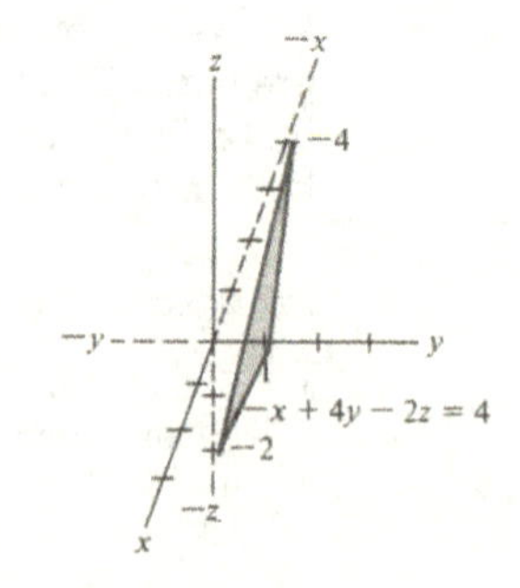

★3.5 Linear Independence and Dependence and Row Rank

When Gauss–Jordan row operations are applied to a matrix, sometimes a whole row "disappears"—that is, it becomes all zeros. Such a row may be thought of as not contributing its full weight to the matrix; it is in some sense dependent on the other rows. Mathematicians have formalized this idea into Definition 3.5.1. Because the definition is so formal, many people at first find it difficult to understand. But all that the following definitions are trying to say is that when the rows of a matrix are considered to be vectors, they are called "dependent" if the reduced form of the matrix has at least one row consisting of all zeros, and they are called "independent" if each row of the reduced form of the matrix contains at least one nonzero number. We shall return to these ideas in Chapter 4.

3.5.1 Definition

A finite set of vectors, $v_1, v_2, \ldots, v_n$, all of the same dimension, is called linearly dependent if there are n scalars (that is, numbers), $c_1, c_2, \ldots, c_n$, not all zero, such that

$$c_1 v_1 + c_2 v_2 + \cdots + c_{n-1} v_{n-1} + c_n v_n = 0$$

The equation is clearly true if all the c's are zero; the definition asks whether some of them can be chosen to be nonzero.

3.5.2 Example

Show that the vectors $v_1 = (0, 1, 0)$, $v_2 = (0, 0, 1)$, and $v_3 = (0, 3, 4)$ are linearly dependent.

Solution:

$$(-3)v_1 + (-4)v_2 + (1)v_3 = 0$$

3.5.3 Definition

A finite set of vectors, $v_1, v_2, \ldots, v_n$, all of the same dimension, is called linearly independent if they are not linearly dependent.

Suppose that we have n linearly dependent row vectors, $v_1, v_2, \ldots, v_n$, and we form a matrix by writing them under each other:

$$A = \begin{bmatrix} v_1 \\ v_2 \\ \vdots \\ v_n \end{bmatrix}$$

Let the notation be as in Definition 3.5.1. Then if $c_n \neq 0$ and we perform the row operations $c_n R_n \longrightarrow R_n$, $R_n + c_1 R_1 \longrightarrow R_n$, $R_n + c_2 R_2 \longrightarrow R_n \cdots R_n + c_{n-1} R_{n-1} \longrightarrow R_n$, we will get the following matrix:

$$\begin{bmatrix} v_1 \\ v_2 \\ \vdots \\ c_n v_n + c_1 v_1 + c_2 v_2 + \cdots + c_{n-1} v_{n-1} \end{bmatrix}$$

Comparing this to Definition 3.5.1, we see that the last row is zero. No matter what row operations are performed on the other rows to change the matrix to reduced form, this row will remain all zeros, so the reduced form of the original matrix will have at least one row of zeros. If $c_n = 0$, then one of the other c's is not 0, and a similar argument will make its row all zeros.

It is harder to show the converse—that the rows of any matrix are linearly dependent if the reduced form of the matrix has at least one row of all zeros. We state the following useful rule without proof.

3.5.4 Rule

A set of row vectors, $v_1, v_2, \ldots, v_n$, with m elements in each vector is linearly dependent if and only if the row-reduced form of the matrix

$$A = \begin{bmatrix} v_1 \\ v_2 \\ \vdots \\ v_n \end{bmatrix}$$

has at least one row consisting entirely of zeros.

We now give some examples showing how this rule is used.

3.5.5 Example

Tell whether the vectors $v_1 = [1, 2, -1]$, $v_2 = [2, -3, -1]$, and $v_3 = [4, 1, -3]$ are linearly dependent or linearly independent.

Solution:

In order to use Rule 3.5.4, we write a matrix whose rows are these three vectors; then we apply Gauss–Jordan row operations to change it to reduced form.

$$\begin{bmatrix} 1 & 2 & -1 \\ 2 & -3 & -1 \\ 4 & 1 & -3 \end{bmatrix} \quad \begin{matrix} R_2 - 2R_1 \longrightarrow R_2 \\ R_3 - 4R_1 \longrightarrow R_3 \end{matrix}$$

$$\begin{bmatrix} 1 & 2 & -1 \\ 0 & -7 & 1 \\ 0 & -7 & 1 \end{bmatrix} \quad R_3 - R_2 \longrightarrow R_3$$

$$\begin{bmatrix} 1 & 2 & -1 \\ 0 & -7 & 1 \\ 0 & 0 & 0 \end{bmatrix}$$

At this point Rule 3.5.4 tells us that the vectors are linearly dependent. To find the c's that determine the dependency, it is necessary to retrace the steps in the Gauss–Jordan row reduction. The rows are dependent because in the second matrix, $R_2 = R_3$. But these rows are equal, respectively, to $R_2 - 2R_1$ and $R_3 - 4R_1$ of the first matrix. Thus in the first matrix, $R_2 - 2R_1 = R_3 - 4R_1$. Collecting all terms on the left of the equal sign, this can be written $2R_1 + R_2 - R_3 = 0$ or, using the original notation,

$$2v_1 + v_2 - v_3 = 0.$$

3.5.6 Example

Tell whether the vectors $e_1 = [1, 0, 0]$, $e_2 = [0, 1, 0]$, and $e_3 = [0, 0, 1]$ are linearly independent or linearly dependent. (The notation using the e's is common for these particular vectors.)

Solution:

Using Rule 3.5.4, we immediately see that the matrix $\begin{bmatrix} 1 & 0 & 0 \\ 0 & 1 & 0 \\ 0 & 0 & 1 \end{bmatrix}$ is already in reduced form, and since no row has all zero entries, the vectors are linearly independent.

Let us examine this example from another angle. If there were c_1, c_2, and c_3 such that

$$c_1[1, 0, 0] + c_2[0, 1, 0] + c_3[0, 0, 1] = [0, 0, 0]$$

then

$$[c_1, 0, 0] + [0, c_2, 0] + [0, 0, c_3] = [0, 0, 0]$$

so

$$[c_1, c_2, c_3] = [0, 0, 0]$$

This implies that $c_1 = 0$, $c_2 = 0$, and $c_3 = 0$.

3.5.7 Example

Tell whether the vectors $v_1 = [1, 1, 0, 0]$, $v_2 = [1, 0, 1, 1]$, $v_3 = [0, 1, 0, 1]$, and $v_4 = [0, 0, 1, 1]$ are linearly dependent or independent.

Solution:

$$\begin{bmatrix} 1 & 1 & 0 & 0 \\ 1 & 0 & 1 & 1 \\ 0 & 1 & 0 & 1 \\ 0 & 0 & 1 & 1 \end{bmatrix} \qquad R_2 - R_1 \longrightarrow R_2$$

$$\begin{bmatrix} 1 & 1 & 0 & 0 \\ 0 & -1 & 1 & 1 \\ 0 & 1 & 0 & 1 \\ 0 & 0 & 1 & 1 \end{bmatrix} \quad \begin{matrix} R_1 + R_2 \longrightarrow R_1 \\ R_3 + R_2 \longrightarrow R_3 \end{matrix}$$

$$\begin{bmatrix} 1 & 0 & 1 & 1 \\ 0 & -1 & 1 & 1 \\ 0 & 0 & 1 & 2 \\ 0 & 0 & 1 & 1 \end{bmatrix} \quad \begin{matrix} -R_2 \longrightarrow R_2 \\ R_1 - R_3 \longrightarrow R_1 \\ R_4 - R_3 \longrightarrow R_4 \end{matrix}$$

$$\begin{bmatrix} 1 & 0 & 0 & -1 \\ 0 & 1 & -1 & -1 \\ 0 & 0 & 1 & 2 \\ 0 & 0 & 0 & -1 \end{bmatrix} \quad \begin{matrix} R_2 + R_3 \longrightarrow R_2 \\ -R_4 \longrightarrow R_4 \end{matrix}$$

$$\begin{bmatrix} 1 & 0 & 0 & -1 \\ 0 & 1 & 0 & 1 \\ 0 & 0 & 1 & 2 \\ 0 & 0 & 0 & 1 \end{bmatrix} \quad \begin{matrix} R_1 + R_4 \longrightarrow R_1 \\ R_2 - R_4 \longrightarrow R_2 \\ R_3 - 2R_4 \longrightarrow R_3 \end{matrix}$$

It is clear at this point that the row-reduced form is the identity, and the original vectors are linearly independent.

3.5.8 Example

Tell whether the vectors $v_1 = [1, 2, -1, 3]$, $v_2 = [2, -3, 4, 1]$, $v_3 = [-3, 2, -2, 0]$, and $v_4 = [1, 3, 0, 7]$ are linearly dependent or independent.

Solution:

$$\begin{bmatrix} 1 & 2 & -1 & 3 \\ 2 & -3 & 4 & 1 \\ -3 & 2 & -2 & 0 \\ 1 & 3 & 0 & 7 \end{bmatrix} \quad \begin{matrix} R_2 - 2R_1 \longrightarrow R_2 \\ R_3 + 3R_1 \longrightarrow R_3 \\ R_4 - R_1 \longrightarrow R_4 \end{matrix}$$

$$\begin{bmatrix} 1 & 2 & -1 & 3 \\ 0 & -7 & 6 & -5 \\ 0 & 8 & -5 & 9 \\ 0 & 1 & 1 & 4 \end{bmatrix} \quad R_3 + R_2 \longrightarrow R_3$$

$$\begin{bmatrix} 1 & 2 & -1 & 3 \\ 0 & -7 & 6 & -5 \\ 0 & 1 & 1 & 4 \\ 0 & 1 & 1 & 4 \end{bmatrix} \quad R_4 - R_3 \longrightarrow R_4 \text{ etc.}$$

It is clear that Rule 3.5.4 now implies that the vectors are linearly dependent. To find the explicit dependency, we observe that the last two rows of the third matrix are equal. This is because in the second matrix, $R_4 = R_2 + R_3$. And this means that in the first matrix,

$$(v_4 - v_1) = (v_2 - 2v_1) + (v_3 + 3v_1)$$

so

$$-2v_1 - v_2 - v_3 + v_4 = 0$$

3.5.9 Definition

The row rank of a matrix is the maximum number of linearly independent rows in the matrix.

Since it happens that the Gauss–Jordan reduction process does not change the linear dependence of the rows of a matrix (although this is by no means obvious), we can use the following rule to find the row rank of a specified matrix.

3.5.10 Rule

The row rank of a matrix is the number of nonzero rows in the row-reduced form of the matrix.

3.5.11 Example

Find the row rank of each of the following matrices.

(a) $\begin{bmatrix} 3 & 0 & 9 \\ 2 & 0 & 6 \\ -7 & 0 & -21 \end{bmatrix}$ (b) $\begin{bmatrix} 3 & 0 & 7 \\ -1 & 2 & -3 \\ 5 & -4 & 13 \end{bmatrix}$

Solution:

(a) $\begin{bmatrix} 3 & 0 & 9 \\ 2 & 0 & 6 \\ -7 & 0 & -21 \end{bmatrix}$ $R_1 - R_2 \longrightarrow R_1$

$\begin{bmatrix} 1 & 0 & 3 \\ 2 & 0 & 6 \\ -7 & 0 & -21 \end{bmatrix}$ $\begin{matrix} R_2 - 2R_1 \longrightarrow R_2 \\ R_3 + 7R_1 \longrightarrow R_3 \end{matrix}$

$\begin{bmatrix} 1 & 0 & 3 \\ 0 & 0 & 0 \\ 0 & 0 & 0 \end{bmatrix}$

so the row rank of the original matrix is 1.

(b) $\begin{bmatrix} 3 & 0 & 7 \\ -1 & 2 & -3 \\ 5 & -4 & 13 \end{bmatrix}$ $\begin{matrix} R_1 + 3R_2 \longrightarrow R_1 \\ R_3 + 5R_2 \longrightarrow R_3 \end{matrix}$

$\begin{bmatrix} 0 & 6 & -2 \\ -1 & 2 & -3 \\ 0 & 6 & -2 \end{bmatrix}$ $R_3 - R_1 \longrightarrow R_3$

$\begin{bmatrix} 0 & 6 & -2 \\ -1 & 2 & -3 \\ 0 & 0 & 0 \end{bmatrix}$

You may at this point see that the row rank of the matrix is 2; if not, keep going until you see that the row-reduced form of the matrix has two nonzero rows.

SUMMARY

A set of vectors all of the same dimension is either linearly dependent or linearly independent. It is linearly dependent if there is a nontrivial linear combination of the vectors which is zero, or, equivalently, if the row-reduced form of the matrix built from these vectors has at least one row consisting entirely of zeros.

The row rank of a matrix is the maximum number of linearly independent rows in the matrix. It is also the number of nonzero rows in the row-reduced form of the matrix.

EXERCISES 3.5. A

Show that each of the following sets of vectors is linearly dependent by finding the c's required by Definition 3.5.1.

1. $v_1 = [1, 0, 0]$, $v_2 = [0, 1, 0]$, $v_3 = [2, 3, 0]$
2. $v_1 = [1, 0, 1]$, $v_2 = [4, 0, 0]$, $v_3 = [0, 0, -2]$
3. $v_1 = [1, 0, 0]$, $v_2 = [0, 1, 0]$, $v_3 = [0, 0, 1]$, $v_4 = [a, b, c]$
4. $v_1 = [0, 0, 1]$, $v_2 = [0, 1, 1]$, $v_3 = [0, 2, 3]$

Use Rule 3.5.4 to discover whether each of the following sets of vectors is linearly dependent or independent. When the vectors are linearly dependent, find the equation given in Definition 3.5.1. (*Hint:* See Examples 3.5.5 and 3.5.8.)

5. $v_1 = [1, 0, 0, 1]$, $v_2 = [1, 0, 1, 1]$, $v_3 = [1, 1, 0, 1]$, $v_4 = [0, 0, 1, 1]$
6. $v_1 = [2, 0, 1, 1]$, $v_2 = [1, 1, 0, -1]$, $v_3 = [1, 2, -1, 0]$, $v_4 = [2, -1, 2, 0]$
7. $v_1 = [1, 0, 0, 1]$, $v_2 = [0, 1, 1, 0]$, $v_3 = [-1, 0, 0, 1]$, $v_4 = [0, 1, -1, 0]$
8. $v_1 = [1, 0, 0, 1]$, $v_2 = [0, 1, 1, 1]$, $v_3 = [-1, 0, 0, 1]$, $v_4 = [0, -1, -1, 1]$
9. $v_1 = [1, 0, 0]$, $v_2 = [0, 1, 0]$, $v_3 = [0, 0, 1]$, $v_4 = [2, 3, 4]$
10. Find the row rank of all the matrices given in problem 3 of Exercises 3.2.A (page 107).

EXERCISES 3.5. B

Show that each of the following sets of vectors is linearly dependent by finding the c's required by Definition 3.5.1.

1. $v_1 = [1, 0, 0]$, $v_2 = [0, 0, 1]$, $v_3 = [5, 0, -3]$
2. $v_1 = [1, 1, 0]$, $v_2 = [6, 0, 0]$, $v_3 = [0, -3, 0]$
3. $v_1 = [1, 0, 0]$, $v_2 = [0, 1, 0]$, $v_3 = [0, 0, 1]$, $v_4 = [x, y, z]$
4. $v_1 = [0, 1, 0]$, $v_2 = [0, 1, 1]$, $v_3 = [0, 4, 5]$

Use Rule 3.5.4 to discover whether each of the following sets of vectors is linearly dependent or independent. When the vectors are linearly dependent, find the equation given in Definition 3.5.1. (*Hint:* See Examples 3.5.5 and 3.5.8.)

5. $v_1 = [1, 0, 0, 1]$, $v_2 = [1, 0, 1, 1]$, $v_3 = [1, 1, 0, 1]$, $v_4 = [1, -1, 1, 1]$
6. $v_1 = [0, 0, 1, 1]$, $v_2 = [1, 1, 0, -1]$, $v_3 = [1, 2, -1, 0]$, $v_4 = [2, -1, 2, 0]$
7. $v_1 = [1, 0, 0, 1]$, $v_2 = [0, 1, 1, 0]$, $v_3 = [-1, 0, 0, 1]$, $v_4 = [0, -1, -1, 2]$
8. $v_1 = [1, 0, 0, 1]$, $v_2 = [0, 1, 1, 1]$, $v_3 = [-1, 0, 0, 1]$, $v_4 = [0, -1, 0, 1]$
9. $v_1 = [1, 0, 0]$, $v_2 = [0, 1, 0]$, $v_3 = [0, 0, 1]$, $v_4 = [4, 5, 6]$
10. Find the row rank of all the matrices given in problem 3 of Exercises 3.2.B (pages 107–108).

ANSWERS 3.5. A

1. $(-2)v_1 + (-3)v_2 + v_3 = 0$
2. $4v_1 + (-1)v_2 + 2v_3 = 0$
3. $(-a)v_1 + (-b)v_2 + (-c)v_3 + v_4 = 0$
4. $(-1)v_1 + (-2)v_2 + v_3 = 0$

There are other correct answers to the exercises above; the definition of linear dependence promises the existence of the c's, but they are not unique.

5. Independent
6. Dependent: $v_1 + v_2 - v_3 - v_4 = 0$
7. Independent
8. Dependent: $v_1 - v_2 + v_3 - v_4 = 0$
9. If there are more vectors in the set than there are dimensions in each vector, the vectors are linearly dependent. In this case, $2v_1 + 3v_2 + 4v_3 - v_4 = 0$.
10. (a) 2 (b) 2 (c) 3 (d) 2 (e) 2 (f) 3

VOCABULARY

solution, inconsistent, redundant, unique solution, closed Leontief model, reduced form, leading 1's, row-reduced matrix, parameter, particular solution, complete solution, general solution, network analysis, z-axis, 3-space, 4-space, *n*-space, null set, linearly dependent, linearly independent, row rank

SAMPLE TEST Chapter 3

1–2. (40 pts) (a) Tell whether each of the following systems of equations has a unique solution or is inconsistent or redundant.
(b) Where a solution(s) exists, write the complete solution.
(c) Where there is more than one solution, give *two* particular solutions. (Label carefully.)
(d) Give the geometric interpretation of the solution set.

1. $$\begin{aligned} -2x + 3y + 4z &= 1 \\ x - 2y - z &= 0 \\ -x + 5z &= 5 \end{aligned}$$

2. $$\begin{aligned} x_1 + x_3 + 2x_4 &= 0 \\ -x_2 - 2x_3 &= 1 \\ 2x_1 - x_2 + 4x_4 &= 1 \\ 3x_1 - 2x_2 - x_3 + 6x_4 &= 2 \end{aligned}$$

3. (20 pts) Write 3 equations in 3 unknowns whose solution set can be interpreted geometrically as each of the following. There will be 12 equations all together.
(a) null set (b) a point (c) a line (d) a plane
4. (5 pts) Use the definition of linearly dependent to show that the following vectors are linearly dependent: $v_1 = [2, 0, 5]$, $v_2 = [-2, 0, 0]$, and $v_3 = [0, 0, -1]$.
5. (10 pts) Use any method to show whether the following are linearly dependent or linearly independent: $v_1 = [2, 1, 3]$, $v_2 = [1, 3, -2]$, and $v_3 = [3, -1, 8]$.
6. (5 pts) What is the row rank of the matrix $\begin{bmatrix} 2 & 1 & 3 \\ 1 & 3 & -2 \\ 3 & -1 & 8 \end{bmatrix}$?

7. (20 pts)
   (a) (5 pts) Write six equations describing the traffic flow shown in the diagram, one for each intersection.
   (b) (10 pts) Write the amount of traffic flow along the other paths in terms of x_6 and x_7.
   (c) (5 pts) Assuming that the traffic moves in the indicated directions, what is the least possible value for x_6?

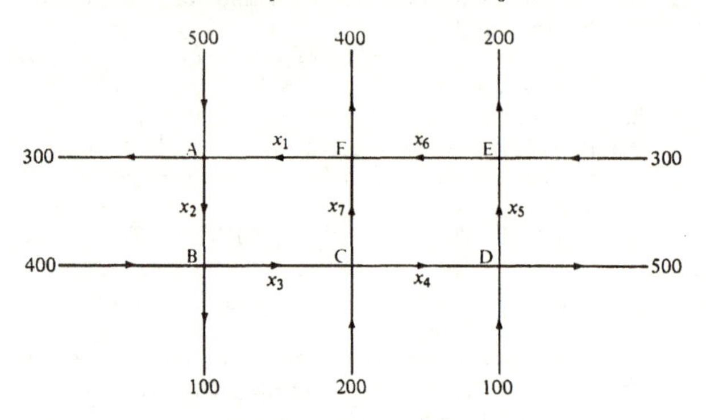

The answers are at the back of the book.

★Chapter 4

DETERMINANTS

4.1 Classical Expansion of Determinants

It is customary to associate with each square matrix a number called the determinant of the matrix. Determinants have a variety of purposes, some of which will be explained in the next section. They can also be used to solve systems of linear equations and to find inverses of square matrices, but the Gauss–Jordan method explained in the previous two chapters is usually more efficient. Gauss–Jordan row operations can also be used to compute determinants; this will be described in Section 4.3.

This section is devoted to explaining the method of computing determinants that has been employed for centuries, beginning with 2 by 2 determinants.

The determinant of any square matrix A is written $|A|$—the matrix with two straight lines down its sides.

4.1.1 Rule

Let $A = \begin{bmatrix} a & b \\ c & d \end{bmatrix}$. Then $|A| = \begin{vmatrix} a & b \\ c & d \end{vmatrix} = ad - bc.$

4.1.2 Example

Compute:

(a) $\begin{vmatrix} 3 & 4 \\ 2 & 1 \end{vmatrix}$ (b) $\begin{vmatrix} 1 & 0 \\ 0 & 1 \end{vmatrix}$ (c) $\begin{vmatrix} -2 & 0 \\ 3 & -5 \end{vmatrix}$

Solution:

(a) $3 \cdot 1 - 4 \cdot 2 = -5$ (b) $1 \cdot 1 - 0 \cdot 0 = 1$

(c) $(-2)(-5) - 0 \cdot 3 = 10$

One way to motivate this curious definition is to observe that if a system of two linear equations in two unknowns,

$$ax + by = e$$
$$cx + dy = f$$

has a unique solution, then this unique solution is

$$x = \frac{ed - bf}{ad - bc} = \frac{\begin{vmatrix} e & b \\ f & d \end{vmatrix}}{\begin{vmatrix} a & b \\ c & d \end{vmatrix}} \qquad y = \frac{af - ec}{ad - bc} = \frac{\begin{vmatrix} a & e \\ c & f \end{vmatrix}}{\begin{vmatrix} a & b \\ c & d \end{vmatrix}}$$

This is a summary of Cramer's rule in two variables. Cramer's rule has traditionally been a primary purpose of determinants. We present the three-variable statement in problem 5 of Exercises 4.1.C, and from it you can probably guess the generalization to more than three variables. Although the formal statement of Cramer's rule is complicated, it is not difficult to use—just tedious! In the next section some other uses of determinants are presented.

The determinant of a 3 by 3 matrix can be computed by using certain 2 by 2 determinants called "minors."

4.1.3 Definition

The minor corresponding to the element a_{ij} of a 3 by 3 matrix is that 2 by 2 determinant which is obtained by crossing out the ith row and the jth column of the original matrix. We shall call the minor corresponding to a_{ij} "minor (a_{ij})."

4.1.4 Example

Find the minors corresponding to a_{11}, a_{12}, a_{13}, and a_{21} in the matrix $\begin{bmatrix} 1 & 2 & 3 \\ 4 & 5 & 6 \\ 7 & 8 & 9 \end{bmatrix}$.

Solution:

(a) $\text{minor }(a_{11}) = \begin{vmatrix} 1 & 2 & 3 \\ 4 & 5 & 6 \\ 7 & 8 & 9 \end{vmatrix} = \begin{vmatrix} 5 & 6 \\ 8 & 9 \end{vmatrix} = 45 - 48 = -3$

(b) $\text{minor }(a_{12}) = \begin{vmatrix} 1 & 2 & 3 \\ 4 & 5 & 6 \\ 7 & 8 & 9 \end{vmatrix} = \begin{vmatrix} 4 & 6 \\ 7 & 9 \end{vmatrix} = 36 - 42 = -6$

(c) $\text{minor }(a_{13}) = \begin{vmatrix} 1 & 2 & 3 \\ 4 & 5 & 6 \\ 7 & 8 & 9 \end{vmatrix} = \begin{vmatrix} 4 & 5 \\ 7 & 8 \end{vmatrix} = 32 - 35 = -3$

(d) $\text{minor }(a_{21}) = \begin{vmatrix} 1 & 2 & 3 \\ 4 & 5 & 6 \\ 7 & 8 & 9 \end{vmatrix} = \begin{vmatrix} 2 & 3 \\ 8 & 9 \end{vmatrix} = 18 - 24 = -6$

To find the determinant of a 3 by 3 matrix, we assign each position in the matrix a sign, checkerboard fashion, starting with a plus sign in the upper left.

$$\begin{matrix} + & - & + \\ - & + & - \\ + & - & + \end{matrix}$$

Then we can select *any* row or *any* column and multiply each of its elements by the minor of that element. We add or subtract these three products according to the checkerboard signs given above to obtain the value of the determinant.

4.1.5 Example

Compute $\begin{vmatrix} 1 & 2 & 3 \\ 4 & 5 & 6 \\ 7 & 8 & 9 \end{vmatrix}$ using:

(a) The first row.
(b) The second row.
(c) The third column.

Solution:

(a) $$\begin{vmatrix} 1 & 2 & 3 \\ 4 & 5 & 6 \\ 7 & 8 & 9 \end{vmatrix} = 1\begin{vmatrix} 5 & 6 \\ 8 & 9 \end{vmatrix} - 2\begin{vmatrix} 4 & 6 \\ 7 & 9 \end{vmatrix} + 3\begin{vmatrix} 4 & 5 \\ 7 & 8 \end{vmatrix}$$

$$= [1(45 - 48)] - [2(36 - 42)] + [3(32 - 35)]$$

$$= (-3) - [2(-6)] + [3(-3)] = 0$$

(b) $$\begin{vmatrix} 1 & 2 & 3 \\ 4 & 5 & 6 \\ 7 & 8 & 9 \end{vmatrix} = -4\begin{vmatrix} 2 & 3 \\ 8 & 9 \end{vmatrix} + 5\begin{vmatrix} 1 & 3 \\ 7 & 9 \end{vmatrix} - 6\begin{vmatrix} 1 & 2 \\ 7 & 8 \end{vmatrix}$$

$$= -[4(18 - 24)] + [5(9 - 21)] - [6(8 - 14)]$$

$$= -[4(-6)] + [5(-12)] - [6(-6)] = 0$$

(c) $$\begin{vmatrix} 1 & 2 & 3 \\ 4 & 5 & 6 \\ 7 & 8 & 9 \end{vmatrix} = 3\begin{vmatrix} 4 & 5 \\ 7 & 8 \end{vmatrix} - 6\begin{vmatrix} 1 & 2 \\ 7 & 8 \end{vmatrix} + 9\begin{vmatrix} 1 & 2 \\ 4 & 5 \end{vmatrix}$$

$$= [3(32 - 35)] - [6(8 - 14)] + [9(5 - 8)]$$

$$= 3(-3) - 6(-6) + 9(-3) = -9 + 36 - 27 = 0$$

No matter which row or column we use, we will always get the same value for the determinant of any given matrix.

4.1.6 Example

Compute $\begin{vmatrix} 6 & -2 & 3 \\ 4 & -3 & 1 \\ 0 & 2 & -7 \end{vmatrix}$.

Solution:

We shall use the third row since the zero in this row makes part of the computation trivial.

$$\begin{vmatrix} 6 & -2 & 3 \\ 4 & -3 & 1 \\ 0 & 2 & -7 \end{vmatrix} = 0\begin{vmatrix} -2 & 3 \\ -3 & 1 \end{vmatrix} - 2\begin{vmatrix} 6 & 3 \\ 4 & 1 \end{vmatrix} + (-7)\begin{vmatrix} 6 & -2 \\ 4 & -3 \end{vmatrix}$$

$$= 0 - 2(6 - 12) - 7(-18 + 8)$$

$$= -2(-6) - 7(-10) = 12 + 70 = 82$$

4.1.7 Example

Compute $\begin{vmatrix} 0 & -1 & 6 \\ 0 & 2 & 8 \\ 5 & -4 & 6 \end{vmatrix}$.

Solution:

This time we shall use the first column, because there are two zeros there.

$$\begin{vmatrix} 0 & -1 & 6 \\ 0 & 2 & 8 \\ 5 & -4 & 6 \end{vmatrix} = 0\begin{vmatrix} 2 & 8 \\ -4 & 6 \end{vmatrix} - 0\begin{vmatrix} -1 & 6 \\ -4 & 6 \end{vmatrix} + 5\begin{vmatrix} -1 & 6 \\ 2 & 8 \end{vmatrix}$$

$$= 0 - 0 + 5(-8 - 12) = 5(-20) = -100$$

Clearly, we would not have had to write the minors that will be multiplied by zero; we did so here only to give you a complete demonstration of what is going on.

We can use 3 by 3 determinants to compute 4 by 4 determinants in a fashion similar to the way in which we used 2 by 2 determinants to compute 3 by 3 determinants. The checkerboard sign arrangement is

$$\begin{matrix} + & - & + & - \\ - & + & - & + \\ + & - & + & - \\ - & + & - & + \end{matrix}$$

4.1.8 Example

Compute $\begin{vmatrix} 1 & -1 & 2 & 0 \\ 2 & -2 & 1 & -1 \\ 1 & -3 & 4 & 3 \\ 0 & 1 & -1 & 0 \end{vmatrix}$.

Solution:

This time we use the last row to expand, taking advantage of the two zeros.

$$-0 + 1\begin{vmatrix} 1 & 2 & 0 \\ 2 & 1 & -1 \\ 1 & 4 & 3 \end{vmatrix} - (-1)\begin{vmatrix} 1 & -1 & 0 \\ 2 & -2 & -1 \\ 1 & -3 & 3 \end{vmatrix} + 0$$

$$= 1\begin{vmatrix} 1 & -1 \\ 4 & 3 \end{vmatrix} - 2\begin{vmatrix} 2 & -1 \\ 1 & 3 \end{vmatrix} + 1\begin{vmatrix} -2 & -1 \\ -3 & 3 \end{vmatrix} + 1\begin{vmatrix} 2 & -1 \\ 1 & 3 \end{vmatrix}$$

$$= (3 + 4) - 2(6 + 1) + 1(-6 - 3) + 1(6 + 1)$$

$$= 7 - 14 - 9 + 7 = -9$$

The same process can be used indefinitely to evaluate large determinants by using smaller ones. We always put a plus sign in the upper left of the checkerboard.

4.1.9 Example

Evaluate the following 5 by 5 determinant (which is comparatively easy since there are three zeros in the second column).

Solution:

$$\begin{vmatrix} 2 & 0 & 1 & 1 & 3 \\ -2 & 1 & 0 & 1 & -1 \\ 3 & 0 & -1 & 0 & 4 \\ 4 & 0 & 2 & -2 & 3 \\ -1 & 1 & 0 & 2 & 3 \end{vmatrix}$$

$$= -0 + 1\begin{vmatrix} 2 & 1 & 1 & 3 \\ 3 & -1 & 0 & 4 \\ 4 & 2 & -2 & 3 \\ -1 & 0 & 2 & 3 \end{vmatrix}$$

$$-0 + 0 - 1\begin{vmatrix} 2 & 1 & 1 & 3 \\ -2 & 0 & 1 & -1 \\ 3 & -1 & 0 & 4 \\ 4 & 2 & -2 & 3 \end{vmatrix}$$

(We expand using the last row of the first 4 by 4 determinant and the second row of the second.)

$$= -(-1)\begin{vmatrix} 1 & 1 & 3 \\ -1 & 0 & 4 \\ 2 & -2 & 3 \end{vmatrix} - 2\begin{vmatrix} 2 & 1 & 3 \\ 3 & -1 & 4 \\ 4 & 2 & 3 \end{vmatrix}$$

$$+ 3\begin{vmatrix} 2 & 1 & 1 \\ 3 & -1 & 0 \\ 4 & 2 & -2 \end{vmatrix} - \left(-(-2)\begin{vmatrix} 1 & 1 & 3 \\ -1 & 0 & 4 \\ 2 & -2 & 3 \end{vmatrix} \right.$$

$$\left. -1\begin{vmatrix} 2 & 1 & 3 \\ 3 & -1 & 4 \\ 4 & 2 & 3 \end{vmatrix} + (-1)\begin{vmatrix} 2 & 1 & 1 \\ 3 & -1 & 0 \\ 4 & 2 & -2 \end{vmatrix} \right)$$

There are now fourteen 2 by 2 matrices just over the horizon. The author pleads the growing paper shortage as a reason for not completing the problem just now; the reader should be able to finish the example if such diversions appeal. In Section 4.3 we shall use Gauss–Jordan row operations to show that the determinant is 40.

SUMMARY

This section was devoted to explaining the classical method for evaluating determinants with a passing mention of Cramer's rule. The next section will consider more ways that determinants can be used.

EXERCISES 4.1. A

1. Evaluate the following determinants.

(a) $\begin{vmatrix} 3 & 5 \\ -6 & 2 \end{vmatrix}$ (b) $\begin{vmatrix} 2 & -4 \\ 3 & 0 \end{vmatrix}$ (c) $\begin{vmatrix} 3 & 0 & 6 \\ 2 & -1 & 3 \\ 4 & 2 & -5 \end{vmatrix}$

(d) $\begin{vmatrix} 2 & 1 & 4 \\ 0 & -2 & 5 \\ 4 & 8 & -7 \end{vmatrix}$ (e) $\begin{vmatrix} 4 & 0 & 2 \\ -2 & 1 & 2 \\ 3 & -1 & 3 \end{vmatrix}$

2. Look at the answers to Exercises 2.3.A (the set where you were finding elementary matrices). Find the determinants of each of those matrices. Look for patterns that enable you to do this quickly. (See page 71.)
3. Tell how to find quickly the determinant of each of the three types of elementary matrices.

EXERCISES 4.1. B

1. Evaluate the following determinants.

(a) $\begin{vmatrix} 2 & 4 \\ -3 & 7 \end{vmatrix}$ (b) $\begin{vmatrix} 0 & 3 \\ -6 & 4 \end{vmatrix}$ (c) $\begin{vmatrix} 3 & 2 & -1 \\ 4 & 0 & 3 \\ -5 & 1 & -4 \end{vmatrix}$

(d) $\begin{vmatrix} 2 & 4 & -4 \\ 3 & 5 & -1 \\ 5 & 10 & 1 \end{vmatrix}$ (e) $\begin{vmatrix} 2 & 3 & 4 \\ 0 & -1 & -2 \\ 1 & 2 & 5 \end{vmatrix}$

2. Evaluate each of the following determinants. Look for patterns that enable you to do this quickly.

(a) $\begin{vmatrix} 1 & 0 & 0 \\ 0 & 2 & 0 \\ 0 & 0 & 1 \end{vmatrix}$ (b) $\begin{vmatrix} 1 & 0 & 0 \\ 0 & \frac{1}{2} & 0 \\ 0 & 0 & 1 \end{vmatrix}$ (c) $\begin{vmatrix} 1 & 0 & 0 \\ 0 & 1 & 0 \\ 0 & 0 & \pi \end{vmatrix}$

(d) $\begin{vmatrix} 0 & 0 & 1 \\ 0 & 1 & 0 \\ 1 & 0 & 0 \end{vmatrix}$ (e) $\begin{vmatrix} 1 & 0 & 0 \\ 0 & 0 & 1 \\ 0 & 1 & 0 \end{vmatrix}$ (f) $\begin{vmatrix} 1 & 0 & 0 \\ 0 & 1 & 0 \\ 1 & 0 & 1 \end{vmatrix}$

(g) $\begin{vmatrix} 1 & 1 & 0 \\ 0 & 1 & 0 \\ 0 & 0 & 1 \end{vmatrix}$ (h) $\begin{vmatrix} 1 & 0 & 0 \\ 0 & 1 & 5 \\ 0 & 0 & 1 \end{vmatrix}$ (i) $\begin{vmatrix} 1 & 0 & 0 \\ 0 & \frac{1}{2} & 0 \\ 0 & 0 & 1 \end{vmatrix}$

(j) $\begin{vmatrix} 1 & 0 & 0 \\ 0 & 1 & 0 \\ 0 & 0 & 1/\pi \end{vmatrix}$ (k) $\begin{vmatrix} 1 & 0 & 1 \\ 0 & 1 & 0 \\ 0 & 0 & 1 \end{vmatrix}$ (l) $\begin{vmatrix} 0 & 1 & 0 \\ 1 & 0 & 0 \\ 0 & 0 & 1 \end{vmatrix}$

(m) $\begin{vmatrix} 1 & 0 & 0 \\ 0 & 1 & 0 \\ -1 & 0 & 1 \end{vmatrix}$ (n) $\begin{vmatrix} 1 & -1 & 0 \\ 0 & 1 & 0 \\ 0 & 0 & 1 \end{vmatrix}$ (o) $\begin{vmatrix} 1 & 0 & -1 \\ 0 & 1 & 0 \\ 0 & 0 & 1 \end{vmatrix}$

(p) $\begin{vmatrix} 1 & 0 & -2 \\ 0 & 1 & 0 \\ 0 & 0 & 1 \end{vmatrix}$ (q) $\begin{vmatrix} 0 & 1 & 0 & 0 \\ 1 & 0 & 0 & 0 \\ 0 & 0 & 1 & 0 \\ 0 & 0 & 0 & 1 \end{vmatrix}$

(r) $\begin{vmatrix} 1 & 0 & 0 & 0 \\ 0 & 1 & 0 & 0 \\ 0 & 0 & 0 & 1 \\ 0 & 0 & 1 & 0 \end{vmatrix}$ (s) $\begin{vmatrix} 1 & 3 & 0 & 0 \\ 0 & 1 & 0 & 0 \\ 0 & 0 & 1 & 0 \\ 0 & 0 & 0 & 1 \end{vmatrix}$

(t) $\begin{vmatrix} 4 & 0 & 0 & 0 \\ 0 & 1 & 0 & 0 \\ 0 & 0 & 1 & 0 \\ 0 & 0 & 0 & 1 \end{vmatrix}$ (u) $\begin{vmatrix} 1 & 0 & 0 & 0 \\ 1 & 1 & 0 & 0 \\ 0 & 0 & 1 & 0 \\ 0 & 0 & 0 & 1 \end{vmatrix}$

(v) $\begin{vmatrix} 0 & 0 & 1 & 0 \\ 0 & 1 & 0 & 0 \\ 1 & 0 & 0 & 0 \\ 0 & 0 & 0 & 1 \end{vmatrix}$ (w) $\begin{vmatrix} 1 & 0 & 0 & 0 \\ 0 & 0 & 0 & 1 \\ 0 & 0 & 1 & 0 \\ 0 & 1 & 0 & 0 \end{vmatrix}$

(x) $\begin{vmatrix} 0 & 1 & 0 & 0 \\ 1 & 0 & 0 & 0 \\ 0 & 0 & 0 & 1 \\ 0 & 0 & 1 & 0 \end{vmatrix}$ (y) $\begin{vmatrix} \frac{1}{4} & 0 & 0 & 0 \\ 0 & 1 & 0 & 0 \\ 0 & 0 & 1 & 0 \\ 0 & 0 & 0 & 1 \end{vmatrix}$

(z) $\begin{vmatrix} 1 & 0 & 0 & 0 \\ -1 & 1 & 0 & 0 \\ 0 & 0 & 1 & 0 \\ 0 & 0 & 0 & 1 \end{vmatrix}$

3. Tell how to find quickly the determinant of each of the three types of elementary matrices.

EXERCISES 4.1. C

Evaluate the following determinants.

1. $\begin{vmatrix} 0 & 1 & 3 & 2 \\ 4 & 0 & -5 & 6 \\ 3 & 7 & -2 & 1 \\ 0 & -1 & 3 & 2 \end{vmatrix}$

2. $\begin{vmatrix} 1 & 0 & -2 & 3 \\ 4 & 2 & 1 & 4 \\ -5 & 3 & -1 & -7 \\ 1 & 5 & -4 & 3 \end{vmatrix}$

3. $\begin{vmatrix} 0 & 2 & 1 & 0 \\ 5 & 0 & 1 & 3 \\ 2 & 0 & 3 & 1 \\ -2 & -1 & 2 & 3 \end{vmatrix}$

4. $\begin{vmatrix} -2 & 3 & -8 & 9 \\ 3 & -6 & 2 & 0 \\ 2 & 5 & 0 & 2 \\ 1 & -2 & 4 & 3 \end{vmatrix}$

5. Cramer's rule for three variables says that if the system

$$ax + by + cz = j$$
$$dx + ey + fz = k$$
$$gx + hy + iz = m$$

has a unique solution, that solution is

$$x = \frac{\begin{vmatrix} j & b & c \\ k & e & f \\ m & h & i \end{vmatrix}}{\begin{vmatrix} a & b & c \\ d & e & f \\ g & h & i \end{vmatrix}} \qquad y = \frac{\begin{vmatrix} a & j & c \\ d & k & f \\ g & m & i \end{vmatrix}}{\begin{vmatrix} a & b & c \\ d & e & f \\ g & h & i \end{vmatrix}}$$

$$z = \frac{\begin{vmatrix} a & b & j \\ d & e & k \\ g & h & m \end{vmatrix}}{\begin{vmatrix} a & b & c \\ d & e & f \\ g & h & i \end{vmatrix}}$$

Use these formulas to solve the problems in Exercises 2.1.A. Start at the "Hints" on page 58; you should get the same answers as those given on page 59.

ANSWERS 4.1. A

1. (a) 36 (b) 12 (c) 45 (d) 0 (e) 18
2. 1. 3 2. $\frac{1}{3}$ 3. π 4. -1 5. -1 6. 1 7. 1 8. 1 9. 1 10. 1
   11. 1 12. 1 13. 1 14. $\frac{1}{3}$ 15. 3 16. $1/\pi$ 17. -1 18. -1 19. 1
   20. 1 21. 1 22. 1 23. 1 24. 1 25. 1 26. 1 27. -1 28. -1
   29. 1 30. 5 31. 1 32. -1 33. -1 34. 1 35. $\frac{1}{5}$ 36. 1
3. An elementary matrix that interchanges two rows when it premultiplies another matrix has determinant -1.
   An elementary matrix that adds a multiple of one row to another row when it premultiplies another matrix has determinant 1.
   Any diagonal matrix has a determinant equal to the product of the elements on its diagonal.

ANSWERS 4.1. C

1. -112 2. 0 3. -115 4. -1382

4.2 Uses of Determinants

In this section we discuss several uses of determinants. We omit providing drill on the two that have classically been most common—inverting matrices and solving simultaneous equations—because the methods presented in Chapters 2 and 3 are much more efficient. (Notice, however, that Cramer's rule for solving simultaneous equations in two and three unknowns was included on pages 131–132 and 138.) Our applications concentrate on the following rule.

4.2.1 Rule

If any one of the following statements is true about a matrix A, then the other statements are all true also.

(a) $|A| = 0$.
(b) A has no multiplicative inverse.
(c) The rows of A are linearly dependent.
(d) The columns of A are linearly dependent.
(e) The equation $AX = 0$ has a nonzero solution.

The proof of this rule is beyond the scope of this text, but we shall verify that it is true in several examples.

4.2.2 Example

Show that all five statements in Rule 4.2.1 are true if $A = \begin{bmatrix} 1 & 2 \\ 3 & 6 \end{bmatrix}$.

(a) $\begin{vmatrix} 1 & 2 \\ 3 & 6 \end{vmatrix} = 1 \cdot 6 - 2 \cdot 3 = 0$

(b) If $\begin{bmatrix} 1 & 2 \\ 3 & 6 \end{bmatrix} \begin{bmatrix} a & b \\ c & d \end{bmatrix} = \begin{bmatrix} 1 & 0 \\ 0 & 1 \end{bmatrix}$, then $\begin{bmatrix} a + 2c & b + 2d \\ 3a + 6c & 3b + 6d \end{bmatrix} = \begin{bmatrix} 1 & 0 \\ 0 & 1 \end{bmatrix}$. But if $a + 2c = 1$, it is impossible to have also $3a + 6c = 0$, because $3a + 6c = 3(a + 2c) = 3(1) \neq 0$. Thus there is no possible choice for a and c; therefore, the matrix has no inverse.

(c) $-3[1, 2] + [3, 6] = 0$

(d) $-2\begin{bmatrix} 1 \\ 3 \end{bmatrix} + \begin{bmatrix} 2 \\ 6 \end{bmatrix} = 0$

(e) $AX = 0$ can also be written $\begin{matrix} x + 2y = 0 \\ 3x + 6y = 0 \end{matrix}$. It is clear that $\begin{bmatrix} 0 \\ 0 \end{bmatrix}$ satisfies $AX = 0$. $\begin{bmatrix} 0 \\ 0 \end{bmatrix}$ is called the <u>zero solution</u>. But $\begin{bmatrix} -2 \\ 1 \end{bmatrix}$ (that is, $x = -2$, $y = 1$) is also a solution; it is a <u>nonzero solution</u>. Can you find another nonzero solution? There are many of them.

4.2.3 Example

Show that all five statements in Rule 4.2.1 are false if $A = \begin{bmatrix} 1 & 2 \\ 3 & 5 \end{bmatrix}$.

(a) $\begin{vmatrix} 1 & 2 \\ 3 & 5 \end{vmatrix} = 1 \cdot 5 - 2 \cdot 3 = -1 \neq 0$

(b) $\begin{bmatrix} 1 & 2 \\ 3 & 5 \end{bmatrix} \begin{bmatrix} -5 & 2 \\ 3 & -1 \end{bmatrix} = \begin{bmatrix} 1 & 0 \\ 0 & 1 \end{bmatrix}$

(c) If $a[1, 2] + b[3, 5] = 0$, then $a + 3b = 0$ and $2a + 5b = 0$. The first equation gives $a = -3b$; substituting this in the second equation, we obtain $2(-3b) + 5b = -6b + 5b = -b = 0$. This implies that $b = 0$, and therefore $a = -3b = 0$. Thus the two vectors are linearly independent.

(d) If $a\begin{bmatrix} 1 \\ 3 \end{bmatrix} + b\begin{bmatrix} 2 \\ 5 \end{bmatrix} = 0$, then $a + 2b = 0$ and $3a + 5b = 0$. The first equation gives $a = -2b$; substituting this in the second equation, we obtain $3(-2b) + 5b = -b = 0$. This implies that $b = 0$, which implies that $a = 0$, so the vectors are linearly independent.

(e) $AX = 0$ can also be written $\begin{array}{r} x + 2y = 0 \\ 3x + 5y = 0. \end{array}$ Part (d) shows that $\begin{bmatrix} 0 \\ 0 \end{bmatrix}$ is the only solution.

4.2.4 Example

Show that statements (a), (c), and (e) in Rule 4.2.1 are true if

$$A = \begin{bmatrix} 1 & 2 & -1 \\ 2 & -3 & -1 \\ 4 & 1 & -3 \end{bmatrix}$$

(a) Using the first row,

$$\begin{vmatrix} 1 & 2 & -1 \\ 2 & -3 & -1 \\ 4 & 1 & -3 \end{vmatrix} = \begin{vmatrix} -3 & -1 \\ 1 & -3 \end{vmatrix} - 2\begin{vmatrix} 2 & -1 \\ 4 & -3 \end{vmatrix} - \begin{vmatrix} 2 & -3 \\ 4 & 1 \end{vmatrix}$$

$$= 9 + 1 - 2(-6 + 4) - (2 + 12)$$

$$= 10 - 2(-2) - 14 = 0$$

(c) This was proved in Example 3.5.5, page 124.

(e) $\begin{bmatrix} 1 & 2 & -1 \\ 2 & -3 & -1 \\ 4 & 1 & -3 \end{bmatrix} \begin{bmatrix} 5 \\ 1 \\ 7 \end{bmatrix} = \begin{bmatrix} 0 \\ 0 \\ 0 \end{bmatrix}$

4.2.5 Example

Show that all five statements of Rule 4.2.1 are true if

$$A = \begin{bmatrix} 1 & 0 & 1 \\ 0 & 0 & 1 \\ 1 & 0 & 0 \end{bmatrix}$$

(a) Using the second column, it is clear that $|A| = 0$.

(b) If $\begin{bmatrix} a & b & c \\ d & e & f \\ g & h & i \end{bmatrix} \begin{bmatrix} 1 & 0 & 1 \\ 0 & 0 & 1 \\ 1 & 0 & 0 \end{bmatrix} = \begin{bmatrix} 1 & 0 & 0 \\ 0 & 1 & 0 \\ 0 & 0 & 1 \end{bmatrix}$, then

$$\begin{bmatrix} a+c & 0 & a+b \\ d+f & 0 & d+e \\ g+i & 0 & g+h \end{bmatrix} = \begin{bmatrix} 1 & 0 & 0 \\ 0 & 1 & 0 \\ 0 & 0 & 1 \end{bmatrix}$$

and the middle element is clearly in trouble.

(c) $[1, 0, 1] - [0, 0, 1] - [1, 0, 0] = 0$

(d) $0 \begin{bmatrix} 1 \\ 0 \\ 1 \end{bmatrix} + 1 \begin{bmatrix} 0 \\ 0 \\ 0 \end{bmatrix} + 0 \begin{bmatrix} 1 \\ 1 \\ 0 \end{bmatrix} = 0$ (Check Definition 3.5.1, page 123.)

(e) $\begin{bmatrix} 1 & 0 & 1 \\ 0 & 0 & 1 \\ 1 & 0 & 0 \end{bmatrix} \begin{bmatrix} 0 \\ 3 \\ 0 \end{bmatrix} = \begin{bmatrix} 0 \\ 0 \\ 0 \end{bmatrix}$

4.2.6 Example

Show that all five statements of Rule 4.2.1 are false if

$$A = \begin{bmatrix} 1 & 0 & 0 \\ 0 & 1 & 0 \\ 0 & 0 & 1 \end{bmatrix} = I$$

(a) $\begin{vmatrix} 1 & 0 & 0 \\ 0 & 1 & 0 \\ 0 & 0 & 1 \end{vmatrix} = 1 \begin{vmatrix} 1 & 0 \\ 0 & 1 \end{vmatrix} = 1(1 - 0) \neq 0$

(b) $I \cdot I = I$; the matrix is its own inverse.

(c) See Example 3.5.6, page 125.

(d) Same reasoning as part (c).

(e) $\begin{bmatrix} 1 & 0 & 0 \\ 0 & 1 & 0 \\ 0 & 0 & 1 \end{bmatrix} \begin{bmatrix} x \\ y \\ z \end{bmatrix} = \begin{bmatrix} 0 \\ 0 \\ 0 \end{bmatrix}$ implies that $\begin{matrix} x = 0 \\ y = 0 \\ z = 0, \end{matrix}$ so there can be no nonzero solution to $AX = 0$.

4.2.7 Example

We return now to the closed Leontief model introduced in Example 3.2.5 on page 104. You may remember the oversized matrix on page 105 that suggested why a 3 by 3 closed Leontief model was such that all its elements were positive and the sum of the elements in each column was 1. Thus, for example, the elements in the left column represent, respectively, the "fraction of the government budget that is consumed by government," the "fraction of the government budget that is invested in industry," and the "fraction of the government budget given to households." The sum of these fractions is 1. Thus the set of equations can be written

$$\begin{aligned} ax + by + cz &= x \\ dx + ey + fz &= y \\ (1 - a - d)x + (1 - b - e)y + (1 - c - f)z &= z \end{aligned}$$

These three equations can be summarized in the matrix expression $AX = X$. You will notice that this is similar to the open Leontief model discussed in Sections 1.4 and 2.5; only the final demands vector D is missing.

Again we solve for X to find what gross outputs vector is suitable to a given technological matrix A:

$$\begin{aligned} AX &= X & & \\ 0 &= X - AX & &\text{(subtracting)} \\ 0 &= IX - AX & &\text{(because } I \text{ is the identity)} \\ 0 &= (I - A)X & &\text{(distributive law)} \end{aligned}$$

All this has a familiar ring. But now if we suppose that $(I - A)$ has an inverse $(I - A)^{-1}$ as we did in Section 2.5, we get

$$(I - A)^{-1}0 = (I - A)^{-1}(I - A)X \qquad \text{[premultiplying both sides by } (I - A)^{-1}]$$

and so

$$0 = X \qquad \text{(by the definition of inverse, and the fact that anything multiplied by 0 gives 0)}$$

Surprise! This time if $(I - A)$ has an inverse, X must be the zero vector—the economy has zero gross outputs! Thus, if a closed Leontief model is to be at all realistic, $(I - A)$ cannot have an inverse.

By Rule 4.2.1 this will happen exactly when the determinant of $(I - A)$ is zero, when $(I - A)X = 0$ has more than one solution, and when the rows of $(I - A)$ are linearly dependent. Since

$$I - A = \begin{bmatrix} 1 - a & -b & -c \\ -d & 1 - e & -f \\ -1 + a + d & -1 + b + e & c + f \end{bmatrix}$$

when we use the notation above, it is clear that the rows are linearly dependent ($R_1 + R_2 + R_3 = 0$). Thus $(I - A)$ does not have an inverse, and we cannot solve to get 0 as we attempted above. There will be many solutions, and these solutions can be found for any specific set of numbers in a fashion similar to that used in Example 3.2.5. As in that example, the solutions will always turn out to be multiples of each other; the economy must contract or expand proportionally if there are no outside influences and the technological matrix remains constant.

SUMMARY

A matrix A has determinant zero if and only if (a) it has no inverse, (b) its rows are linearly dependent, (c) its columns are linearly dependent, and/or (d) $AX = 0$ has a nonzero solution.

EXERCISES 4.2. A

1. Show that if $A = \begin{bmatrix} 2 & 3 \\ 4 & 6 \end{bmatrix}$, all the statements in Rule 4.2.1 are true.
2. Show that if $A = \begin{bmatrix} 4 & 3 \\ 5 & 4 \end{bmatrix}$, all five statements in Rule 4.2.1 are false.
3. Evaluate the determinants to discover which of the following matrices has an inverse.
   (a) $\begin{bmatrix} 0 & 2 & -3 \\ 2 & -1 & 4 \\ 4 & 4 & -1 \end{bmatrix}$ (b) $\begin{bmatrix} 2 & 2 & 0 \\ 1 & 0 & 0 \\ 0 & -10 & 1 \end{bmatrix}$ (c) $\begin{bmatrix} 3 & -2 & 1 \\ 1 & 0 & 2 \\ 1 & 2 & 7 \end{bmatrix}$
4. In which matrices of problem 3 are the rows linearly independent?
5. *Without* performing the Gauss–Jordan reduction process, what can you say about the row rank of each of the matrices in problem 3?
6. Show that for any 2 by 2 matrix of the form $\begin{bmatrix} a & b \\ ma & mb \end{bmatrix}$, the five statements in Rule 4.2.1 are true. [*Hint:* (b) is hardest; do it last.]
7. If $A = \begin{bmatrix} a & b \\ c & d \end{bmatrix}$ does not have a zero determinant, then $ad - bc \neq 0$.

   Show that in this case the inverse of A is $\begin{bmatrix} \dfrac{d}{ad - bc} & \dfrac{-b}{ad - bc} \\ \dfrac{-c}{ad - bc} & \dfrac{a}{ad - bc} \end{bmatrix}$.

EXERCISES 4.2. B

1. Show that if $A = \begin{bmatrix} 3 & 1 \\ 6 & 2 \end{bmatrix}$, all the statements in Rule 4.2.1 are true.
2. Show that if $A = \begin{bmatrix} 3 & 2 \\ 4 & 3 \end{bmatrix}$, all five statements in Rule 4.2.1 are false.

3. Evaluate the determinants to discover which of the following matrices has an inverse.

(a) $\begin{bmatrix} -3 & 2 & 9 \\ 0 & 1 & 4 \\ 1 & 0 & 0 \end{bmatrix}$ (b) $\begin{bmatrix} 4 & -3 & 5 \\ 0 & 1 & -2 \\ 4 & -1 & 1 \end{bmatrix}$

(c) $\begin{bmatrix} 2 & 0 & -1 \\ 3 & 1 & 2 \\ -4 & -2 & -5 \end{bmatrix}$

4. In which matrices of problem 3 are the rows linearly independent?
5. *Without* performing the Gauss–Jordan reduction process, what can you say about the row rank of each of the matrices in problem 3?

6–7. Same as in Exercises 4.2.A.

ANSWERS 4.2. A

1. Mimic Example 4.2.2. One nonzero solution for part (e) is $\begin{bmatrix} -3 \\ 2 \end{bmatrix}$. There are many others.
2. Mimic Example 4.2.3. (b) $A^{-1} = \begin{bmatrix} 4 & -3 \\ -5 & 4 \end{bmatrix}$.
3. Only part (b) has an inverse; the others have determinant 0.
4. Part (b) by Rule 4.2.1.
5. Matrices (a) and (c) have row rank 2; matrix (b) has row rank 3. [Since the rows are linearly dependent in (a) and (c), they cannot have row rank 3. Since the rows are not multiples of each other, they cannot have row rank 1.]
6. (a) $a(mb) - b(ma) = 0$

(b) If $\begin{bmatrix} a & b \\ ma & mb \end{bmatrix}\begin{bmatrix} c & d \\ e & f \end{bmatrix} = \begin{bmatrix} 1 & 0 \\ 0 & 1 \end{bmatrix}$, then $ac + be = 1$ and $mac + mbe = 0$. But $mac + mbe = m(ac + be) = m(1) = m$. Thus $m = 0$.

But $\begin{bmatrix} a & b \\ 0 & 0 \end{bmatrix}\begin{bmatrix} c & d \\ e & f \end{bmatrix} = \begin{bmatrix} ac + be & ad + bf \\ 0 & 0 \end{bmatrix} \neq \begin{bmatrix} 1 & 0 \\ 0 & 1 \end{bmatrix}$.

This contradiction shows that no inverse $\begin{bmatrix} c & d \\ e & f \end{bmatrix}$ can exist.

(c) $-m[a, b] + [ma, mb] = 0$

(d) $\begin{bmatrix} a \\ ma \end{bmatrix} - \dfrac{a}{b}\begin{bmatrix} b \\ mb \end{bmatrix} = 0$, if $b \neq 0$. If $b = 0$, $0\begin{bmatrix} a \\ ma \end{bmatrix} + 1\begin{bmatrix} 0 \\ 0 \end{bmatrix} = 0$.

(e) $\begin{bmatrix} a & b \\ ma & mb \end{bmatrix}\begin{bmatrix} -b \\ a \end{bmatrix} = \begin{bmatrix} 0 \\ 0 \end{bmatrix}$, if $a \neq 0$ or $b \neq 0$. If $a = b = 0$, part (e) is obvious.

7. $$\begin{bmatrix} \dfrac{d}{ad - bc} & \dfrac{-b}{ad - bc} \\ \dfrac{-c}{ad - bc} & \dfrac{a}{ad - bc} \end{bmatrix}\begin{bmatrix} a & b \\ c & d \end{bmatrix} = \begin{bmatrix} 1 & 0 \\ 0 & 1 \end{bmatrix}$$

4.3 The Gauss–Jordan Method Applied to Determinants

The following rule, which we state here without proof, combined with the facts about elementary matrices which you discovered while doing the exercises in Section 4.1, is the key to a much faster way of computing the determinant of large matrices.

4.3.1 Rule

If A and B are two matrices, then the determinant of their product is the product of their determinants. Symbolically,

$$|AB| = |A| \cdot |B|$$

Since performing a Gauss–Jordan row operation on a matrix is the same as premultiplying it by an elementary matrix, we can use elementary row operations to reduce any matrix to a simpler form as long as we keep track of how the determinant has been changed; to do this we use the rules derived in problem 3 of Exercises 4.1.A and 4.1.B. When the matrix has been changed to upper triangular form, it is easy to calculate the determinant using the following rule.

4.3.2 Rule

The determinant of an upper triangular matrix is the product of its elements on the major diagonal.

It is not hard to see why this is so. Consider the following arbitrary 4 by 4 triangular matrix, and see what happens if we evaluate its determinant by always expanding using the first column.

$$\begin{vmatrix} a & b & c & d \\ 0 & e & f & g \\ 0 & 0 & h & i \\ 0 & 0 & 0 & j \end{vmatrix} = a \begin{vmatrix} e & f & g \\ 0 & h & i \\ 0 & 0 & j \end{vmatrix} - 0 + 0 - 0$$

$$= a\left[e \begin{vmatrix} h & i \\ 0 & j \end{vmatrix} - 0 + 0\right]$$

$$= a \cdot e \cdot (hj - i0) = aehj$$

The same argument works for a lower triangular matrix (using the first rows instead of first columns) and for triangular matrices of other dimensions.

Our method, therefore, for evaluating matrices in this section will be to change a matrix to upper triangular form (not necessarily reduced form, which is harder) by using the Gauss–Jordan row-reduction method and the following three facts.

4.3.3 Rule

1. If two rows of a matrix are interchanged, the determinant of the matrix is multiplied by -1.
2. If one row of a matrix is multiplied by a constant, the determinant of that matrix is multiplied by that constant. (Thus if a row is multiplied by a number, a compensating reciprocal must appear outside the determinant to maintain equality, as in Example 4.3.5.)
3. If a multiple of one row of a matrix is added to (or subtracted from) another row of the same matrix, the value of its determinant does not change.

4.3.4 Example

Give examples of each part of Rule 4.3.3.

Solution:

1. $\begin{vmatrix} 0 & 1 & 0 \\ 1 & 0 & 0 \\ 0 & 0 & 1 \end{vmatrix} = -1$ 2. $\begin{vmatrix} 1 & 0 & 0 \\ 0 & 5 & 0 \\ 0 & 0 & 1 \end{vmatrix} = 5$

3. $\begin{vmatrix} 1 & 0 & 4 \\ 0 & 1 & 0 \\ 0 & 0 & 1 \end{vmatrix} = 1$

4.3.5 Example

Evaluate $\begin{vmatrix} 5 & 2 & 8 \\ 7 & 3 & 0 \\ 0 & 0 & 1 \end{vmatrix}$

Solution:

To solve this using (2) of Rule 4.3.3, we might want to make the non-zero elements in the first column equal so that we can subtract one from the other and get zero. Thus we multiply the first row by 7 and the second by 5. To keep the value of the new determinant the same as before, we must multiply the new determinant by $\frac{1}{7}$ and then by $\frac{1}{5}$, as follows.

$$\begin{vmatrix} 5 & 2 & 8 \\ 7 & 3 & 0 \\ 0 & 0 & 1 \end{vmatrix} = \frac{1}{7}\begin{vmatrix} 35 & 14 & 56 \\ 7 & 3 & 0 \\ 0 & 0 & 1 \end{vmatrix} = \frac{1}{5}\cdot\frac{1}{7}\begin{vmatrix} 35 & 14 & 56 \\ 35 & 15 & 0 \\ 0 & 0 & 1 \end{vmatrix}$$

$$= \frac{1}{35}\begin{vmatrix} 35 & 14 & 56 \\ 0 & 1 & -56 \\ 0 & 0 & 1 \end{vmatrix} = \frac{1}{35}\cdot 35 = 1$$

4.3.6 Example

Evaluate the following determinants using the row-reduction method.

(a) $\begin{vmatrix} 3 & 2 & 1 \\ -1 & 2 & 4 \\ 2 & -4 & 2 \end{vmatrix}$ (b) $\begin{vmatrix} 0 & 2 & 1 \\ 1 & 3 & -1 \\ -2 & -4 & 3 \end{vmatrix}$

Solution:

(a) $\begin{vmatrix} 3 & 2 & 1 \\ -1 & 2 & 4 \\ 2 & -4 & 2 \end{vmatrix}$ $\quad R_1 + 3R_2 \longrightarrow R_1$, $R_3 + 2R_2 \longrightarrow R_3$

$$= \begin{vmatrix} 0 & 8 & 13 \\ -1 & 2 & 4 \\ 0 & 0 & 10 \end{vmatrix} \quad R_1 \longleftrightarrow R_2$$

$$= -\begin{vmatrix} -1 & 2 & 4 \\ 0 & 8 & 13 \\ 0 & 0 & 10 \end{vmatrix}$$

$$= -(-1)(8)(10) = 80$$

(b) $\begin{vmatrix} 0 & 2 & 1 \\ 1 & 3 & -1 \\ -2 & -4 & 3 \end{vmatrix}$ $\quad R_3 + 2R_2 \longrightarrow R_3$

$$= \begin{vmatrix} 0 & 2 & 1 \\ 1 & 3 & -1 \\ 0 & 2 & 1 \end{vmatrix} \quad R_3 - R_1 \longrightarrow R_3$$

$$= \begin{vmatrix} 0 & 2 & 1 \\ 1 & 3 & -1 \\ 0 & 0 & 0 \end{vmatrix} \quad R_1 \longleftrightarrow R_2$$

$$= -\begin{vmatrix} 1 & 3 & -1 \\ 0 & 2 & 1 \\ 0 & 0 & 0 \end{vmatrix}$$

$$= -(1)(2)(0) = 0$$

Part (b) of this example suggests two facts which are indeed true.

4.3.7 Rule

If any row of a matrix is all zeros, the determinant of that matrix is zero.

4.3.8 Rule

If any two rows of a matrix are equal, the determinant of that matrix is zero.

Rule 4.3.8 follows immediately from Rules 4.3.7 and 4.3.3 (part 3). Rule 4.3.7 can be deduced from the fact that if any row is all zeros, it can be interchanged with the last row, and then no matter what other row operations are performed to change the upper rows to triangular form, the zero in the lower right of the matrix will be there when the elements of the diagonal are multiplied together. And one zero factor in a product is enough to assure that the product is zero.

Since Rule 4.2.1 requires only the knowledge of whether a determinant is zero to find out whether its matrix has an inverse or whether its rows are linearly independent, Rules 4.3.7 and 4.3.8 are often effective time-savers. We need only to reduce the associated determinant of a matrix to one where two rows are equal or one row is entirely zeros in order to show that the given matrix has no inverse or that its rows are linearly dependent.

We now show how, for large matrices, the technique of this section is remarkably more efficient than the (better known) method of Section 4.1.

4.3.9 Example

Evaluate the determinant of Example 4.1.9 using the row-reduction method.

Solution:

$$\begin{vmatrix} 2 & 0 & 1 & 1 & 3 \\ -2 & 1 & 0 & 1 & -1 \\ 3 & 0 & -1 & 0 & 4 \\ 4 & 0 & 2 & -2 & 3 \\ -1 & 1 & 0 & 2 & 3 \end{vmatrix}$$

$$= \begin{vmatrix} 0 & 2 & 1 & 5 & 9 \\ 0 & -1 & 0 & -3 & -7 \\ 0 & 3 & -1 & 6 & 13 \\ 0 & 4 & 2 & 6 & 15 \\ -1 & 1 & 0 & 2 & 3 \end{vmatrix}$$

$$= -\begin{vmatrix} -1 & 1 & 0 & 2 & 3 \\ 0 & -1 & 0 & -3 & -7 \\ 0 & 3 & -1 & 6 & 13 \\ 0 & 4 & 2 & 6 & 15 \\ 0 & 2 & 1 & 5 & 9 \end{vmatrix}$$

$$= -\begin{vmatrix} -1 & 1 & 0 & 2 & 3 \\ 0 & -1 & 0 & -3 & -7 \\ 0 & 0 & -1 & -3 & -8 \\ 0 & 0 & 2 & -6 & -13 \\ 0 & 0 & 1 & -1 & -5 \end{vmatrix}$$

$$= -\begin{vmatrix} -1 & 1 & 0 & 2 & 3 \\ 0 & -1 & 0 & -3 & -7 \\ 0 & 0 & -1 & -3 & -8 \\ 0 & 0 & 0 & -12 & -29 \\ 0 & 0 & 0 & -4 & -13 \end{vmatrix}$$

$$= -\begin{vmatrix} -1 & 1 & 0 & 2 & 3 \\ 0 & -1 & 0 & -3 & -7 \\ 0 & 0 & -1 & -3 & -8 \\ 0 & 0 & 0 & 0 & 10 \\ 0 & 0 & 0 & -4 & -13 \end{vmatrix}$$

$$= \begin{vmatrix} -1 & 1 & 0 & 2 & 3 \\ 0 & -1 & 0 & -3 & -7 \\ 0 & 0 & -1 & -3 & -8 \\ 0 & 0 & 0 & -4 & -13 \\ 0 & 0 & 0 & 0 & 10 \end{vmatrix}$$

$$= (-1)(-1)(-1)(-4)(10) = 40$$

SUMMARY

The Gauss–Jordan reduction process can be used to evaluate determinants by noticing that the determinant of every upper (and lower) triangular matrix is the product of the elements on its major diagonal. Interchanging two rows of a determinant changes its signs; multiplying one row of a determinant by any number multiplies the whole determinant by the same number; and adding a multiple of one row of the matrix to another row does not change the value of the matrix at all.

Any determinant is zero if it has a row of all zeros or if two rows are equal.

EXERCISES 4.3. A

Evaluate the following determinants using Gauss–Jordan row reduction.

1. $\begin{vmatrix} 2 & 1 & 4 & 2 \\ 1 & 0 & 2 & 3 \\ 4 & 2 & 2 & 5 \\ 2 & 2 & 2 & 6 \end{vmatrix}$ 2. $\begin{vmatrix} 3 & 1 & 0 & 5 \\ 2 & 4 & -2 & 6 \\ 7 & -1 & 3 & 9 \\ -2 & 6 & -5 & 2 \end{vmatrix}$

3. $\begin{vmatrix} -1 & 0 & 1 & 3 \\ 3 & 1 & 0 & -1 \\ 1 & 0 & -1 & 4 \\ -3 & 1 & 2 & 3 \end{vmatrix}$ 4. $\begin{vmatrix} 2 & 4 & 6 & 2 \\ 0 & 2 & 3 & 1 \\ 1 & 1 & 2 & 2 \\ 4 & 2 & 2 & 1 \end{vmatrix}$

EXERCISES 4.3. B

Evaluate the determinants of problems 1–4 in Exercises 4.1.C (page 138) using Gauss–Jordan row reduction.

EXERCISES 4.3. C

Evaluate the following determinants using Gauss–Jordan row reduction.

1. $\begin{vmatrix} 3 & 6 & 9 & 12 & 15 \\ 1 & 2 & 3 & 4 & 3 \\ 2 & 3 & 1 & 6 & 3 \\ 4 & 1 & 3 & 3 & 5 \\ 2 & 2 & 1 & 4 & 3 \end{vmatrix}$ 2. $\begin{vmatrix} 1 & 1 & 2 & 1 & 0 \\ 2 & 0 & 6 & 4 & 1 \\ 4 & 0 & 3 & 1 & 2 \\ 0 & 1 & 2 & 1 & 3 \\ 1 & 0 & 4 & 0 & 1 \end{vmatrix}$

ANSWERS 4.3. A

1. 46 2. 0 3. −28 4. 6

ANSWERS 4.3. C

1. 30 2. 172

VOCABULARY

determinant, minor, checkerboard sign pattern, zero solution, nonzero solution

SAMPLE TEST

Chapter 4

1. (a) $\begin{vmatrix} 1 & 2 & 1 \\ 1 & 0 & -1 \\ 2 & -1 & 1 \end{vmatrix} =$ (b) Does $\begin{bmatrix} 1 & 2 & 1 \\ 1 & 0 & -1 \\ 2 & -1 & 1 \end{bmatrix}$ have an inverse?

   (c) Does the following set of equations have more than one solution?

$$\begin{aligned} x + 2y + z &= 0 \\ x - \quad\;\; z &= 0 \\ 2x - y + z &= 0 \end{aligned}$$

2. Use a determinant to find out whether the following set of vectors is linearly dependent or independent: [2, 3, −1], [5, 0, 2], and [1, 9, −5].
3. Complete: A matrix has determinant 0 if and only if
   (a) (b) (c) (d)
4. Does the set of equations $\begin{aligned} 2x + 6y &= 0 \\ x + 3y &= 0 \end{aligned}$ have a nonzero solution? How do you know? If "yes," give one such solution.
5. Use the Gauss–Jordan technique to evaluate the determinant

$$\begin{vmatrix} 0 & 2 & 1 & 0 \\ 1 & 0 & -2 & 3 \\ 0 & 1 & 3 & 4 \\ 3 & 2 & 4 & -5 \end{vmatrix}$$

The answers are at the back of the book.

Chapter 5

INTRODUCTION TO LINEAR PROGRAMMING

5.1 Graphing Linear Inequalities

This section discusses how to graph a linear inequality in two variables, because this skill will be needed in the next section. First we review numerical inequalities.

If you add the same number to both sides of an inequality, the inequality remains true (Figure 5.1–1). Similarly, if you multiply both

Figure 5.1–1

sides of an inequality by the same positive number, the inequality remains true (Figure 5.1–2). On the other hand, if you multiply an inequality by a negative number, the inequality is reversed (Figure 5.1–3). These facts are used in manipulating expressions such as the following:

$2x + y \leq 4$	is the same as	$y \leq 4 - 2x$
$5x + 10y + 15z \leq 60$	is the same as	$x + 2y + 3z \leq 12$
$-x + y - z \leq 8$	is the same as	$x + y + z \geq -8$
$6x - 3y \geq 9$	is the same as	$-2x + y \leq -3$

Figure 5.1–2

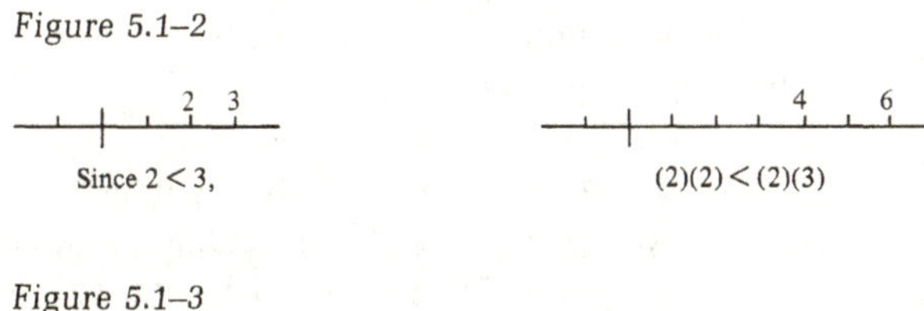

Figure 5.1–3

When variables appear in an inequality, there will usually be some values of the variable(s) that make it true and others that make it false. When the values that make it true are pictured on a graph, we say we have "graphed" that inequality. You have had some experience in graphing linear equations (if you want more, consult Appendix 2), and you may remember that the graph of a linear equation is always a straight line. By contrast, the graph of a linear inequality is a whole region.

5.1.1 Example

Graph $2x + y \geq 4$.

Solution:

It is helpful to subtract 2x from each side of this inequality and write it in the form

$$y \geq 4 - 2x$$

Now consider the equation $y = 4 - 2x$. It is the straight line indicated in Figure 5.1–4 with y-intercept of 4 and x-intercept of 2. For each x_1, the corresponding y_1 such that $y_1 = 4 - 2x_1$ will determine the point (x_1, y_1) lying on the line. If x_1 remains the same, but y grows larger than $y_1 = 4 - 2x_1$, the point (x_1, y) will be above the line. This is true no matter how much y may grow greater than y_1. Thus *the points on and above the line* $y = 4 - 2x$ form the graph of $y \geq 4 - 2x$.

Figure 5.1–4

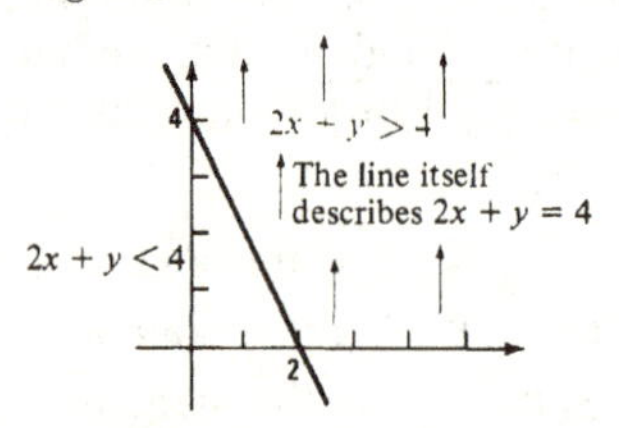

By similar reasoning, the graph of $2x + y \leq 4$ is the set of points on or below the line $2x + y = 4$.

The equation $2x + y = 4$ can be changed into an inequality in any one of four standard ways:

(1) $2x + y < 4$ "less than"

(2) $2x + y > 4$ "greater than"

(3) $2x + y \leq 4$ "less than or equal to"

(4) $2x + y \geq 4$ "greater than or equal to"

The graphs of the first two inequalities are the points strictly below and strictly above, respectively, the line $2x + y = 4$. The graphs of the last two inequalities are similar but include the line $2x + y = 4$.

Similarly, the graph of any linear inequality in two unknowns will be the points on one side of the line that graphs the corresponding equality.

The new skill required in graphing inequalities is deciding which inequality describes which side of the line. Probably the easiest way to do this is to pick a point (preferably the origin) not on the line. If the corresponding ordered pair (x, y) satisfies the inequality, its side of the line is the side we seek. If not, it is the opposite side.

5.1.2 Example

Graph $2x + 3y \leq 12$.

Solution:

First we graph the corresponding straight line, $2x + 3y = 12$, by using the facts that $(0, 4)$ and $(6, 0)$ lie on the line. Then we can test the origin $(0, 0)$ and discover that it *does* satisfy the inequality. Thus the side of the line containing the origin should be shaded (Figure 5.1–5).

Figure 5.1–5

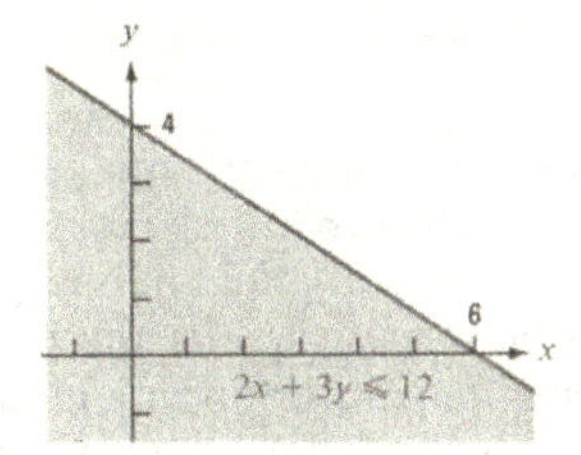

5.1.3 Example

Graph $2x - 3y \geq 12$.

Solution:

First we plot the intercepts $(0, -4)$ and $(6, 0)$ to identify the corresponding line. Then we note that $(0, 0)$ does *not* satisfy the inequality, so we shade the side of the line opposite the origin (Figure 5.1–6).

Figure 5.1–6

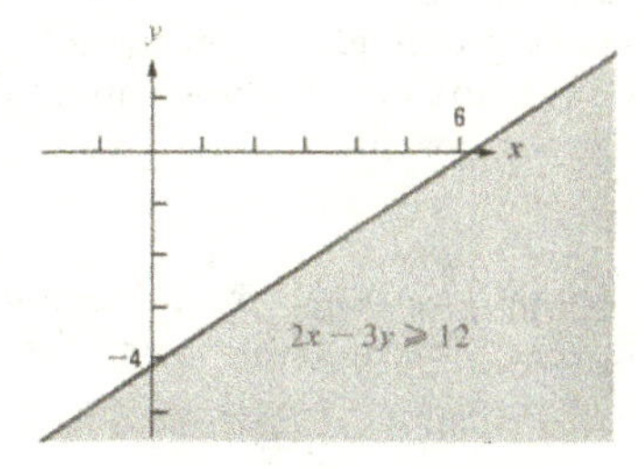

5.1.4 Example

Graph $y \leq 2x$.

Solution:

First we graph $y = 2x$. The origin cannot be used as a test point because it lies on the line. In such cases (unless the line is the x-axis) an easy point to use is (1, 0). Since $0 \leq 2(1)$, the point (1, 0) *does* satisfy the inequality, so the right side of the line should be shaded (Figure 5.1–7).

Figure 5.1–7

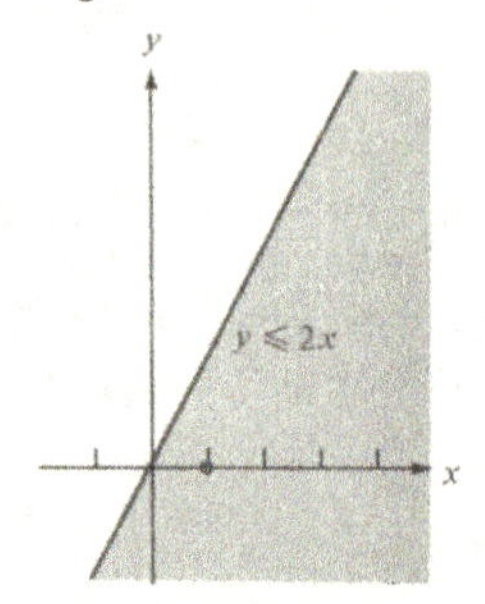

If the inequality does not include an equal sign (that is, if it is of the form $<$ or $>$), the line itself is omitted from the graph. These latter cases are often indicated by making the boundary line dotted or dashed, but in this course we need never consider these cases. We do, however, need to use simultaneous inequalities in this text, and now we turn to a problem that indicates why and how they will be used. We postpone the complete solution of the problem, however, until the next section.

5.1.5 Example

Suppose that a grain wholesaler regularly receives 3000 lb of wheat and 4000 lb of barley, which is sold in two types of retail bags. "Premium" feed bags contain 1 lb of wheat and 2 lb of barley, and "Regular" feed bags contain 1 lb each of wheat and barley; the wholesaler gets a profit of 30 cents on each Premium bag sold and 20 cents on each Regular bag sold. The amount of each type of bag that is sold will depend to a considerable extent on the advertising and marketing techniques used. How many of each type should the wholesaler attempt to sell to get a maximum profit?

Partial Solution:

When we ask, "What quantities am I looking for?" as we did in Chapter 2, the answer is the number of bags of each retail type. Thus we can set

x = number of Premium bags sold

y = number of Regular bags sold

But we are also looking for the profit, P, and espe ially the maximum possible profit, P_{max}. The number of each type of bag sold will determine the profit in a predictable way, and this relationship is called the objective function:

$$P = 30x + 20y$$

This objective function says that the profit will be 30 times the number of Premium bags sold plus 20 times the number of Regular bags sold. The objective function of any linear programming problem describes the objective, or goal, of that problem.

If there were no restrictions on the number of bags sold, the profit would be unbounded, but this is rarely the case in real life. The mathematical expressions of the restrictions are inequalities called the constraints. In this example, there is one constraint resulting from a limited supply of barley and another resulting from a limited supply of wheat.

Each bag—both Premium and Regular—contains 1 lb of wheat, so the total number of bags cannot exceed the supply of wheat, which is 3000 lb.

$$x + y \leq 3000 \qquad \text{(wheat constraint)}$$

Furthermore, x Premium bags use $2x$ lb of barley and y Regular bags require y lb of barley, so

$$2x + y \leq 4000 \qquad \text{(barley constraint)}$$

Since it would make no sense to order a negative number of either Premium or Regular bags, we must also have $x \geq 0$ and $y \geq 0$. These last constraints are called the nonnegativity constraints.

This whole problem can be summarized as:

$$\text{Maximize } P = 30x + 20y \qquad \text{subject to} \qquad \begin{array}{ll} x + y \leq 3000 & x \geq 0 \\ 2x + y \leq 4000 & y \geq 0 \end{array}$$

In order to be considered as a solution to this problem, any point (x, y) must satisfy all these constraints. The set of points satisfying all the constraints is called the feasible region; if the problem has only two independent variables, we can picture the feasible region by graphing the simultaneous solution of all the constraints.

We do this now. Each constraint is a linear inequality, so it can be graphed using the method explained earlier in this section. The feasible points of this problem will be the set of points that lie simultaneously in *all* the shaded regions of Figure 5.1–8. To picture this set, you might combine them gradually. For example, the constraints $x \geq 0$ and $y \geq 0$ have the upper right quadrant as their simultaneous solution.

Figure 5.1–8

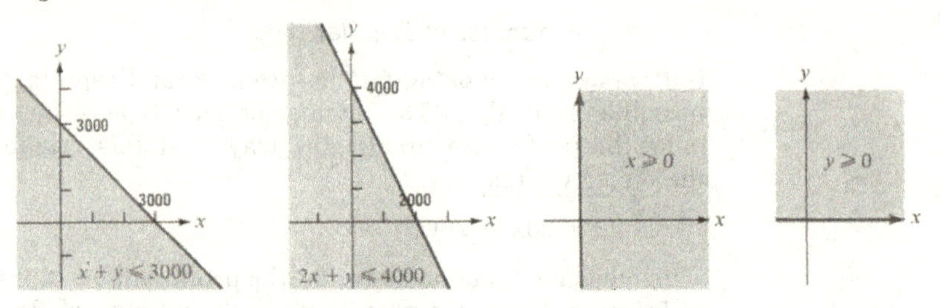

The constraints $x + y \leq 3000$ and $2x + y \leq 4000$ are both graphed by the set of points lying *under* the corresponding lines, so the simultaneous solution of all four inequalities will be the points in the upper right quadrant lying on and under *both* the lines $x + y = 3000$ and $2x + y = 4000$, as pictured in Figure 5.1–9. To find the coordinates of the vertex (corner) in the upper right, we simultaneously solve $x + y = 3000$ and $2x + y = 4000$:

$$
\begin{aligned}
2x + y &= 4000 \\
x + y &= 3000 \\
x \quad &= 1000 && \text{(subtracting the second from the first)} \\
1000 + y &= 3000 && \text{(substituting } x = 1000 \text{ into the second)} \\
y &= 2000 && \text{(subtracting 1000 from both sides)}
\end{aligned}
$$

Thus the vertex in the upper right of the feasible region is the point (1000, 2000).

Figure 5.1–9

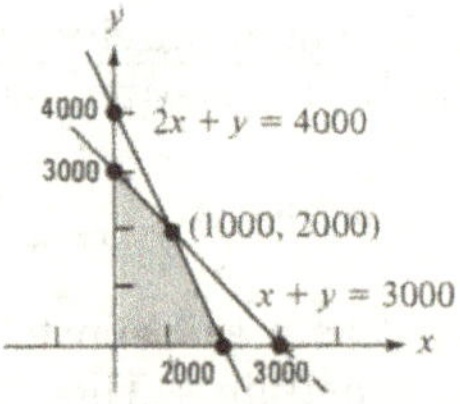

The solution for this example—that is, the point that satisfies all the given constraints and for which the profit is a maximum—is among those shaded in Figure 5.1–9 and the boundary of that region. The identification of which point is "best" is postponed until the next section.

We conclude this section with two more examples of the solution of simultaneous inequalities.

5.1.6 Example

Find the simultaneous solution to the inequalities

$$x \geq 0 \qquad 2x + 5y \leq 3500$$
$$y \geq 0 \qquad x + 10y \leq 4000$$

Solution:

The four given inequalities have the graphs shown in Figure 5.1–10. Thus the simultaneous solution will be those points that lie in the upper right quadrant below the lines $2x + 5y = 3500$ and $x + 10y = 4000$, as shown in Figure 5.1–11. (If you have trouble simultaneously solving $2x + 5y = 3500$ and $x + 10y = 4000$ to find the upper right vertex, consult Appendix 3.)

Figure 5.1–10

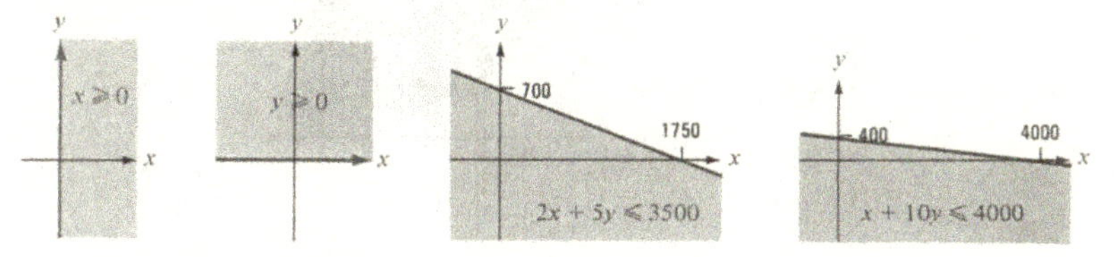

Figure 5.1–11

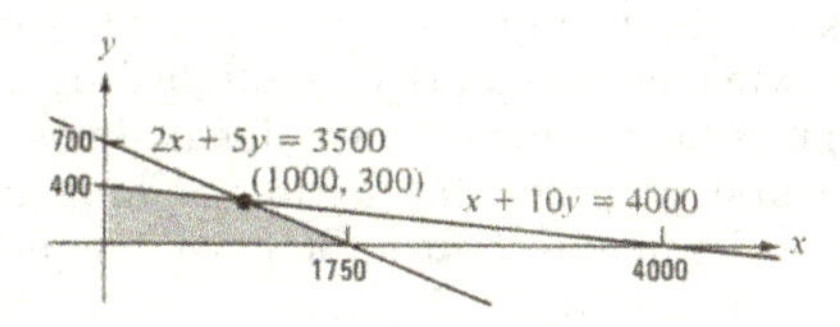

5.1.7 Example

Find the simultaneous solution to

$$x \geq 0 \qquad 3x + 2y \geq 12$$
$$y \geq 0 \qquad x + 2y \geq 8$$

Solution:

The four given inequalities have the graphs shown in Figure 5.1–12. Thus the simultaneous solution will be those points lying in the upper right quadrant *above* both the lines $3x + 2y = 12$ and $x + 2y = 8$, as shown in Figure 5.1–13. (Notice that the region is unbounded.)

Figure 5.1–12

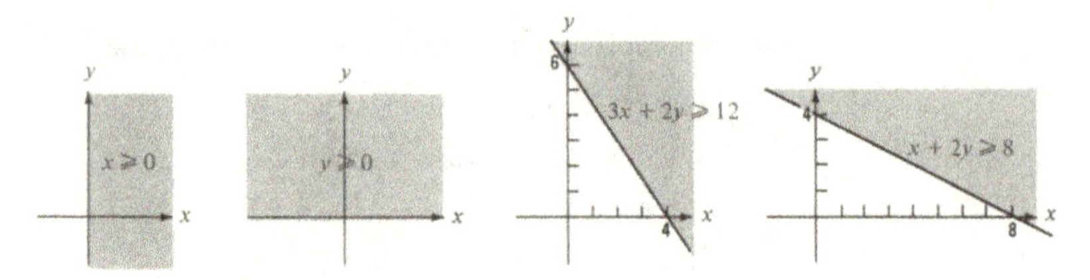

Figure 5.1–13

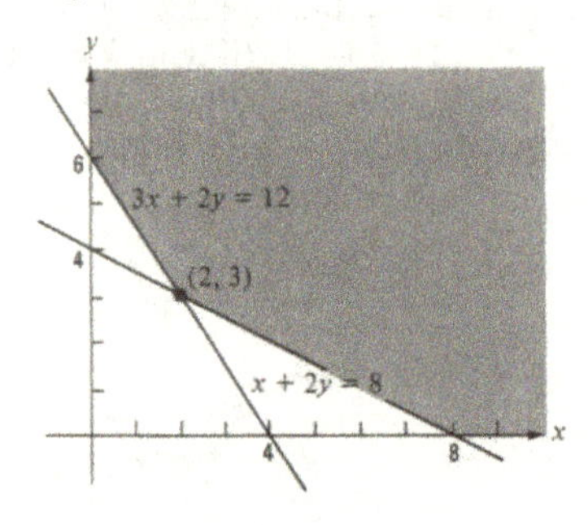

SUMMARY

Inequalities can be manipulated in much the same way as equations, except that when an inequality is multiplied by a negative number, the inequality must be reversed. Linear inequalities in two unknowns are graphed by first graphing the corresponding line and then deciding which side of the line describes the given inequality; this can be done by testing one point not on the line, often the origin.

Simultaneous linear inequalities will be used to picture the feasible regions of linear programming problems. One such problem was introduced in this section; the solution of this problem will be completed in the next section. Two other sets of simultaneous inequalities were solved to emphasize this essential skill.

EXERCISES 5.1. A

Graph the following inequalities:

1. $5x + 3y \leq 30$
2. $5x - 3y \leq 30$
3. $5y - 3x \geq 30$
4. $2x - y \geq 6$
5. $x - 2y \leq 6$
6. $4x + y \geq 8$
7. $x + 5y \leq 5$
8. $x + 2y \leq 0$
9. $x \geq y$
10. $3x + y \geq 0$

11. Tell whether each of the given inequalities means the same as $x - y + 2z \geq 3$.
   (a) $y \geq x + 2z - 3$
   (b) $5x - 5y + 10z \geq 15$
   (c) $3x + 6z \geq 3y + 9$
   (d) $y - x - 2z \geq -3$
   (e) $0.4x - 0.4y + 0.8z \leq 1.2$

Graph the simultaneous solution of the set of inequalities given in each of the next five problems. Save your answers; they will be useful for Exercises 5.2.A.

12. $x \geq 0,\ y \geq 0,\ x + 3y \leq 15,\ 8x + 4y \leq 40$
13. $x \geq 0,\ y \geq 0,\ 5x + 3y \leq 30,\ x + y \leq 8$
14. $x \geq 0,\ y \geq 0,\ 2x + y \leq 6,\ x \leq y$
15. $x \geq 0,\ y \geq 0,\ 5x + 4y \leq 360,\ 4x + 5y \leq 360$
16. $x \geq 0,\ y \geq 0,\ 2x + 3y \leq 6,\ 3x + 4y \geq 12$

EXERCISES 5.1. B

Graph the following inequalities:

1. $3x + 4y \leq 12$
2. $3x - 4y \leq 12$
3. $3x - 4y \geq 12$
4. $3x - y \geq 6$
5. $x - 3y \leq 6$
6. $3x + y \geq 9$
7. $x + 4y \leq 4$
8. $x + 3y \leq 0$
9. $2x \geq y$
10. $4x + y \geq 0$
11. Tell whether each of the given inequalities means the same as $x - y - 3z \leq 2$.
   (a) $6x - 6y - 18z \leq 12$
   (b) $y + 3z \geq x - 2$
   (c) $y - x + 3z \leq -2$
   (d) $y \leq x - 3z - 2$
   (e) $0.2x - 0.2y - 0.6z \leq 0.4$

Graph the simultaneous solution of the set of inequalities given in each of the next five problems. Save your answers; they will be useful for Exercises 5.2.B.

12. $x \geq 0,\ y \geq 0,\ 3x + 4y \leq 36,\ x + 2y \leq 14$
13. $x \geq 0,\ y \geq 0,\ 3x + 4y \leq 60,\ x + 2y \leq 26$
14. $x \geq 0,\ y \geq 0,\ 3x + 2y \leq 1200,\ x + 2y \leq 800$
15. $x \geq 0,\ y \geq 0,\ x + y \leq 10,\ 3x + y \leq 12$
16. $x \geq 0,\ y \geq 0,\ x + y \leq 2,\ 3x + 5y \geq 15$

ANSWERS 5.1. A

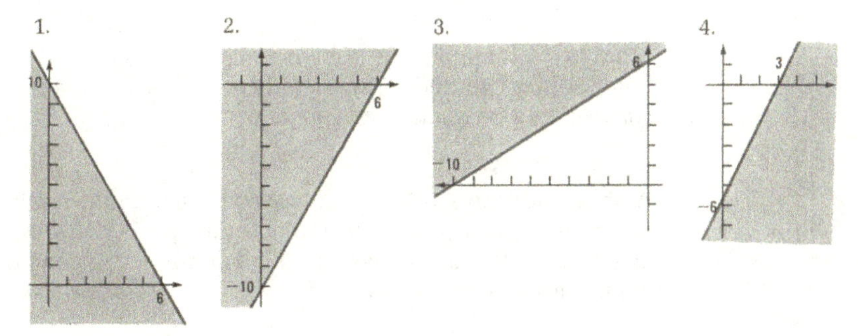

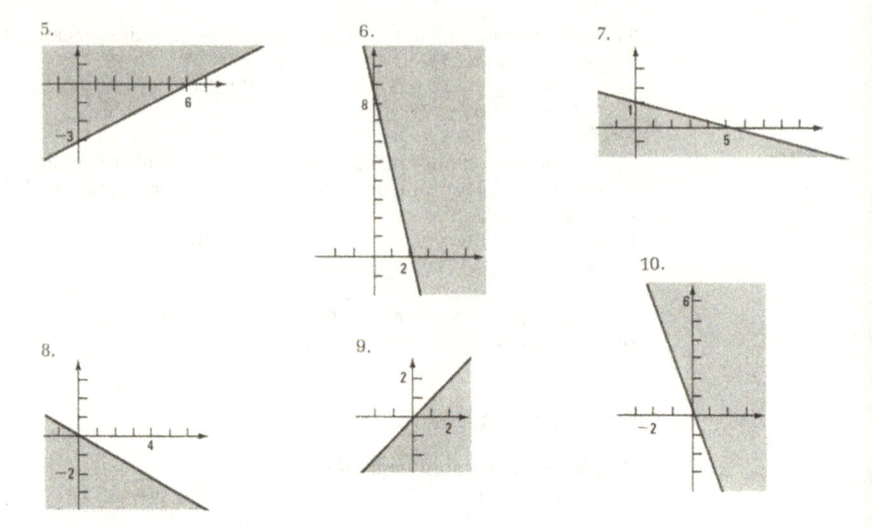

11. (a) no (b) yes (c) yes (d) no (e) no
12–15. See the answers to problems 2–5 in Exercises 5.2.A.
16. There are no common points; the inequalities are inconsistent.

5.2 Setting Up Linear Programming and the Graphical Approach

Linear programming (LP) is concerned with the efficient allocation of resources in order to achieve some clearly defined objective, such as maximizing profits or minimizing costs. It is one of the mathematical techniques most commonly applied to managerial decision making and other planning.

Linear programming always involves at least two independent variables (typically many more), and at no time is any variable raised to any power (other than 1, of course; $x^1 = x$). There is always an objective equation (called an objective function), such as

$$P = 30x + 20y$$

that gives the quantity to be maximized or minimized in terms of the independent variables. There is also a set of linear inequalities called constraints (or conditions), such as

$$x + y \leq 3000 \qquad 2x + y \leq 4000 \qquad x \geq 0 \quad \text{and} \quad y \geq 0$$

Among these are often nonnegativity constraints (either written or assumed), which guarantee that the independent variables are never negative; in the list above, the inequalities $x \geq 0$ and $y \geq 0$ are the nonnegativity constraints.

5.2.1 Definition

In a linear programming problem with two independent variables, x and y, the pairs (x, y) that satisfy all the constraints are called feasible points for this problem. The set of all feasible points is called the feasible region.

If the constraining inequalities are inconsistent, there will be no points in the feasible region, and the problem will have no solution. In this text the feasible regions will always be nonempty.

Section 5.1 provided practice in graphing the feasible region of LP problems, although we explicitly stated the problem in only one example. We return now to the solution of this problem.

5.2.2 Example

Complete the solution to Example 5.1.5 about the feed grain. That is: Maximize $P = 30x + 20y$ subject to

$$x + y \leq 3000 \qquad x \geq 0$$
$$2x + y \leq 4000 \qquad y \geq 0$$

Solution:

In Section 5.1 we graphed the feasible region for this problem; the graph is repeated in Figure 5.2–2. To solve the problem, we must choose the "best" point in the feasible region—where "best" means that point (x, y) which gives the biggest value for $30x + 20y$.

To see how to do this, it is convenient to solve first the objective function, $P = 30x + 20y$, for y:

$$-20y = 30x - P \qquad \text{(subtracting } 20y \text{ and } P \text{ from each side)}$$
$$y = -\tfrac{3}{2}x + \tfrac{1}{20}P \qquad \text{(dividing each side by } -20\text{)}$$

This computation reveals that for each fixed profit P the objective function may be viewed as a straight line with slope $-\frac{3}{2}$ and y-intercept $\frac{1}{20}P$. (If these ideas are foggy, consult Appendix 2.) Figure 5.2–1 shows some lines with slope $-\frac{3}{2}$. It is not hard to see that the higher the line on this graph, the larger the corresponding value of P. Conversely,

Figure 5.2–1

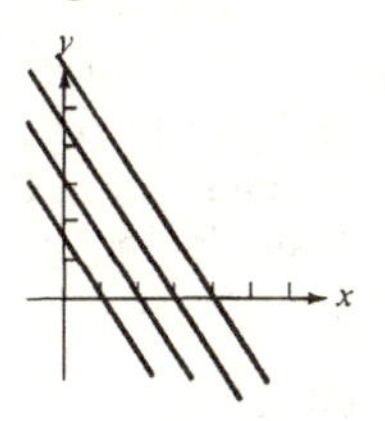

the larger the value of P, the higher the line. (Note that since the lines have equal slope, they must all be parallel.)

The question in the example can now be expressed: What is the largest P for which at least one point of the corresponding line is inside the feasible area? The answer is to pick the highest line on the graph that touches the feasible area, because it has the largest y-intercept and therefore the largest P.

It is clear from Figure 5.2–2 that the desired line is the one that passes through the point (1000, 2000). Then the corresponding P is

$$P = 30(1000) + 20(2000) = 70{,}000$$

Thus the grain wholesaler in Example 5.1.5 will have a maximum profit of 70,000 cents = $700 by selling 1000 Premium bags and 2000 Regular bags.

Figure 5.2–2

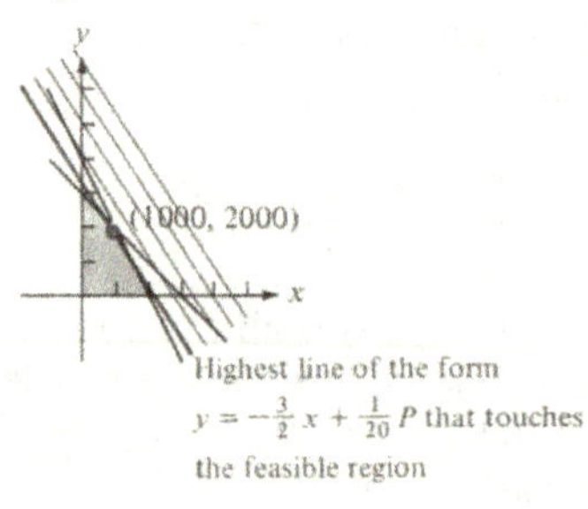

5.2.3 Example

Maximize $P = 20x + 30y$ subject to the same conditions as those of Example 5.2.2: $x + y \leq 3000$, $2x + y \leq 4000$, $x \geq 0$, and $y \geq 0$.

Solution:

Since the constraints are the same, the feasible region will be the same. We need consider only the new objective function, $P = 20x + 30y$:

$$-30y = 20x - P \qquad \text{(subtracting } P \text{ and } 30y \text{ from each side)}$$
$$y = -\tfrac{2}{3}x + \tfrac{1}{30}P \qquad \text{(dividing each side by } -30)$$

This time the new set of equations, for varying P, can be graphed by parallel lines with slope $-\frac{2}{3}$ and y-intercept $\frac{1}{30}P$. When we impose these lines on the same feasible region, as in Figure 5.2–3, we see that the highest one touching the feasible region touches it at (0, 3000). The corresponding P is

$$P = 20(0) + 30(3000) = 90{,}000$$

Figure 5.2–3

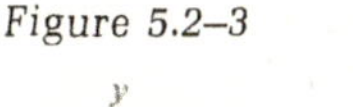

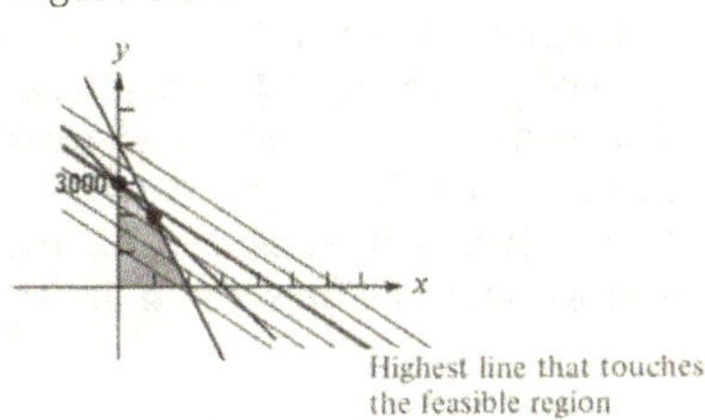

5.2.4 Example

Maximize $P = 20x + 20y$ subject to the same constraints as in Examples 5.2.2 and 5.2.3: $x + y \leq 3000$, $2x + y \leq 4000$, $x \geq 0$, and $y \geq 0$.

Solution:

Since the feasible region is again the same as in Examples 5.2.2 and 5.2.3, we again need to consider only the objective function, $P = 20x + 20y$:

$$-20y = 20x - P$$
$$y = -x + \tfrac{1}{20}P$$

This time the slope of the parallel lines is -1, and a line segment of one of them falls on one of the boundary segments of the feasible region (Figure 5.2–4). This indicates that there will be many points in the feasible region—all those on that line segment—which yield the maximum possible value for P. Three of these points are (0, 3000), (500, 2500), and (1000, 2000). Notice that for all three of these points the corresponding value of P is 60,000:

$$P = 20(0) + 20(3000) = 60{,}000$$
$$P = 20(500) + 20(2500) = 60{,}000$$
$$P = 20(1000) + 20(2000) = 60{,}000$$

Figure 5.2–4

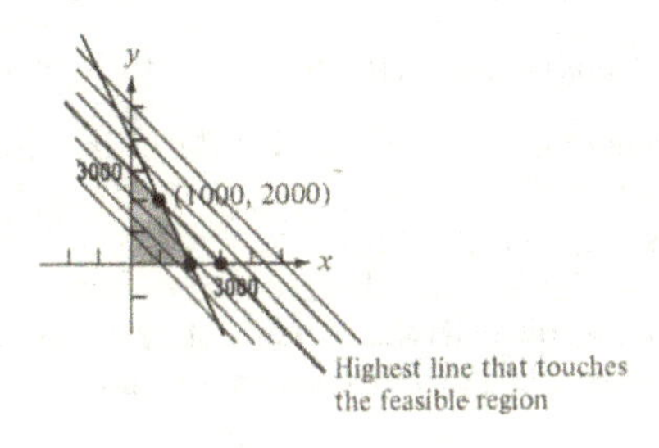

You might conclude—correctly—from these examples that for any linear objective function, the maximum value will be taken at one of the vertices (corners) of the feasible region. Thus we need not draw all the parallel lines describing possible objective functions to solve a linear programming problem; we need only examine the value of the objective function at the vertices of the feasible region. This idea is the basis of the graphical method for solving linear programming problems.

5.2.5 Rule

To solve a linear programming problem in two independent variables:

1. Graph all the constraining inequalities to obtain a picture of the feasible region.
2. Solve the corresponding equations to find the vertices (corners) of the feasible area.
3. Substitute each of these vertex points into the objective function and choose that one for which the resulting value is a maximum (or minimum).

5.2.6 Example in Biology

Suppose that a lake in a national park is being stocked with two species of fish, S_1 and S_2. Both of them feed on two foods, F_1 and F_2. The first species eats 2 units a day of F_1 and 1 unit a day of F_2. The second species consumes 5 units a day of F_1 and 10 units of F_2. If 3500 units of F_1 and 4000 units of F_2 grow in the lake each day, and the first species weighs 1 lb per fish and the second weighs 2 lb per fish, how should the lake be stocked so that it supports the maximum total weight of these two species of fish?

Solution:

The quantities we seek are

$$x = \text{number of species } S_1 \qquad y = \text{number of species } S_2$$

Together x and y determine the total weight of these two species of fish that live in the lake:

$$x \text{ lb} = \text{total weight of species } S_1 \qquad 2y \text{ lb} = \text{total weight of species } S_2$$

Thus the objective function, which describes the total weight of both species of fish, is $W = x + 2y$.

The constraints result from limited amounts of food. There are 3500 units of F_1 available each day, and the first species will consume $2x$ of them, since each individual fish eats 2 units and there are x fish of the first species. The second species will consume $5y$ units, so the total eaten by both species will be $2x + 5y$. This cannot exceed the amount available, so

$$2x + 5y \leq 3500 \qquad (F_1 \text{ constraint})$$

By the same reasoning, these two species will consume $x + 10y$ units of F_2 each day because the S_1 eats x units and the S_2 eats 10y units. This cannot exceed the 4000 available, so another constraint is

$$x + 10y \leq 4000 \qquad (F_2 \text{ constraint})$$

Obviously, we cannot have a negative number of fish, so $x \geq 0$ and $y \geq 0$. Thus this LP problem can be expressed:

$$\text{Maximize } W = x + 2y \qquad \text{subject to} \qquad \begin{matrix} 2x + 5y \leq 3500 & x \geq 0 \\ x + 10y \leq 4000 & y \geq 0 \end{matrix}$$

These inequalities were solved simultaneously in Example 5.1.6. We repeat Figure 5.1–11 in Figure 5.2–5.

Figure 5.2–5

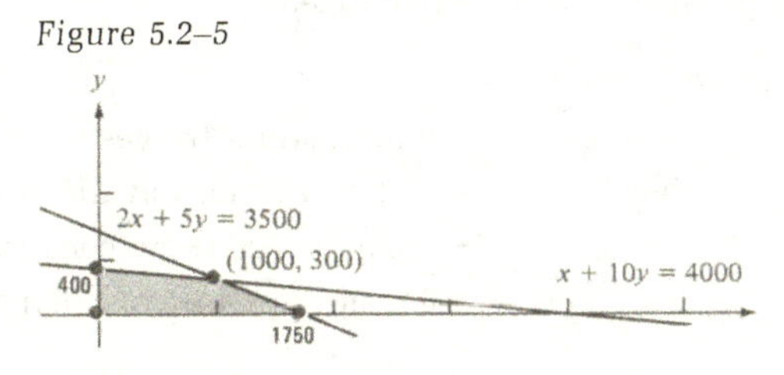

Rule 5.2.5 says that we need only test the vertices of the feasible (shaded) region:

Vertex	*Weight*: $W = x + 2y$
(0, 0)	0
(1750, 0)	1750
(1000, 300)	1600
(0, 400)	800

Since the maximum of the four weights 0, 1750, 1600, and 800 is 1750, the maximum possible weight is 1750 lb. It is attained by keeping 1750 of the first species in the lake and omitting the second altogether.

5.2.7 Example

Stringing tennis rackets is piecework; in a certain shop a worker is paid \$1.20 for each tennis racket with an aluminum frame that she strings and only \$1 for each racket with a steel frame. Suppose that there are only 8 aluminum rackets and 10 steel rackets available for a certain worker but that from these she can choose as many of each as she likes during the 5 hours (300 minutes) that she is in the shop. How many of each should she string if she wants the largest possible paycheck and it takes her 25 minutes to string each aluminum frame and 20 minutes to string each steel frame?

Solution:

The setup is similar to that in Examples 5.1.5 and 5.2.6:

$$\begin{aligned} x &= \text{number of aluminum-framed rackets to string} \\ y &= \text{number of steel-framed rackets to string} \\ 1.2x &= \text{amount earned from aluminum-framed rackets} \\ 1y &= \text{amount earned from steel-framed rackets} \\ 25x &= \text{time used for aluminum-framed rackets} \\ 20y &= \text{time used for steel-framed rackets} \end{aligned}$$

This problem can be summarized:

Maximize $P = 1.2x + y$

$$\text{subject to} \quad \begin{aligned} &x \leq 8 \text{ (aluminum frames available)} \\ &y \leq 10 \text{ (steel frames available)} \\ &25x + 20y \leq 300 \text{ (time constraint)} \\ &x \geq 0 \quad \text{and} \quad y \geq 0 \qquad \text{(nonnegativity constraints)} \end{aligned}$$

The time constraint is easier to manage if we divide it by the positive number 5, obtaining $5x + 4y \leq 60$. Again we begin by graphing the constraints on one graph and finding the vertices of the feasible region

Figure 5.2–6

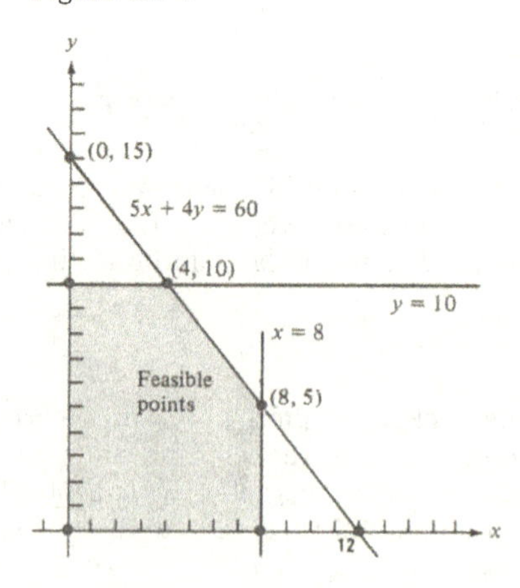

(Figure 5.2–6). Then to find the maximum value of P over the entire feasible region, we need only consider its value at the five vertices:

Vertex	$P = 1.2x + y$
(0, 0)	0
(0, 10)	10
(4, 10)	14.8
(8, 5)	14.6
(8, 0)	9.6

Thus the maximum amount the pieceworker can make is $14.80 by stringing 4 aluminum and 10 steel rackets.

Examples 5.2.2, 5.2.6, and 5.2.7 not only demonstrated how to solve linear programming problems using graphs but also illustrated how linear programming can be used in mixing products, in ecology, and in scheduling, respectively. These are but three of many fields in which linear programming is useful.

Whenever two or more products are mixed in varying amounts to form other products, a possible linear programming problem lurks not far away. This is especially true if large numbers of items are to be sold, or if the products can be mixed in any proportion. Companies distributing such products as drinks, chemicals, coffees, fertilizers, and petroleum products often use linear programming to maximize profits each month; there will, of course, typically be many variables and many constraints.

Ecological problems such as Example 5.2.6 can be much more complicated if there are many species of animals in a game forest or wildlife farm. Linear programming can be used to prevent any one species in such a confined ecological system from disappearing completely or from becoming too numerous.

One of the earliest linear programming applications was to the very complex scheduling problem presented by the Berlin airlift, which began in 1948. There the constraints included the number of runways, crews, and aircraft, and the time and money available; the objective was to maximize the tonnage supplied to West Berlin.

Today linear programming is used by airlines to decide how to schedule and recruit flight crews; each crew member can fly only a limited number of hours per month and there tends to be a high, but statistically predictable, turnover rate. Scheduling problems also occur in the manufacture of a variety of products, each of which requires a different amount of processing time by several machines, departments, or specially skilled workers. (See Examples 5.3.2 and 5.3.3.)

Another early application of linear programming was to diet problems—minimizing the cost of providing a specified amount of various nutrients. Sections 5.4, 5.5, and 6.3 discuss such problems. Similar mathematics is useful when choosing the size of garbage trucks, locating retail stores, or determining the cheapest combination of dyes with which to color a fabric or thread.

Linear programming is also used when a mutual fund is deciding how to distribute its investments (Example 6.3.2), when an advertising agency is planning the best use of available media (Example 6.3.1), and when a river authority is deciding the minimum cost for removing waste from a stream (Section 6.4). It is also used to decide the cheapest shipping pattern (Example 8.1.1), the least expensive way to use machines (Example 8.1.7), and the best way to assign each of a given group of workers or work teams a single job (Example 8.4.4).

SUMMARY

This section introduces the graphical method for solving linear programming problems and emphasizes that the maximum value of the objective function will always occur at a vertex of the feasible region. Thus to solve a linear programming problem in two independent variables: (1) graph the feasible region, (2) find the location of the vertices of the feasible region by solving simultaneous equations, and (3) evaluate the objective function at each vertex. The maximum value of the objective function at the vertices will be its maximum value throughout the feasible region.

Real-world linear programming problems involve many variables (usually called x_1, x_2, x_3, etc., so there is an abundant supply of names) and many constraints. Any one such problem could require days to solve, so the author has chosen to use artificial examples in this text which require only a realistic amount of student time.

The exercises of this section have two aspects—setting up applications as linear programming problems and using the graphical method to solve these problems. Thus each exercise is really two exercises; the "Hints" give the answers to the setups, and the "Answers" tell the results of using the graphical method to find the optimum solutions.

EXERCISES 5.2. A

1. (a) Sketch the feasible points satisfying the constraints $x \geq 0$, $y \geq 0$, $3x + 2y \leq 12$, and $x + 2y \leq 8$. Then find the maximum profit, P, for each of the following objective functions where the constraints above are given:
   (b) $P = 3x + y$ (c) $P = x + 3y$ (d) $P = 3x + 3y$
   (e) Where is P a minimum for each of parts (b), (c), and (d)?
2. A woodworker makes fancy coffee tables and simpler dining tables, which are three times as large as the coffee tables but take only half as long to make. Each week a truck arrives to take away his accumulated work. Meanwhile he has to store what he makes in a room just big enough to accommodate 15 coffee tables. Assuming that each coffee table occupies 1 unit of space, a dining table requires 3 such units

of space. If it takes the worker 8 hours to make a coffee table and 4 hours to make a dining table and he works 40 hours each week, how many of each should he make:

(a) If his profit on each coffee table is \$100 and on each dining table is \$200?

(b) If his profit on every table is \$100?

(c) If the interesting hand crafting on his coffee tables becomes so popular that he can get \$300 for each of them and still only \$100 for the plain dining tables?

3. A small company makes basketballs and footballs. Each type of ball requires both machine time and hand labor. Each gross of basketballs requires 5 hours of machine time and 1 hour of worker's time, and each gross of footballs requires only 3 hours of machine time but also 1 hour of the worker's labor. The worker has 8 hours available during the day, but since he has several machines, there is 30 hours of machine time available.

   (a) If the company makes \$150 profit for each gross of basketballs and \$100 profit for each gross of footballs, how many of each should be produced for maximum profit?

   (b) If the profit is \$100 for each gross of basketballs and \$150 for each gross of footballs, how does this affect your answer?

   (c) Give an objective function for which there is more than one optimal solution. Find two optimal solutions for the objective function you have just given.

4. A woman decides to convert her knitting hobby into a small business by devoting 18 hours a week to knitting ponchos and matching hats. She can sell the hats without the ponchos, but every poncho must be accompanied by a hat; thus she must make at least as many hats as ponchos. It takes her 6 hours to make a poncho and 3 hours to make a hat.

   (a) If she profits \$10 on each poncho and \$2 on each hat, how many of each should she make for maximum profit?

   (b) If she profits only \$7 on each poncho but \$4 on each hat, how many of each should she make for maximum profit?

5. Suppose that a company makes tennis balls both with seams and without and that both types require time on two machines. The balls with seams require 50 seconds on machine I and 40 seconds on machine II. The seamless balls require 40 seconds on machine I and 50 seconds on machine II. If the machines can be used for 1 hour each and the company makes 60 cents on each seamless ball and 40 cents on each ball with seams, how many of each should be made to maximize profits?

EXERCISES 5.2. B

1. (a) Sketch the feasible points satisfying the constraints $x \geq 0$, $y \geq 0$, $x + 2y \leq 40$, and $9x + 4y \leq 108$. Then find the maximum profit, P, for each of the following objective equations where the constraints above are given:

   (b) $P = x + y$ (c) $P = 3x + y$ (d) $P = x + 3y$

2. An independent craftsman produces wooden cabinets for both television and stereo sets. He has only 14 units of space in which to store his finished products and the TV cabinets occupy 1 unit each; the stereo cabinets take up twice as much space. If it takes him 3 hours to make the TV cabinets and 4 hours to make each stereo cabinet and he has 36 hours to work between pickups, how many of each should he make if:
   (a) The TV cabinets bring $15 profit and the stereos $17.50?
   (b) The TV cabinets bring $20 profit and the stereos $45?
   (c) The TV cabinets bring $17.50 profit and the stereos $25?
3. A shoemaker became famous for his oxford-type shoes called "the strider" and his knee-high boots called "the hiker." He stores them in 26 units of space; the strider takes up 1 such unit and the hiker occupies 2 such units. It takes him 3 hours to make a strider and 4 hours to make a hiker, and his family complains if he spends more than 60 hours a week at his shoemaking. How many of each should he make if the profits are:
   (a) $20 each on both the strider and hiker?
   (b) $20 each on the strider and $30 each on the hiker?
   (c) Give an objective function for which there is more than one optimal solution. What are the optimal solutions for the objective function you have just written?
4. Suppose that a lake is being stocked with two species of fish, S_1 and S_2. They both feed on two foods, F_1 and F_2, of which there is a daily supply of 1200 units and 1600 units, respectively. If the first species eats 3 units of F_1 and 2 units of F_2 daily, and the second species eats 2 units of F_1 and 4 units of F_2 daily, how should the lake be stocked to maintain a maximum number of total fish?
5. A boy decides to spend Saturday washing and waxing cars; he can either wash a car, or both wash and wax it. There are 10 cars available to him.
   (a) It takes him $2\frac{1}{4}$ hours to wash and wax a car; for this he gets $2.50. He charges $1 to wash each car and it takes him $\frac{3}{4}$ hour. If he works a total of 9 hours, how many jobs of each type should he do to make the most money?
   (b) Suppose that the cars are exceptionally dirty and his customers are especially anxious, so he can raise his rates to $5 for washing and waxing and $1.50 for washing. How should he arrange his work schedule now for maximum earnings?

HINTS 5.2. A

2. x = number of coffee tables; y = number of dining tables; constraints: $x \geq 0$, $y \geq 0$, $x + 3y \leq 15$ (space), $8x + 4y \leq 40$ or $2x + y \leq 10$ (time).
   (a) $P = 100x + 200y$ (b) $P = 100x + 100y$ (c) $P = 300x + 100y$
3. x = number of gross of basketballs; y = number of gross of footballs; constraints: $x \geq 0$, $y \geq 0$, $5x + 3y \leq 30$ (machine time), $x + y \leq 8$ (worker's time).
   (a) $P = 150x + 100y$ (b) $P = 100x + 150y$
4. x = number of ponchos; y = number of hats; constraints: $x \geq 0$, $y \geq 0$, $6x + 3y \leq 18$ or $2x + y \leq 6$ (time), $x \leq y$ (sets).
   (a) $P = 10x + 2y$ (b) $P = 7x + 4y$

5. x = number of tennis balls with seams; y = number of seamless tennis balls; constraints: $x \geq 0$, $y \geq 0$, $50x + 40y \leq 3600$ (machine I), $40x + 50y \leq 3600$ (machine II) because there are $60 \cdot 60 = 3600$ seconds in 1 hour. $P = 0.4x + 0.6y$.

HINTS 5.2. B

2. x = number of TV cabinets; y = number of stereo cabinets; constraints: $x \geq 0$, $y \geq 0$, $3x + 4y \leq 36$ (time), $x + 2y \leq 14$ (space).
   (a) $P = 15x + 17.5y$ (b) $P = 20x + 45y$ (c) $P = 17.5x + 25y$
3. x = number of striders, y = number of hikers; constraints: $x \geq 0$, $y \geq 0$, $3x + 4y \leq 60$ (time), $x + 2y \leq 26$ (space).
   (a) $P = 20x + 20y$ (b) $P = 20x + 30y$
4. x = number of species S_1; y = number of species S_2; constraints: $x \geq 0$, $y \geq 0$, $3x + 2y \leq 1200$ (F_1), $2x + 4y \leq 1600$ (F_2); objective equation: $P = x + y$.
5. x = number of cars both washed and waxed, y = number of cars washed (if you let x be the number of cars waxed, the computations are different but the answer will be the same); constraints: $x \geq 0$, $y \geq 0$, $x + y \leq 10$ (cars available), $\frac{9}{4}x + \frac{3}{4}y \leq 9$ (time).
   (a) $P = 2.5x + y$ (b) $P = 5x + 1.5y$

ANSWERS 5.2. A

1. (a)

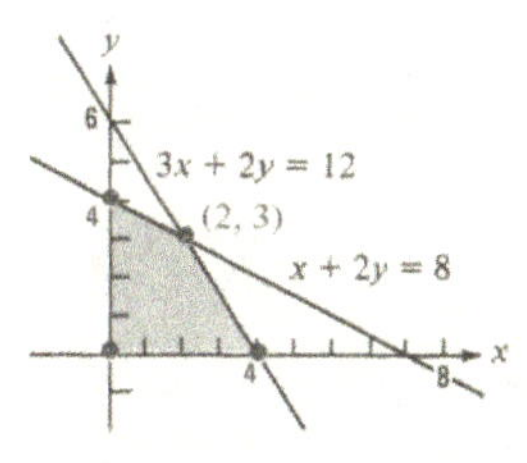

(b) $x = 4, y = 0$ $P_{max} = 12$
(c) $x = 0, y = 4$ $P_{max} = 12$
(d) $x = 2, y = 3$ $P_{max} = 15$
(e) $x = y = 0$

2. (a) 3 coffee tables and 4 dining tables, for a profit of \$1100
   (b) 3 coffee tables and 4 dining tables, for a profit of \$700
   (c) 5 coffee tables and no dining tables, for a profit of \$1500

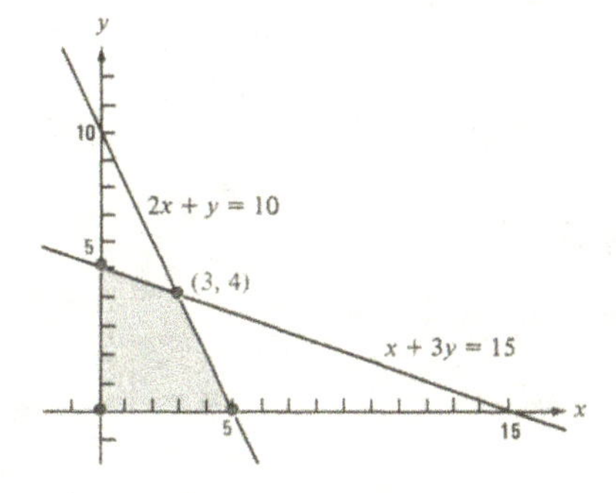

3. (a) 3 gross of basketballs and 5 gross of footballs
   (b) no basketballs and 8 gross of footballs
   (c) There are many answers to this problem, but basically the answers are of two types. One possibility is an objective function of $P = 150x + 90y$ or some multiple of these coefficients. Then (6, 0) and (3, 5) both give the maximum value of P. The second possibility is an objective function of $P = 100x + 100y$ (or some other with the same coefficients for both variables); then (0, 8) and (3, 5) both give the maximum.

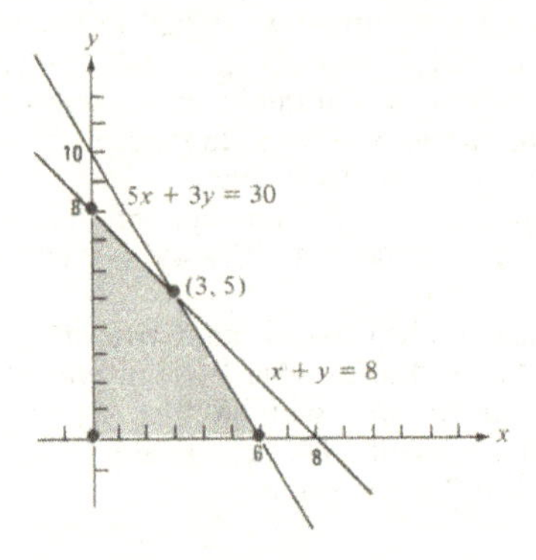

4. (a) 2 each for $P_{max} = \$24$
   (b) 6 hats and no ponchos for $P_{max} = \$24$

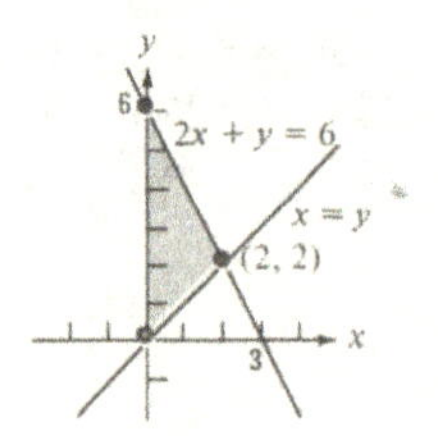

5. $P_{max} = \$43.20$ when $x = 0$ and $y = 72$

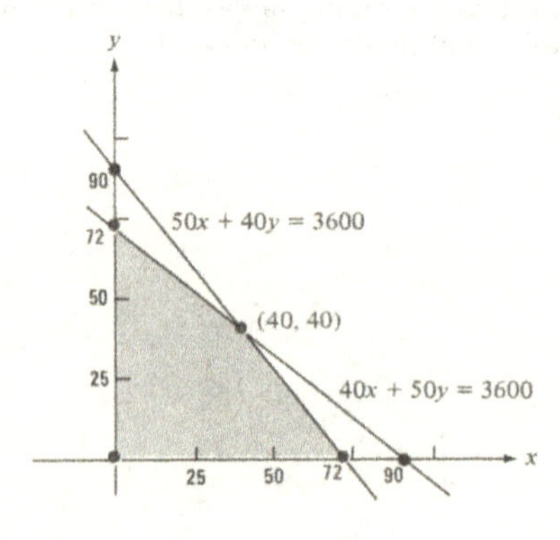

5.3 Tabular Solutions of Linear Programming Problems

A major purpose of the previous section was to convince you that the objective function in any linear programming problem always attains its maximum and minimum (if there is one) at a vertex of the feasible area described by the constraining inequalities. Once you realize this, you may also suspect that it is often easier to solve the problem by inspecting the vertices directly without actually graphing the inequalities. This *tabular method* is investigated in this section.

The first step in the tabular method is to locate the vertices of the feasible region. How is this done without graphing? By *solving simultaneous equations*!

Notice that in the case of two independent variables the vertices of the feasible region always occur at the intersection of two lines describing two equations corresponding to the constraining inequalities. If you solve all pairs of equations matching the inequalities in any one of the examples or exercises in Section 5.2, you will obtain all the corner points of the feasible region; you will also obtain some other points not in the feasible region, but it is easy to throw them away. A point is not in the feasible region if it does not satisfy all the given constraints.

Thus we can make a table of all the vertex points, decide which are feasible, and then evaluate the objective function for all the feasible vertex points. Finally, by inspection, we can pick the maximum or minimum as required. This is called the tabular method of solving a linear programming problem.

5.3.1 Example

Repeat Example 5.2.2 using the tabular method.

Solution:

It is convenient to label the given inequalities, which correspond to the equations on the right:

(1)	$x \geq 0$	(1)	$x = 0$
(2)	$y \geq 0$	(2)	$y = 0$
(3)	$x + y \leq 3000$	(3)	$x + y = 3000$
(4)	$2x + y \leq 4000$	(4)	$2x + y = 4000$

Each pair of equations on the right can be solved for x and y and will thus determine the intersection of two lines; we test all such intersection points to see if they are feasible.

Equations 1 and 2 intersect at $(0, 0)$, which is a feasible point.

Equations 1 and 3 intersect at $(0, 3000)$, which is also feasible.

Equations 1 and 4 intersect at (0, 4000), which is not feasible because it does not satisfy constraint 3.

Equations 2 and 3 have solution (3000, 0), which is not feasible because it does not satisfy constraint 4.

Equations 2 and 4 intersect at (2000, 0), which is feasible.

Equations 3 and 4 intersect at (1000, 2000), which is feasible.

We make a table of these facts as follows, including the value of the objective function *only* for the feasible points:

Equations	*Point*	*Feasible?*	$P = 30x + 20y$
1, 2	(0, 0)	yes	0
1, 3	(0, 3000)	yes	60,000
1, 4	(0, 4000)	no	—
2, 3	(3000, 0)	no	—
2, 4	(2000, 0)	yes	60,000
3, 4	(1000, 2000)	yes	70,000 ⟵ maximum

Figure 5.3–1 reveals the placement of the two nonfeasible vertices; they are the intersections of two of the lines *outside* the feasible area.

Figure 5.3–1

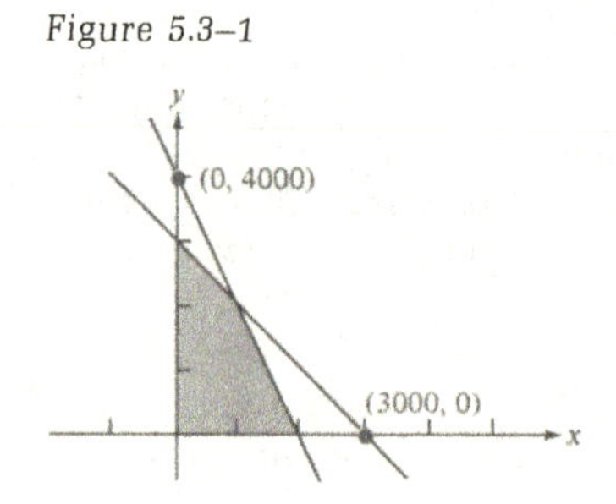

The graphical method is unappealing for solving linear programming problems with more than two independent variables because even with three it is difficult to draw and read the graphs (see Section 3.4). With more than three it is virtually impossible. Thus as we introduce linear programming problems with more than two independent variables in this section, it is necessary to turn to the tabular method.

If we have three independent variables, a linear equation describes a plane in 3-space. Three such equations are needed to determine one point (Section 3.4). A given set of three equations may not intersect in a point, but if they are inconsistent or redundant, they can be merely ignored in the table. Every vertex of a feasible area bounded by a set of linear inequalities in 3-space will be the solution of some trio of corresponding equations.

5.3.2 Example

A company is making lawn chairs, lawn rocking chairs, and chaise lounges; each is made using a tube-bending machine and a webbing machine. The lawn chair requires 2 minutes on each machine. The rocking chair requires 3 minutes on the tube-bending machine and 2 minutes on the webbing machine. The chaise requires 4 minutes on the tube-bending machine and 3 minutes on the webbing machine. Suppose that 60 minutes are available in each hour for the tube-bending machine and 48 minutes for the webbing machine. If the profit on the lawn chair is \$2, on the rocking chair is \$2.50, and on the chaise is \$3, how many of each should be produced each hour for maximum profit?

Solution:

x = number of lawn chairs produced per hour

y = number of lawn rocking chairs produced per hour

z = number of chaise lounges produced per hour

Maximize $P = 2x + 2.5y + 3z$ subject to

$x \geq 0 \qquad y \geq 0 \qquad z \geq 0$

$2x + 3y + 4z \leq 60$ (tube-bending constraint)

$2x + 2y + 3z \leq 48$ (webbing machine constraint)

This time we list all possible combinations of *three* equations in the left column of the table. We list again the inequalities here for your convenience, but we leave the corresponding equalities (listed explicitly in the previous example) to your imagination.

(1) $x \geq 0$

(2) $y \geq 0$

(3) $z \geq 0$

(4) $2x + 3y + 4z \leq 60$ (tube-bending machine)

(5) $2x + 2y + 3z \leq 48$ (webbing machine)

Equations	*Point*	*Feasible?*	$P = 2x + 2.5y + 3z$
1, 2, 3	(0, 0, 0)	yes	0
1, 2, 4	(0, 0, 15)	yes	45
1, 2, 5	(0, 0, 16)	no (does not satisfy 4)	—
1, 3, 4	(0, 20, 0)	yes	50
1, 3, 5	(0, 24, 0)	no (does not satisfy 4)	—
1, 4, 5	(0, −12, 24)	no (does not satisfy 2)	—
2, 3, 4	(30, 0, 0)	no (does not satisfy 5)	—
2, 3, 5	(24, 0, 0)	yes	48
2, 4, 5	(6, 0, 12)	yes	48
3, 4, 5	(12, 12, 0)	yes	54 ⟵ maximum

Thus the maximum of \$54 is attained when 12 lawn chairs and 12 lawn rocking chairs are produced.

The parenthetical comments telling why a point is not feasible are not part of the solution; they are there merely to help the reader. Figure 5.3–2 indicates how this example might be graphed; the axes and the solid lines bound the feasible region.

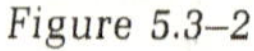

Figure 5.3–2

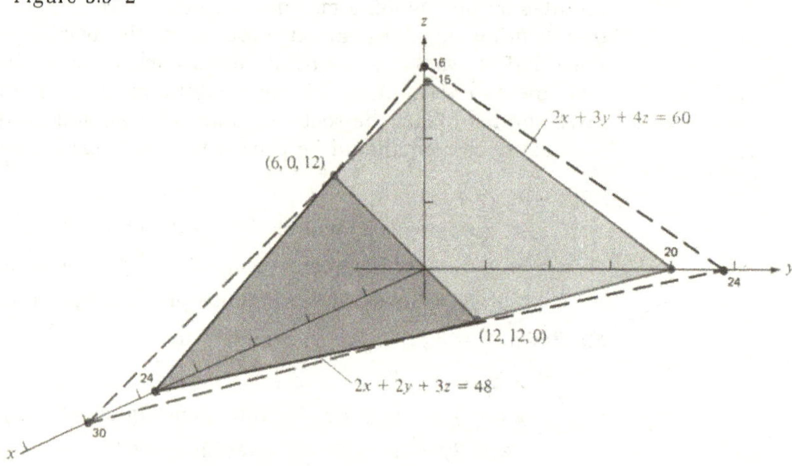

In the next, more complicated, example we find a trio of equations which are inconsistent. If either an inconsistency or a redundancy occurs when solving simultaneous equations in a linear programming problem, it should be clear that we can merely ignore those equations. The vertex points of the feasible region will all appear by examining the other sets of equations—in this case the other trios of equations.

5.3.3 Example

Suppose that a predelivery service center for cars prepares its midget cars, its subcompact cars, and its compact cars in three departments. The midget cars require 10 minutes in department I, 20 minutes in department II, and 30 minutes in department III. The subcompact cars require 20 minutes in department I, 20 minutes in department II, and 30 minutes in department III. The compact cars require 20 minutes in department I, 30 minutes in department II, and 40 minutes in department III. If department I has 200 minutes of operating time available during a certain period, department II has 300 minutes available, and department III has 360 minutes available, and the company profits \$48 on each midget car, \$70 on each subcompact car, and \$96 on each compact car, how many of each should be prepared for maximum profit?

Solution:

Letting x = number of midget cars, y = number of subcompact cars, and z = number of compact cars, the problems can be stated: Maximize $P = 48x + 70y + 96z$ subject to

(1) $x \geq 0$

(2) $y \geq 0$

(3) $z \geq 0$

(4) $10x + 20y + 20z \leq 200$ or $x + 2y + 2z \leq 20$ (department I)

(5) $20x + 20y + 30z \leq 300$ or $2x + 2y + 3z \leq 30$ (department II)

(6) $30x + 30y + 40z \leq 360$ or $3x + 3y + 4z \leq 36$ (department III)

Equations	*Point*	*Feasible?*	$P = 48x + 70y + 96z$
1, 2, 3	(0, 0, 0)	yes	0
1, 2, 4	(0, 0, 10)	no	—
1, 2, 5	(0, 0, 10)	no	—
1, 2, 6	(0, 0, 9)	yes	864
1, 3, 4	(0, 10, 0)	yes	700
1, 3, 5	(0, 15, 0)	no	—
1, 3, 6	(0, 12, 0)	no	—
1, 4, 5	(0, 0, 10)	no	—
1, 4, 6	(0, 4, 6)	yes	856
1, 5, 6	(0, −12, 18)	no	—
2, 3, 4	(20, 0, 0)	no	—
2, 3, 5	(15, 0, 0)	no	—
2, 3, 6	(12, 0, 0)	yes	576
2, 4, 5	(0, 0, 10)	no	—
2, 4, 6	(−4, 0, 12)	no	—
2, 5, 6	(−12, 0, 18)	no	—
3, 4, 5	(10, 5, 0)	no	—
3, 4, 6	(4, 8, 0)	yes	752
3, 5, 6	inconsistent	—	—
4, 5, 6	(−8, −4, 18)	no	—

Thus the maximum profit of \$864 is obtained by preparing 9 compact cars and no midgets or subcompacts.

You may wonder how many points (that is, sets of equations) must be examined as the number of variables and the number of inequalities grows larger. The remainder of this section briefly states the facts about this "counting problem," without stating why they are true. If this interests you, the reasons can be found in Section 9.2. (Your professor may choose to insert that section at this point.) Or the next paragraphs can be omitted altogether without interrupting the continuity of this book.

5.3.4 Definition

$n! = n(n-1)(n-2)\cdots 3\cdot 2\cdot 1$ is called n *factorial*. For example,

$$3! = 3\cdot 2\cdot 1 = 6 \qquad 4! = 4\cdot 3\cdot 2\cdot 1 = 24$$

$$5! = 5\cdot 4\cdot 3\cdot 2\cdot 1 = 120 \qquad 7! = 7\cdot 6\cdot 5\cdot 4\cdot 3\cdot 2\cdot 1 = 5040$$

$$10! = 3{,}628{,}800$$

5.3.5 Rule

If there are n inequalities in k unknowns, there are $\dfrac{n!}{k!(n-k)!}$ combinations of k equations each, and therefore this is the number of potential points to examine.

In Example 5.3.1 there were 4 equations (inequalities) in 2 unknowns, so there were $\dfrac{4!}{2!(4-2)!} = \dfrac{4\cdot 3\cdot 2}{2(2)} = 6$ vertex points. In Example 5.3.2 there were 5 equations in 3 unknowns, so there were $\dfrac{5!}{3!(5-3)!} = \dfrac{5\cdot 4\cdot 3\cdot 2}{3\cdot 2(2)} = 10$ vertex points. In Example 5.3.3 there were 6 equations in 3 unknowns, so there were $\dfrac{6!}{3!(6-3)!} = \dfrac{6\cdot 5\cdot 4\cdot 3\cdot 2}{3\cdot 2(3\cdot 2)} =$ 20 potential vertex points.

As the number of equations grows, the number of potential vertex points to be checked grows rapidly. The simplex method, which is the subject matter of Chapter 6, enables us to find the desired vertex point without checking all the others.

SUMMARY

The tabular method of solving linear programming problems uses the fact that the maximum or minimum of a linear programming problem must be attained at some vertex of the feasible region. If there are n independent variables in the linear programming problem, all possible combinations of n of the constraints are examined in turn. Each such set is solved (if possible) for the values of the n variables, and these values are then substituted into the other constraints to see if they define a feasible point. If so, they are substituted into the objective equation. The results are compiled into a table so that the maximum (or minimum) value can easily be seen.

EXERCISES 5.3. A

Do each of the following linear programming problems using the tabular method.

1. Two college students decide to make bookcases together; one will do the carpentry and the other the painting and decorating. The plain bookcases require 2 hours of the carpenter's time and 2 hours of the painter's time. The decorated bookcases require 5 hours of the painter's time and 1 hour of the carpenter's time. If the painter has 20 hours a week available and the carpenter has 22 hours a week available and the profit for each bookcase is $15, how many of each type should they make for maximum profit?
2. A baker is selling chocolate and vanilla cookies. There is a deluxe pack containing 2 lb of chocolate cookies and 1 lb of vanilla, gaining the baker a profit of $4; an assorted pack containing 1 lb each of the chocolate and vanilla cookies and netting the baker $2; and a plain box containing 1 lb of vanilla (only) cookies and netting $1. If the baker has 8 lb of vanilla cookies and 10 lb of chocolate cookies available, how many of each type of pack should be sold for maximum profit?
3. A gasoline station sells tanks of Skytop, Extra, and Good Gas for $60, $60, and $40, respectively. (To make the numbers easy to handle, we shall assume that the gasoline is sold in tanks; a genuine model of this type of situation involves fiendish arithmetic, which we assume that our readers would rather avoid.) Each tank of Skytop gasoline contains 3 units of high-octane gasoline and 2 units of medium-octane gasoline. Each tank of Extra contains 3 units of high octane, 1 unit of medium octane, and 2 units of lower grade. Each tank of Good Gas contains 1 unit of medium octane and 2 units of lower grade. If each shipment contains 18 units of high-octane gas and 12 units each of medium- and lower-octane gas, what is the greatest possible income the station can get from one shipping of ingredients?
4. A candy maker has made 6 lb of caramels and 12 lb of cream candies. If 1 lb of caramels and 2 lb of creams are put into a type A box, 1 lb of caramels are put into a type B box, 3 lb of creams are put into a type C box, and 2 lb of caramels and 1 lb of creams are put into each type D box, how many of each type of box should be sold to maximize the profit when each box brings a profit of $1? (*Suggestion*: When there are more than three independent variables, use subscripts—x_1, x_2, x_3, x_4, etc.)

EXERCISES 5.3. B

Save your answers because you will need them for Exercises 7.1.B.

Do each of the following LP problems using the tabular method.

1. A manufacturer of bicycles makes two-wheelers and three-wheelers. The two-wheelers require 30 minutes in department I and 20 minutes in department II. The three-wheelers require 20 minutes in department I and 10 minutes in department II. If department I is available for 300 minutes during the day and department II is available for 240 minutes and the company makes $100 and $75 on two-wheelers and three-wheelers, respectively, how many of each should the company make for maximum profit?
2. A company makes three types of footballs—cowhide, horsehide, and pigskin. Each football has to go through two machines. The cowhide footballs require 1 minute on machine I and 2 minutes on machine II.

The horsehide balls require 2 minutes on machine I and 2 minutes on machine II. The pigskin balls require 2 minutes on machine I and 3 minutes on machine II. Machine I is available (on the average) to footballs for a total of 20 minutes per hour and machine II is available for 30 minutes per hour. If the cowhide balls sell for $3, the horsehide balls for $3.50, and the pigskin balls for $4, how many of each should be produced per hour for maximum revenue?

3. Suppose that a large hospital classifies its surgical operations into three categories according to their length and charges a fee of $150, $225, and $300, respectively, for each of the categories. The average time of the operations in the three categories is 30 minutes, 1 hour, and 2 hours, respectively, and the hospital has four operating rooms, each of which can be used for 10 hours per day. If the total number of operations cannot exceed 60, how many of each type should the hospital schedule to maximize its revenues?
4. Suppose that the profits from a certain brand of single, double, and queen-sized beds are $50, $75, and $90, respectively, and they are made in three departments of a factory that have 100, 200, and 300 hours of time available, respectively. Suppose further that each single bed spends 1 hour in each of the three departments; each double bed spends 1 hour each in departments I and III and 2 hours in department II; and each queen-sized bed spends 1 hour each in departments I and II and 3 hours in department III. How many of each should be made to maximize total profits?

EXERCISES 5.3. C

1. If there are 5 equations in 2 unknowns in a linear programming problem, how many potential vertex points are there to test?
2. If there are 7 equations in 3 unknowns in a linear programming problem, how many potential vertex points are there to test?
3. If there are 8 equations in 5 unknowns in a linear programming problem, how many potential vertex points are there to test?
4. The day of a grazing animal can be divided into time spent (a) grazing, (b) moving, and (c) resting. The net energy gain during the grazing time is about 200 calories per hour and the net energy loss during the moving and resting time is 150 and 50 calories per hour, respectively. Suppose that the animal must spend at least as much time moving as grazing (to keep away from predators and to find new sources of food) and it must rest at least 6 hours per day. How would an animal trained in linear programming divide its time when it wants to maximize net energy gain?

HINTS 5.3. A

Henceforth all variables in a linear programming problem will be assumed to be nonnegative.

1. x = number of plain bookcases; y = number of decorated bookcases; (3) $2x + 5y \le 20$, (4) $2x + y \le 22$; $P = 15x + 15y$.
2. x = number of deluxe packs; y = number of assorted packs; z = number of plain packs; (4) $x + y + z \le 8$, (5) $2x + y \le 10$; $P = 4x + 2y + z$.

3. x = number of tanks of Skytop; y = number of tanks of Extra; z = number of tanks of Good Gas; (4) $3x + 3y \leq 18$, (5) $2x + y + z \leq 12$; (6) $2y + 2z \leq 12$; $P = 60x + 60y + 40z$.
4. x_1 = number of type A boxes; x_2 = number of type B boxes; x_3 = number of type C boxes; x_4 = number of type D boxes; (5) $x_1 + x_2 + 2x_4 \leq 6$, (6) $2x_1 + 3x_3 + x_4 \leq 12$; $P = x_1 + x_2 + x_3 + x_4$.

HINTS 5.3. B

1. x = number of two-wheelers; y = number of three-wheelers; (3) $30x + 20y \leq 300$, (4) $20x + 10y \leq 240$; $P = 100x + 75y$.
2. x = number of cowhide footballs; y = number of horsehide footballs; z = number of pigskin footballs; (4) $x + 2y + 2z \leq 20$, (5) $2x + 2y + 3z \leq 30$; $R = 3x + 3.5y + 4z$.
3. x = number of 30-minute operations; y = number of hour-long operations; z = number of 2-hour operations; (4) $x + y + z \leq 60$, (5) $30x + 60y + 120z \leq 2400$ (the number of minutes in 40 hours); $R = 150x + 225y + 300z$.
4. x = number of single beds; y = number of double beds; z = number of queen-sized beds; (4) $x + y + z \leq 100$, (5) $x + 2y + z \leq 200$, (6) $x + y + 3z \leq 300$; $P = 50x + 75y + 90z$.

HINTS 5.3. C

4. x = number of hours per day spent grazing; y = number of hours per day moving; z = number of hours per day resting; (4) $x \leq y$, (5) $z \geq 6$, (6) $x + y + z = 24$; $E = 200x - 150y - 50z$.

ANSWERS 5.3. A

1. P_{max} = \$150 when 10 plain and 0 decorated bookcases are made.
2. P_{max} = \$23 for 5 deluxe and 3 plain boxes.
3. P_{max} = \$480 when 3 tanks of each are sold.
4. P_{max} = \$10 when 6 type B and 4 type C are sold.

ANSWERS 5.3. C

1. $\frac{5!}{2!3!} = 10$ 2. $\frac{7!}{3!4!} = 35$ 3. $\frac{8!}{5!3!} = 56$
4. E_{max} = 150 for 9 hours each of moving and grazing and 6 hours of resting. Notice that there is only one feasible point, so whatever value it gives to the objective equation is both the minimum and the maximum.

5.4 Minimum Problems

The skills presented in the previous two sections can be easily applied to minimization problems with virtually no change.

5.4.1 Example

Suppose that we want to obtain at least 4800 International Units of vitamin A and at least 80 mg of vitamin C from eating two kinds of vitamin pills. Each type I pill contains 1200 I.U. of vitamin A and 10 mg of vitamin C. Each type II pill contains 800 I.U. of vitamin A and 20 mg of vitamin C. If bottles containing the same number of type I and type II pills cost $1 each, how many of each type should be taken to minimize the cost?

Solution:

This time the quantities we seek are the number of each type of pill and their total cost. We can write:

x = number of type I pills

$1200x$ = amount of vitamin A in x type I pills

$10x$ = amount of vitamin C in x type I pills

y = number of type II pills

$800y$ = amount of vitamin A in y type II pills

$20y$ = amount of vitamin C in y type II pills

Since the amount of vitamin A must be *greater than* 4800 I.U., we have

$$1200x + 800y \geq 4800 \qquad \text{(vitamin A constraint)}$$

Similarly, since the amount of vitamin C must exceed 80, we also must have

$$10x + 20y \geq 80 \qquad \text{(vitamin C constraint)}$$

Notice that these inequalities are in the opposite direction from the corresponding inequalities in the previous examples. But the nonnegativity inequalities are identical; we cannot have negative quantities of either type of pill:

$$x \geq 0 \qquad y \geq 0 \qquad \text{(nonnegativity constraints)}$$

When writing the objective function, you may be bothered by the fact that the cost per pill is not given. But there are the same number of pills per bottle for each type, so the cost per pill will be proportional to the cost per bottle. Thus it is sufficient to find the minimum of

$$C = x + y$$

This problem can be summarized: Minimize $C = x + y$ subject to $1200x + 800y \geq 4800$, $10x + 20y \geq 80$, $x \geq 0$, and $y \geq 0$.

To simplify the arithmetic we divide the first inequality by 400 and the second by 10. We then graph these two inequalities as shown in Figure 5.4–1. The green shaded area shows where $3x + 2y \geq 12$, and the gray shaded area indicates where $x + 2y \geq 8$. Points satisfying $x \geq 0$ and $y \geq 0$ must lie in the first quadrant. Thus the feasible region is that part of the first quadrant bounded by the thick lines.

Figure 5.4–1

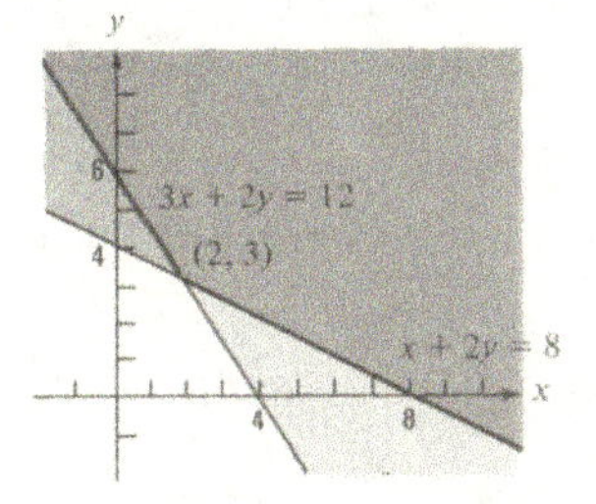

For each constant C, the objective function can be written $y = -x + C$, and thus is a straight line with slope -1 and y-intercept C as shown in Figure 5.4–2. To get a minimum cost, therefore, we want the lowest such line that intersects the feasible region. The vertex point $(2, 3)$ will be touched by the lowest line intersecting the feasible region and for this point $C = 2 + 3$. Therefore, $C_{\min} = 2 + 3 = 5$.

Figure 5.4–2

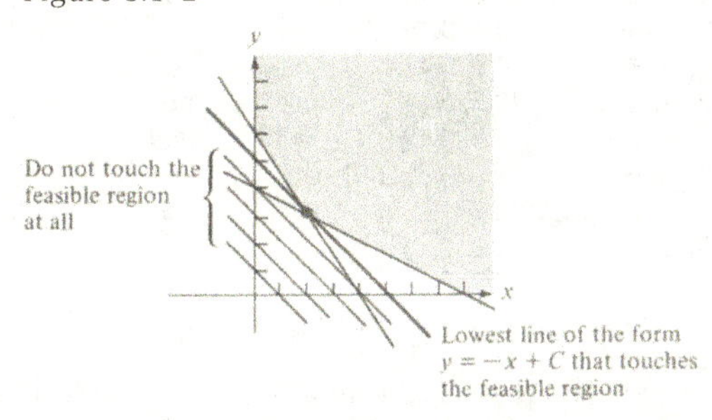

There are some minimization linear programming problems with no minimum, either because the objective function can become arbitrarily small or because the feasible region is empty. Or it is possible to have the minimum taken at more than one point of the feasible region if the slope of the lines describing the objective function is the same as the slope of one of the segments bounding the region. But if there is a minimum in a minimization linear programming problem, it will be taken at (at least) one of the vertices of the feasible region.

The tabular method can also be used for minimization problems.

5.4.2 Example

Minimize $C = 200u + 300v + 360w$ subject to

(1) $u \geq 0$

(2) $v \geq 0$

(3) $w \geq 0$

(4) $10u + 20v + 30w \geq 48$

(5) $20u + 20v + 30w \geq 70$

(6) $20u + 30v + 40w \geq 96$

Solution:

Since there are three variables, we use the tabular method for this problem.

Equations	*Point*	*Feasible?*	$C = 200u + 300v + 360w$
1, 2, 3	(0, 0, 0)	no	—
1, 2, 4	(0, 0, 1.6)	no	—
1, 2, 5	$(0, 0, \frac{7}{3})$	no	—
1, 2, 6	(0, 0, 2.4)	yes	864 ⟵ minimum
1, 3, 4	(0, 2.4, 0)	no	—
1, 3, 5	(0, 3.5, 0)	yes	1050
1, 3, 6	(0, 3.2, 0)	no	—
1, 4, 5	inconsistency	—	—
1, 4, 6	(0, 9.6, −4.8)	no	—
1, 5, 6	(0, 0.8, 1.8)	yes	888
2, 3, 4	(4.8, 0, 0)	yes	960
2, 3, 5	(3.5, 0, 0)	no	—
2, 3, 6	(4.8, 0, 0)	yes	960
2, 4, 5	$(2.2, 0, \frac{13}{15})$	no	—
2, 4, 6	(4.8, 0, 0)	yes	960
2, 5, 6	(−0.4, 0, 2.6)	no	—
3, 4, 5	(2.2, 1.3, 0)	no	—
3, 4, 6	(4.8, 0, 0)	yes	960
3, 5, 6	(0.9, 2.6, 0)	yes	960
4, 5, 6	(2.2, 5.2, −2.6)	no	—

Thus we see that $C_{min} = 864$ at the vertex (0, 0, 2.4).

If the numbers in this example look vaguely familiar, you might want to skip ahead and briefly read the beginning of Chapter 7, where dual problems are introduced and the relationship of the example above to Example 5.3.3 is discussed.

SUMMARY

Graphs and tables can be used to solve linear programming problems in which the minimum value of the objective function is wanted in almost the same way as they are used to solve maximum linear programming problems.

EXERCISES 5.4. A

Do the first two problems using the graphing method.

1. Suppose that a special baby formula is supposed to supply at least 80 g of protein and 2 mg of iron. The skim milk available costs 40 cents a quart and the evaporated milk costs 18 cents a can. If each quart of skim milk contains 32 g of protein and 0.5 mg of iron, and each can of evaporated milk contains 16 g of protein and 1 mg of iron, what formula made with these ingredients costs least?
2. Suppose that a nutritionist wants to supply at least 15 mg of iron and 20 mg of niacin from two breakfast cereals. Morning Joy provides 1 mg of iron and 2 mg of niacin in each serving, and Naturally Yours provides 5 mg each of iron and niacin. If Morning Joy costs 5 cents per serving and Naturally Yours costs 10 cents per serving, what is the minimum cost that can be paid to meet the requirements?
3. Suppose that 12 mg of iron and 48 g of protein are to be provided through milk, oatmeal, and dark molasses. The milk costs 5 cents per glass, each serving of oatmeal is 3 cents, and a giant spoonful of dark molasses is 2 cents. If the milk provides about 8 g of protein and 1 mg of iron per glass, the oatmeal provides about 4 g of protein and 2 mg of iron per serving, and the dark molasses provides 3 mg of iron but negligible protein per giant spoonful, how much of each food gives the desired quantities of iron and protein at minimum cost?
4. Suppose that a combination of wheat cereal, cornmeal, and rolled oats is supposed to supply 60 units of nutrient I and 30 units of nutrient II. Wheat cereal costs 4 cents per cup, cornmeal costs 3 cents per cup, and rolled oats costs 3 cents per cup. How much of each must be used to make a satisfactory combination at minimum cost if each cup of wheat cereal provides 1 unit of nutrient I and 2 units of nutrient II, each cup of cornmeal provides 2 units of nutrient I and 1 unit of nutrient II, and each cup of rolled oats supplies 3 units of nutrient I and 1 unit of nutrient II?

EXERCISES 5.4. B

Save your answers because you will need them for Exercises 7.1.B.
Use graphing to solve the first two problems.

1. Suppose that we are making a fruit punch which we want to contain at least 5000 International Units of vitamin A and 750 mg of vitamin C. Our red juice contains 1000 I.U. of vitamin A and 200 mg of vitamin C per can. The purple juice contains 2000 I.U. of vitamin A and 150 mg of vitamin C per can. If each can costs 50 cents, how much of each should be used in the punch?
2. Suppose that eggs, costing 6 cents each, and blackstrap molasses, costing 2 cents per spoonful, are to be used to supply 15 mg of iron and 25 grams of protein. If each egg has about 1.5 mg of iron and 6 g of protein and each spoonful of blackstrap molasses contains about 9 mg of iron and 1 g of protein, how can the required amounts of iron and protein be consumed most cheaply?
3. Suppose that your pet goldfish needs at least 4 mg of crude protein and 12 mg of vitamin B_6. It is to be supplied by oat gruel, which costs 4 cents per teaspoon, soybean meal, costing 2 cents per teaspoon, and

dayfly egg, which costs only 1 cent per teaspoon. If the oat gruel provides about 6 mg of B_6 and 1 mg of crude protein, the soybean meal provides 2 mg of each, and the dayfly egg provides 3 mg of crude protein and only 1 mg of B_6, how much of each food should be provided if you are to feed your pet adequately as cheaply as possible?

4. Suppose that a combination of wheat cereal, cornmeal, and rolled oats is supposed to supply your family with 18 units of nutrient I, 24 units of nutrient II, and 16 units of nutrient III. Wheat cereal costs 2 cents per cup and provides 1 unit of nutrient I, 2 units of nutrient II, and 1 unit of nutrient III. Cornmeal costs 3 cents per cup and provides 2 units of nutrient I, 3 units of nutrient II, and 1 unit of nutrient III. Rolled oats costs 4 cents per cup and provides 1 unit each of nutrients I and II and 4 units of nutrient III. How much of each should you serve for a minimum cost? What is the minimum cost?

HINTS 5.4. A

1. Minimize $C = 40x + 18y$ when $32x + 16y \geq 80$ and $0.5x + y \geq 2$.
2. Minimize $C = 5x + 10y$ if $x + 5y \geq 15$ and $2x + 5y \geq 20$.
3. Minimize $C = 5x + 3y + 2z$ if $x + 2y + 3z \geq 12$ and $8x + 4y \geq 48$.
4. Minimize $C = 4x + 3y + 3z$ for $x + 2y + 3z \geq 60$ and $2x + y + z \geq 30$.

HINTS 5.4. B

1. Minimize $C = 0.5x + 0.5y$ when $1000x + 2000y \geq 5000$ and $200x + 150y \geq 750$.
2. Minimize $C = 6x + 2y$ if $1.5x + 9y \geq 15$ and $6x + y \geq 25$.
3. Minimize $C = 4x + 2y + z$ when (4) $x + 2y + 3z \geq 4$ and (5) $6x + 2y + z \geq 12$.
4. Minimize $C = 2x + 3y + 4z$ when (4) $x + 2y + z \geq 18$, (5) $2x + 3y + z \geq 24$, and (6) $x + y + 4z \geq 16$.

ANSWERS 5.4. A

1. 5 cans of evaporated milk yields $C_{min} = 90$ cents.

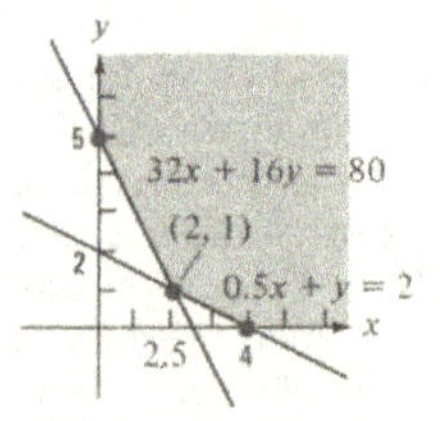

2. $C_{min} = 40$ cents for no servings of Morning Joy and 4 servings of Naturally Yours.

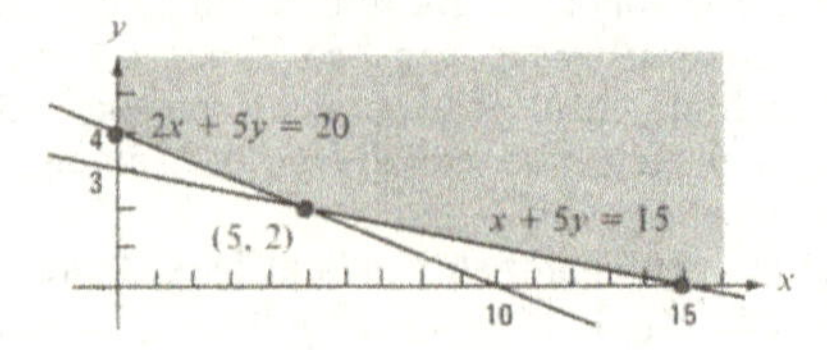

3. C_{min} = 32 cents for 4 glasses of milk, 4 servings of oatmeal, and no molasses.
4. C_{min} = 78 cents is attained for 6 cups of wheat cereal and 18 cups of rolled oats.

★5.5 The Classic Diet Problem

This section summarizes "The Cost of Subsistence" by George J. Stigler, which appeared in Volume 27 of the *Journal of Farm Economics*. The paper is interesting for several reasons, one of which is its statement: "Thereafter the procedure is experimental because there does not appear to be any direct method of finding the minimum of a linear function subject to linear conditions." This appeal for help in solving an LP problem was soon to be answered with the invention of the simplex method, the subject matter of Chapter 6.

In the early 1940s the problem of supplying people with adequate nutrition at minimum cost was especially urgent. In his paper Stigler quotes "minimum-cost" diets recommended by several nutritionists in the late 1930s, all of which amount to about $100 per year for food for an average man, or about $400 in 1974 dollars. Then, using careful mathematical analysis (basically the tabular method), he showed how nutrition needs could be supplied for a mere $39.93 in August, 1939—or about $160 in 1974 dollars! The minimum-cost diet, naturally, is hardly tempting to the taste buds, but Stigler points out that the delicacies needed to make it so vary greatly from one individual to another and should be acknowledged to be personal options.

The paper begins with a discussion of the current incomplete knowledge of nutrition. There are several essential nutrients for which minimum daily allowances have not yet been established, and there are probably others whose existence has not even been discovered. But adequate quantities of other nutrients are believed to be supplied if the following nine are eaten in natural foods (as opposed to vitamin pills).

Daily Allowances of Nutrients for a Moderately Active Man Weighing 154 Pounds

Nutrient	Allowance
Calories	3000 calories
Protein	70 grams
Calcium	0.8 gram
Iron	12 milligrams
Vitamin A	5000 International Units
Thiamine (B_1)	1.8 milligrams
Riboflavin (B_2 or G)	2.7 milligrams
Niacin	18 milligrams
Ascorbic acid (C)	75 milligrams

Next Stigler's paper discusses which foods will be used in the analysis. The figures are those reported by the Bureau of Labor Statistics, data averaged from 51 large cities in 1939 and 56 cities in 1944. Eighty foods were listed, each with a reasonable unit in which to measure that particular food, the price per unit, and the quantity of each of the nine nutrients per unit. Stigler points out that the list is by no means exhaustive; seasonal foods and those fluctuating widely in price are omitted from the national survey, so the inclusion of some cheap fresh fruits and vegetables might considerably lower his estimate of minimum cost.

After some interesting comments regarding preparation, preservation, and waste of various foods, the paper begins the mathematical analysis. First, any commodity is discarded if all its nutrients (per dollar of expenditure) are less than those of some other commodity. The list is then further reduced by eliminating any food which is notably inferior to some other in its important nutrients and only slightly superior in the remaining nutrients. For example, except for calcium, white bread has less than half the nutrients of white flour. But neither white bread nor white flour is an economical source for calcium, so white bread is eliminated. This procedure excludes all but 15 of the original 80 foods; these 15 are starred in the list below. The new list contains no meat except liver, and no sugars or beverages.

By taking linear combinations of various foods (that is, mixing them as suggested in the examples of Chapter 2), it is possible to find combination foods which are better in all respects than some food previously on the list. By this means the list was reduced from 15 to 9, which are double-starred in the list below.

Readers who are interested in the entire table can consult the original paper. Here we settle for merely listing the 80 foods included. They are wheat flour**, macaroni, wheat cereal, cornflakes, cornmeal, hominy grits, rice, rolled oats, white bread, whole wheat bread, rye bread, pound cake, soda crackers, milk, evaporated milk**, butter, oleomargarine*, eggs, cheddar cheese**, cream, peanut butter, mayonnaise, Crisco, lard, sirloin steak, round steak, rib roast, chuck roast, plate, beef liver**, leg of lamb, rib lamb chops, pork chops, pork loin roast, bacon, smoked ham, salt pork, roasting chicken, veal cutlets, canned pink salmon, apples, bananas, lemons, oranges, green beans*, cabbage**, carrots, celery, lettuce, onions*, potatoes*, spinach**, sweet potatoes**, canned peaches, canned pears, canned pineapple, canned asparagus, canned green beans, canned pork and beans, canned corn, canned peas, canned tomatoes, canned tomato soup, dried peaches*, dried prunes*, dried raisins, dried peas, dried lima beans**, dried navy beans**, coffee, tea, cocoa, chocolate, sugar, corn syrup, molasses, strawberry preserves, pancake flour, beets, pork liver.

After stating his regret that there was no "direct method of finding the minimum of a linear function subject to linear conditions," Stigler resorted to trial and error, computing the cost of diets for various combinations of the remaining 9 foods. He examined only a few of the many possible combinations and points out that there might be some more economical diet than the one he presents.

But it could not be much cheaper, by the following simple argument. "The nutrient with the highest cost (when secured from its most economical source) is calories; it would require $24.50 to supply for a year the calories from flour. But then only 61 days' calcium would be provided, and the most efficient source (cheese) could meet the deficiency only at a cost of $14.90, and the contribution to calories would be relatively small. The requirements for vitamin A and ascorbic acid would still be unfilled. Use of other commodities for calories yields a similar conclusion."

Thus the following diet, adequate for human subsistence, has been shown to be near minimum costs.

Minimum-Cost Annual Diets, August, 1939 and 1944

	August, 1939		August, 1944	
Commodity	Quantity	Cost	Quantity	Cost
Wheat flour	370 lb	$13.33	535 lb	$34.53
Evaporated milk	57 cans	3.84	—	—
Cabbage	111 lb	4.11	107 lb	5.23
Spinach	23 lb	1.85	13 lb	1.56
Dried navy beans	285 lb	16.80	—	—
Pancake flour	—	—	134 lb	13.08
Pork liver	—	—	25 lb	5.48
		$39.93		$59.88

Stigler was careful to emphasize that he does *not* recommend such a diet; he only points out that variations are a matter of taste. Clearly, most people prefer more variety in their diet than is afforded by five foods. But they might like to be aware that there are many ways to alter this rock-bottom diet.

The 1944 analysis was done using a similar technique, but the prices had changed so much in 5 years that now the minimum-cost diet consisted of wheat flour, cabbage, spinach, pancake flour, and pork liver. Although the price index rose only 47 percent from 1939 to 1945, the minimum-cost diet rose 50 percent. This is because of the large increase in the price of efficient food sources, especially wheat flour.

Notice that although the price of wheat flour rose 79 percent, far more than the price index rise of 47 percent, the amount of wheat flour included in the 1944 minimum diet is *greater* than that in the 1939 diet.

This is reminiscent of the "Giffin paradox," which states that a rise in the price of the cheapest foods increases poor families' consumption of those same foods, because they can no longer afford as much of the more expensive foods as they had bought before.

Although there is no sophisticated mathematics in Stigler's paper, it does show how mathematical techniques can be applied to a problem of broad social concern with strikingly unexpected results. The paper also mentions the need for a more efficient method of solving linear programming problems just a few years before the simplex method was invented.

SUMMARY

This section tells about a paper that seeks to find the minimum costs of a year's diet for an average man given nine constraints requiring a certain amount of each of nine known essential nutrients. The paper shows how the tabular method of solving a linear programming problem has been applied to an important practical problem. The exercises of this section provide a review of Sections 5.1–5.4.

EXERCISES 5.5. A

Solve the first two problems using the graphing method and the last two using the tabular method.

1. In her spare time an elderly woman makes machine-embroidered tablecloths in 2 sizes. The large tablecloths take 16 minutes to sew and 2 minutes to embroider, and the small tablecloths take 8 minutes to sew and 6 minutes to embroider. She decides to spend 80 minutes sewing and 30 minutes embroidering. What is her maximum profit if she makes $10 on each large and $5 on each small tablecloth?
2. Suppose that we want to make a fruit punch containing 4000 I.U. of vitamin A per can and 1000 mg of vitamin C per can. There is a yellow drink containing 1500 I.U. of vitamin A and 300 mg of vitamin C per can and there is a red drink containing 500 I.U. of vitamin A and 200 mg of vitamin C per can. If the yellow drink costs $2 per can and the red drink costs $1 per can, what is the minimum-cost punch that fulfills the requirements, and how should it be made?
3. Suppose that a combination of dried fruit is to provide 60 units of nutrient I and 36 units of nutrient II. Each box of prunes provides 2 units of nutrient II but negligible quantities of nutrient I. Each box of dried peaches provides 3 units of nutrient I and 1 unit of nutrient II, and each box of dried pears supplies 2 units of nutrient I and 1 of nutrient II. If the prunes cost $1 per box and the others each cost $2 per box, how many boxes of each provide the required nutrients at minimum cost?
4. The good doctor thinks it would help his elderly patient to take a supplement providing 6 units of nutrient I and 12 units of nutrient II each day. Red pills cost 5 cents each and provide 3 units of nutrient I and 4 units of nutrient II. Blue pills cost 2 cents each and contain 1

unit of nutrient I and 3 units of nutrient II. Green pills cost 3 cents each and have 2 units of nutrient I and 3 units of nutrient II. With kind consideration for the patient's budget, how much of each type of pill should the doctor prescribe?

EXERCISES 5.5. B

Do the first two problems using the graphing method and the last two using the tabular method.

1. The photographer has a problem. He has postponed his project until 9:00 P.M. the night before it is due, and he discovers that he has only a small amount of his processing chemical left, enough to process only 200 5 by 7 prints. Or he could use the same amount of chemical needed to process two 5 by 7 prints to make instead just one 8 by 10 print. It takes 1.5 minutes to process each print, regardless of size, and the chemical is depleted after 4 hours' exposure to the air. If his client will pay him $3 for each 8 by 10 and $2 for each 5 by 7, how much of each should he make for a maximum take?
2. Suppose that you want to obtain 10 mg of nutrient I and 15 mg of nutrient II from two kinds of instant breakfast drinks. Type *A* provides 1 mg of nutrient I and 3 mg of nutrient II per glass. Type *B* provides 4 mg of nutrient I and 2 mg of nutrient II per glass. If type *A* costs 20 cents per glass and type *B* costs 25 cents per glass, what is the minimum cost that can be paid to meet the requirement?
3. Suppose that a scalper is to provide at least 36 tickets to an upcoming concert and at least 48 tickets to a popular show. He gets a discount if he buys them in sets. Set *A* contains one ticket each to the concert and show and costs $2. Set *B* contains 2 tickets to the concert and 1 to the show and costs $3. Set *C* contains 3 tickets to the concert and 2 to the show; it costs $5. How many of each set should he buy for the least possible minimum cost?
4. A shirt manufacturer makes dress shirts, sports shirts, and T shirts. His factory has three departments—the cutting department, the sewing department, and the collar and cuff department. In each of these departments he has 5, 3, and 4 workers, respectively; they all work an 8-hour day. A dress shirt needs $\frac{3}{4}$ hour in both the cutting and sewing departments and $\frac{1}{2}$ hour in the collar and cuff department. A sports shirt needs $\frac{1}{4}$ hour each in the cutting and collar and cuff departments and $\frac{1}{2}$ hour in the sewing department. A T shirt requires $\frac{1}{4}$ hour in each of the three departments. If the selling prices of the shirts are $18 for a dress shirt, $10 for a sports shirt, and $6 for a T shirt, how many of each type must be manufactured for a maximum revenue?

HINTS 5.5. A

1. Maximize $P = 10x + 5y$ when $16x + 8y \leq 80$ and $2x + 6y \leq 30$.
2. Minimize $C = 2x + y$ when $15x + 5y \geq 40$ and $3x + 2y \geq 10$.
3. Minimize $C = x + 2y + 2z$ when $3y + 2z \geq 60$ and $2x + y + z \geq 36$.
4. Minimize $C = 5x + 2y + 3z$ when $3x + y + 2z \geq 6$ and $4x + 3y + 3z \geq 12$.

HINTS 5.5. B

1. Maximize $P = 3x + 2y$ when $2x + y \leq 200$ and $1.5x + 1.5y \leq 240$.
2. Minimize $C = 0.2x + 0.25y$ when $x + 4y \geq 10$ and $3x + 2y \geq 15$.
3. Minimize $C = 2x + 3y + 5z$ when $x + 2y + 3z \geq 36$ and $x + y + 2z \geq 48$.
4. Maximize $P = 18x + 10y + 6z$ when $3x + y + z \leq 160$, $3x + 2y + z \leq 96$, and $2x + y + z \leq 128$.

ANSWERS 5.5. A

1.

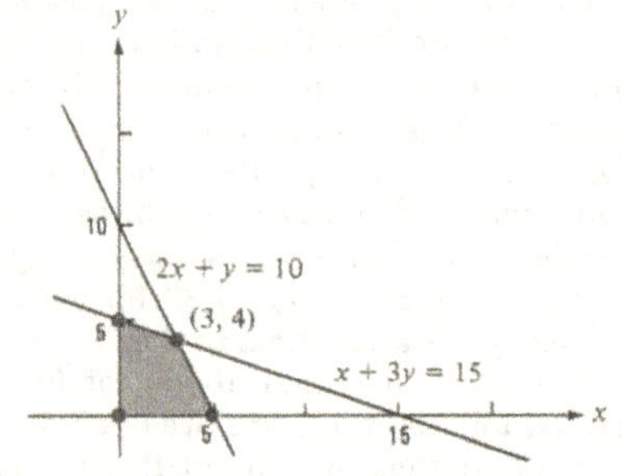

P_{max} = \$50 when 5 of the large tablecloths are made or 3 of the large and 4 of the small are made. Since (4, 2) lies on the line segment connecting (5, 0) and (3, 4), it, too, gives \$50.

2. C_{min} = \$6 when 2 cans of each type of juice are used.

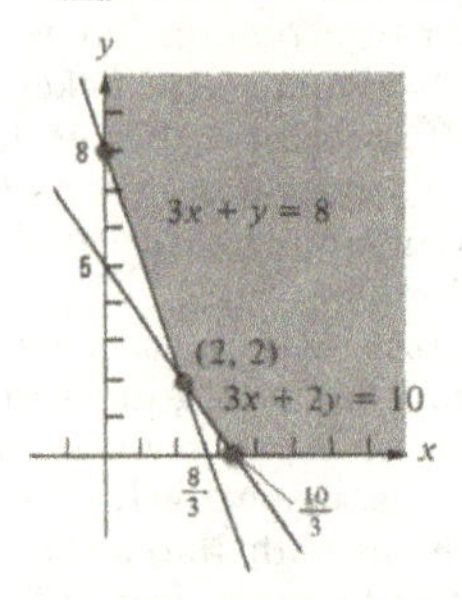

3. C_{min} = \$48 when 8 boxes of prunes and 20 boxes of peaches are used.
4. C_{min} = 10 cents when 2 blue pills and 2 green pills are used.

VOCABULARY

objective equation, constraints, nonnegativity conditions, inequality, feasible points, vertex points, graphing method, tabular method

SAMPLE TEST

Chapter 5

Set the first two problems up in mathematical language as LP problems. Label the constraints and objective equation. *Do not solve.* These problems count 10 points each. The remaining four problems count 20 points each.

1. As a hobby, an accountant spends 4 hours (240 minutes) per week making handmade turquoise bracelets and necklaces. He receives a shipment of 360 turquoise beads a week, of which 40 are needed for each bracelet and 120 for each necklace. It takes him about 30 minutes per bracelet and 1 hour per necklace to string the beads, attach the clasp, and assemble the materials. If he profits \$3 on each bracelet and \$5 on each necklace, how many of each should he make for a maximum profit?
2. A sailing yacht dealer is trying to decide how many of next year's craft he should order. He has three models from which his franchise can choose: the Morgan 55, for which he receives a profit of \$5000; the Columbia 42, for which he receives a profit of \$4600; and the Trident 36, for which he receives a profit of \$3900. Each yacht goes through two phases of preparation: hull/cabin and rigging. The Morgan 55 requires 19 hours in hull/cabin and 16 hours in rigging. The Columbia 42 requires 10 hours in hull/cabin and 20 hours in rigging. The Trident 36 requires 5 hours in hull/cabin and 50 hours in rigging. The hull/cabin crew will be available for work for 1960 hours next year and the rigging crew will be available for 1800 hours. How many yachts of each model type should be ordered for maximum profit?
3. Use the graphing method to maximize $P = 7x + 3y$ when $x \geq 0$, $y \geq 0$, $3x + y \leq 12$, and $3x + 2y \leq 18$.
4. Use the graphing method to minimize $C = 3x + y$ when $x \geq 0$, $y \geq 0$, $2x + y \geq 16$, and $3x + 2y \leq 30$. Note! The inequalities in the last two constraints are in opposite directions.
5. Minimize $C = 6x + 6y + 5z$ when $x \geq 0$, $y \geq 0$, $z \geq 0$, $3x + 5y + 8z \geq 50$, $x + 5y + 4z \geq 40$.
6. Maximize $P = 2x + 4y + 5z$ when $x \geq 0$, $y \geq 0$, $z \geq 0$, $3x + y \leq 13$, $x + 2z \leq 13$, $y + 2z \leq 14$.

The answers are at the back of the book.

Chapter 6

THE SIMPLEX ALGORITHM

6.1 Solving Standard Linear Programming Problems Using the Simplex Algorithm

The linear programming problems in Chapter 5 included both nonnegativity constraints and other constraints. We shall refer to the constraints in any linear programming problem that are not nonnegativity constraints as significant constraints. These will be the constraints of primary interest, since whenever the simplex method is used, all the independent variables are assumed to be nonnegative. Thus the nonnegativity constraints are often not written, but simply understood.

A linear programming problem is said to be standard if the objective function is to be maximized and if all the significant constraints are of the form $a_1x_1 + a_2x_2 + \cdots + a_nx_n \leq b$ where the a_i and b are constants ($b \geq 0$) and the x_i are variables. Section 5.2 and 5.3 discussed only standard linear programming problems, as will this section and the next. In Section 7.1 we shall see that every standard linear programming problem can be expressed in the form "Maximize $P = DX$ when $AX \leq B$," where D is a row vector, B is a column vector, A is a matrix, and all the elements in B are nonnegative.

An algorithm is a method for solving a particular type of routine problem. The simplex algorithm (or method) is a mathematical technique that was developed in the middle of the twentieth century for the purpose of solving linear programming problems. This section shows how to use the simplex algorithm to solve standard linear programming problems. The next section explains more of the reasoning behind the algorithm, and Section 6.3 shows how it can be adapted to linear programming problems that are not in standard form.

The simplex algorithm uses a basic idea from the tabular method–that of examining vertices of the feasible region. But only certain vertices are examined by the simplex algorithm. Each such vertex is described by a simplex tableau; a series of tableaux are written, each one describing a vertex adjacent to the one described by the previous tableau. In each tableau the value of the objective function is increased over the value in the previous tableau. When the objective function can increase no more, we have reached the maximum, and the location of the corresponding vertex can be read off the tableau. A standard linear programming problem always includes the origin as one of its feasible points; thus the first tableau can and will always describe the origin.

To use the simplex method, the constraints, which were first given as inequalities, are now converted to equalities by introducing slack variables. For example, the inequality $2x + 3y \leq 10$ can be written

$$2x + 3y + s = 10 \qquad s \geq 0$$

To say that $(2x + 3y)$ is less than or equal to 10 is the same as to say that there is some nonnegative number, which, when added to $(2x + 3y)$, gives 10. The s in the equation above is called the slack variable; it takes up the "slack" in the inequality. Since all variables in an LP problem are assumed to be nonnegative, we omit writing $s \geq 0$.

6.1.1 Example

Maximize $P = x + 2y + z$ subject to

$$\begin{array}{lll} x + 2z \leq 4 & (\text{or } x + 2z + s_1 & = 4) \\ 2y + z \leq 6 & (\text{or } 2y + z + s_2 & = 6) \\ 3x + y + 2z \leq 12 & (\text{or } 3x + y + 2z + s_3 & = 12) \end{array}$$

Solution:

The constraining inequalities on the left become the constraining equations given on the right. In this case we used three slack variables, a different one for each equation.

The first tableau summarizes the coefficients in the constraining equations in its inner rectangle and the coefficients of the objective function in its last row. Across the top are listed all the independent variables, both those that appeared in the original problem and the slack variables.

The left column of any simplex tableau lists those variables that may be nonzero in that tableau. These are called basic variables for that tableau, and together they form the basis of the point described by that tableau. In standard linear programming problems the basic variables for the first tableau are the slack variables, because we start at the origin, where the original independent variables are all zero. We see that when $x = y = z = 0$, the values of the slack variables are given, respectively, at the right of the tableau.

First tableau

	x	y	z	s_1	s_2	s_3	Solution
s_1	1	0	2	1	0	0	4
s_2	0	②	1	0	1	0	6
s_3	3	1	2	0	0	1	12
P	1	2	1	0	0	0	0

This tableau represents

$$\left.\begin{aligned} &s_1 + x + 2z = 4 \\ &s_2 + 2y + z = 6 \\ &s_3 + 3x + y + 2z = 12 \\ &P = x + 2y + z \end{aligned}\right\} \text{where } x = 0,\ y = 0,\ z = 0$$

For these values of the variables, clearly $P = 0$.

After each tableau is complete, its last row is examined to see if we can find a better tableau. If it contains a positive number, the value of the objective function can be increased by putting the variable above that number into the basis—that is, making it larger than zero. We pick the largest such positive number, which in this case is 2. This indicates that *per unit* we can gain more by increasing y than by increasing x or z. So we decide to put y in the basis, and we call the y-column the pivot column.

Which of the three slack variables shall we remove from the basis? Since y does not appear in the first constraint, there is no need to worry about that constraint. But the second constraint will not be satisfied if y gets any bigger than $3 = \frac{6}{2}$. And the third constraint will not be satisfied if y gets larger than $12 = \frac{12}{1}$. Thus the largest y can get (assuming the other variables remain nonnegative) is 3; in this situation $s_2 = 0$. So y will replace s_2 in the basis and we say the s_2-row is the pivot row. The number 2 which lies in the y-column and the s_2-row is called the pivot, and it is customary to circle it.

Now we are ready to begin writing the second tableau. The top row of variables will not change, and the left column will change *only* in that y will replace s_2. We say that y has "joined the basis" and s_2 has "left the basis."

The pivot row is divided by the pivot. In Example 6.1.1 this means dividing the second constraint (that is, the second row of the tableau) by 2 to show that $y = 3$ at the second tableau.

The other computations involve using Gauss–Jordan row operations in order to reduce the other numbers in the pivot column to zeros. But be careful! The row operations must be used in a special way. *An appropriate multiple of the pivot row* (only!) *must be subtracted from the other rows so as to obtain zeros elsewhere in the pivot column.*

In Example 6.1.1 the number above the pivot is already 0, so that the whole row remains unchanged. To change the 1 below the pivot to 0, we subtract $\frac{1}{2}$ the pivot row from the row just below it. And to change the 2 in the last row to 0, we merely subtract the pivot row from the last.

This is because the last row of the second tableau must represent the original objective function written in terms of the nonbasic variables of the second tableau: x, z, and s_2. It is obtained by subtracting the original second constraint from the original objective function:

$$
\begin{aligned}
P &= x + 2y + z \\
6 &= \quad 2y + z + s_2 \\
P - 6 &= x - s_2
\end{aligned}
$$

Thus the second tableau has been computed:

Second tableau

	x	y	z	s_1	s_2	s_3	Solution
s_1	1	0	2	1	0	0	4
y	0	1	$\frac{1}{2}$	0	$\frac{1}{2}$	0	3
s_3	③	0	$\frac{3}{2}$	0	$-\frac{1}{2}$	1	9
P	1	0	0	0	-1	0	-6

This tableau represents

$$
\left.
\begin{aligned}
&s_1 + x + 2z = 4 \\
&y + \tfrac{1}{2}z + \tfrac{1}{2}s_2 = 3 \\
&s_3 + 3x + \tfrac{3}{2}z - \tfrac{1}{2}s_2 = 9 \\
&P - 6 = x - s_2
\end{aligned}
\right\}
\text{where } x = 0,\ z = 0,\ s_2 = 0
$$

For these values of the variables, clearly $P = 6$. (Yes, $+6$, the negative of the number in the lower right of the tableau.) This tableau represents the point $(0, 3, 0)$.

The expression $P - 6 = x - s_2$, represented by the last row of the second tableau, suggests that we can increase P still more by letting x get larger than zero. Thus x will join the basis in the third tableau. How large can x grow? The s_1-row, representing the first constraint, says that x cannot be larger than $\frac{4}{1} = 4$. Since x has a zero coefficient in the second row, we need not worry about that row. (Similarly, we would not worry about any row where x had a negative coefficient, because such a row would put no restriction on how large x could grow.) The third row says that x dare not grow larger than $\frac{9}{3} = 3$. Since 3 is the smaller of these two quotients, x will become 3 in the next tableau, and s_3 will become 0. Thus s_3 will leave the basis, and the s_3-row will be the pivot row. The 3 in the x-column and the s_3-row will be the new pivot; it is circled.

To write the third tableau, we replace s_3 in the left column by x. Then we divide the pivot row by 3 to get a 1 in the pivot place. The first row of numbers is replaced by itself minus a third the pivot row to obtain a 0 at the upper left of the new inner rectangle. Since there is already a 0 in the middle row of the pivot column, that row is not changed. And since there is a 1 in the lower left, one-third the pivot row is subtracted from the last row.

Third tableau

	x	y	z	s_1	s_2	s_3	Solution
s_1	0	0	$\frac{3}{2}$	1	$\frac{1}{6}$	$-\frac{1}{3}$	1
y	0	1	$\frac{1}{2}$	0	$\frac{1}{2}$	0	3
x	1	0	$\frac{1}{2}$	0	$-\frac{1}{6}$	$\frac{1}{3}$	3
P	0	0	$-\frac{1}{2}$	0	$-\frac{5}{6}$	$-\frac{1}{3}$	-9

This tableau represents

$$\left.\begin{aligned} s_1 + \tfrac{3}{2}z + \tfrac{1}{6}s_2 - \tfrac{1}{3}s_3 &= 1 \\ y + \tfrac{1}{2}z + \tfrac{1}{2}s_2 &= 3 \\ x + \tfrac{1}{2}z - \tfrac{1}{6}s_2 + \tfrac{1}{3}s_3 &= 3 \\ P - 9 &= -\tfrac{1}{2}z - \tfrac{5}{6}s_2 - \tfrac{1}{3}s_3 \end{aligned}\right\} \text{where } z = 0,\ s_2 = 0,\ s_3 = 0$$

This tableau corresponds to the point (3, 3, 0), at which $P = 9$.

The expression describing the objective function in the last line of this tableau contains no positive coefficients. Thus we cannot increase P by making any of the nonbasic variables positive.

If there are no positive elements in the last row of a tableau, it is the last tableau of that problem. The numbers in the solutions column on the right match up with the variables given to their left. The variables that do not appear in the left column all have value 0. The number in the lower right is the negative of the maximum value for P—that is, P_{max}.

Thus the solution to this problem is $P_{max} = 9$ when $x = 3$, $y = 3$, $z = 0$, $s_1 = 1$, $s_2 = 0$, and $s_3 = 0$. We can get a rough check (at least the answers are plausible) by substituting back into the original problem: $P_{max} = 3 + 2(3) + 0 = 9$.

Inequalities	*Equalities*
$3 + 0 + 2(0) \leq 4$	$3 + 0 + 2(0) + 1 + 0 + 0 = 4$
$0 + 2(3) + 0 \leq 6$	$0 + 2(3) + 0 + 0 + 0 + 0 = 6$
$3(3) + 3 + 2(0) \leq 12$	$3(3) + 3 + 2(0) + 0 + 0 + 0 = 12$

Notice that except for locating the pivot, the computations used in the simplex algorithm are all familiar. Once the pivot is found, divide its entire row by the pivot and then add or subtract appropriate multiples of the pivot row to all the other rows so that the rest of the numbers in the pivot column are all zeros. You will thereby represent an adjoining vertex of the feasible region.

The following remarkably uncomplicated problem, devised by Robert Bixby of Northwestern University, clearly shows in three dimensions how the tableaux march around the feasible region, always increasing the value of the objective function.

6.1.2 Example

Maximize $P = 3x + 2y + 2z$ subject to $x + z \leq 8$, $x + y \leq 7$, and $x + 2y \leq 12$. (The nonnegativity constraints are assumed.)

Solution:

First, we must convert the inequalities to equalities using slack variables:

$$\begin{aligned} x \quad + z + s_1 \qquad\qquad &= 8 \\ x + y \qquad\quad + s_2 \qquad &= 7 \\ x + 2y \qquad\qquad\quad + s_3 &= 12 \end{aligned}$$

Then we are ready to set up the first tableau, where s_1, s_2, and s_3 are the basic variables and the vertex represented is the origin.

First tableau

	x	y	z	s_1	s_2	s_3	Solution
s_1	1	0	1	1	0	0	8
s_2	(1)	1	0	0	1	0	7
s_3	1	2	0	0	0	1	12
P	3	2	2	0	0	0	0

Since 3 is the largest number in the last row, the first column will be the pivot column. It is easy to examine the three quotients and discover that $7 \div 1 = 7$ is the smallest. Therefore, it is s_2 that is replaced by x at the left of the next tableau.

Second tableau

	x	y	z	s_1	s_2	s_3	Solution
s_1	0	−1	(1)	1	−1	0	1
x	1	1	0	0	1	0	7
s_3	0	1	0	0	−1	1	5
P	0	−1	2	0	−3	0	−21

Since y, z, and s_2 do not appear at the left of the second tableau, they are all zero at the vertex it represents, which is (7, 0, 0). At this vertex $P = 3(7) + 2(0) + 2(0) = 21$. Since 2 is the only positive number in the last row, it indicates the pivot column; the 1 in the s_1-row must be the pivot.

Third tableau

	x	y	z	s_1	s_2	s_3	Solution
z	0	−1	1	1	−1	0	1
x	1	1	0	0	1	0	7
s_3	0	①	0	0	−1	1	5
P	0	1	0	−2	−1	0	−23

This tableau represents the vertex (7, 0, 1), at which P = 23. Since 1 is the only positive number in the last row, it indicates the pivot column. Pivots are always positive, so we ignore the first row. Since $5 \div 1 = 5$ is less than $7 \div 1 = 7$, the pivot row will be the s_3-row. Since none of the

Fourth tableau

	x	y	z	s_1	s_2	s_3	Solution
z	0	0	1	1	−2	1	6
x	1	0	0	0	2	−1	2
y	0	1	0	0	−1	1	5
P	0	0	0	−2	0	−1	−28

numbers in the last row are positive, this is the final tableau for this problem. We have $P_{max} = 28$ at (2, 5, 6).

Figure 6.1–1 pictures the feasible region of this problem and shows how the tableaux move from one vertex to another, each time increasing the value taken by the objective function.

Figure 6.1–1

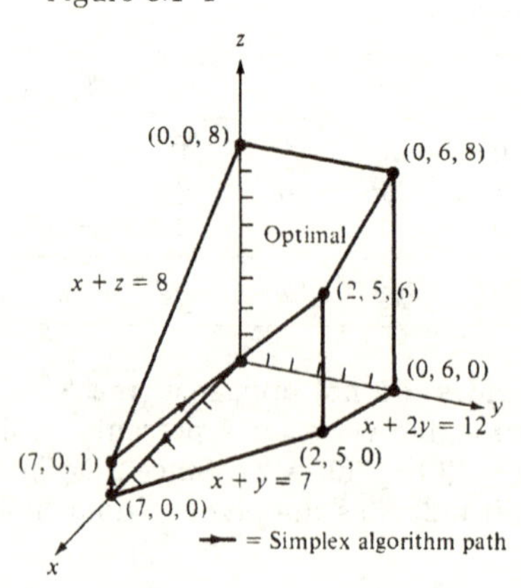

SUMMARY OF THE SIMPLEX METHOD

1. Rewrite the constraints as equalities using slack variables.
2. Construct the first tableau using the coefficients of the constraining equalities and the objective equation.
3. Choose the largest positive number in the last row; this determines the pivot column.
4. Construct the next tableau as follows:
   (a) Divide the positive elements of the inner rectangle in the pivot column into their corresponding elements in the "solutions" column. The one that gives the smallest quotient is the pivot.
   (b) Replace the letter to the left of the pivot row by the letter above the pivot column.
   (c) Divide the pivot row by the pivot.
   (d) Add the appropriate multiple of the pivot row to the other rows of numbers in such a way as to obtain zeros in the rest of the pivot column.
5. Examine the last row.
   (a) If it contains any positive numbers, the largest of them determines the new pivot column. Go back to step 4 and repeat it.
   (b) If all the elements in the last row are negative or zero, read the final solution down the right column. Each number in the solution column gives the value of the corresponding variable named at the left of its row. The number in the lower right is the negative of P_{max}. All independent variables that do not appear in the left column have the value 0.

Notice that the word "pivot" is used both as a verb and as a noun. As a verb, "pivoting" refers to the process by which we transform one tableau into the next. As a noun, the "pivot" is that number in a tableau around which the pivoting revolves.

In each tableau the numbers down the left are called the *basic* variables for that tableau. They are the only independent variables that can have nonzero values. Notice that the number of independent variables with nonzero values can never exceed the number of original significant constraints.

The exercises of this section provide practice in solving linear programming problems using the simplex method. The tableaux of the first problem in each exercise set are given in their entirety in the "Hints"; checking your work as you go may help relieve the normal human fear of failure in the first encounter with such long problems.

EXERCISES 6.1. A

Solve the following linear programming problems using the simplex method. Assume that all variables must be nonnegative.

1. Maximize $P = 4x + 3y + 2z$ given that $2x + 2y + z \leq 40$, $4y + 2z \leq 60$, and $2x + z \leq 20$.

2. Maximize $P = 2x + 3y + 4z$ subject to $x + 2y \leq 3$, $y + z \leq 4$, $x + z \leq 2$.
3. Maximize $P = 3x + 2y + 2z$ when $2x + 3y + 2z \leq 10$, $x + 2y + z \leq 8$, and $4x + 2y + 3z \leq 16$.

EXERCISES 6.1. B

Solve the following linear programming problems using the simplex method. Assume that all variables are nonnegative.

1. Maximize $P = 4x + 3y + z$ subject to $2x + 2y + z \leq 8$ and $4x + 2y + 4z \leq 12$.
2. Maximize $P = x + 3y + 4z$ when $y + 2z \leq 12$, $2x + 3y \leq 10$, and $x + y + 2z \leq 6$.
3. Maximize $P = 2x + 4y + 2z$ given that $2x + 4z \leq 8$, $4y + 2z \leq 12$, and $6x + 2y + 4z \leq 24$.

HINTS FOR PROBLEM 1, EXERCISES 6.1. A

First tableau

	x	y	z	s_1	s_2	s_3	Solution
s_1	2	2	1	1	0	0	40
s_2	0	4	2	0	1	0	60
s_3	(2)	0	1	0	0	1	20
P	4	3	2	0	0	0	0

Second tableau

	x	y	z	s_1	s_2	s_3	Solution
s_1	0	(2)	0	1	0	-1	20
s_2	0	4	2	0	1	0	60
x	1	0	$\frac{1}{2}$	0	0	$\frac{1}{2}$	10
P	0	3	0	0	0	-2	-40

Third tableau

	x	y	z	s_1	s_2	s_3	Solution
y	0	1	0	$\frac{1}{2}$	0	$-\frac{1}{2}$	10
s_2	0	0	2	-2	1	2	20
x	1	0	$\frac{1}{2}$	0	0	$\frac{1}{2}$	10
P	0	0	0	$-\frac{3}{2}$	0	$-\frac{1}{2}$	-70

HINTS FOR PROBLEM 1, EXERCISES 6.1. B

First tableau

	x	y	z	s_1	s_2	Solution
s_1	2	2	1	1	0	8
s_2	④	2	4	0	1	12
P	4	3	1	0	0	0

Second tableau

	x	y	z	s_1	s_2	Solution
s_1	0	①	-1	1	$-\frac{1}{2}$	2
x	1	$\frac{1}{2}$	1	0	$\frac{1}{4}$	3
P	0	1	-3	0	-1	-12

Third tableau

	x	y	z	s_1	s_2	Solution
y	0	1	-1	1	$-\frac{1}{2}$	2
x	1	0	$\frac{3}{2}$	$-\frac{1}{2}$	$\frac{1}{2}$	2
P	0	0	-2	-1	$-\frac{1}{2}$	-14

ANSWERS 6.1. A

1. $P_{max} = 70$ when $x = 10$, $y = 10$, $z = 0$
2. $P_{max} = \frac{25}{2}$ when $x = 0$, $y = \frac{3}{2}$, $z = 2$
3. $P_{max} = 12.5$ when $x = 3.5$, $y = 1$, $z = 0$

6.2 Why the Simplex Algorithm Works

The name of George Dantzig is associated with many important papers about the simplex algorithm that appeared in the 1940s and 1950s. Perhaps his greatest contribution to simplex algorithm theory was that before it became commonly used, he believed in it; he believed that in practical problems there would be few enough tableaux in each problem to make it easier to use than the tabular method. And he was right!

Basically, the simplex algorithm gives a systematic way of moving from one feasible vertex to an adjoining feasible vertex, each time increasing the value of the objective function until the maximum value is reached. To understand the algorithm from another point of view, we return now to Example 5.1.5 (page 154) about the wholesaler with limited quantities of wheat and barley who is deciding how many Premium and Regular bags to try to sell. (This problem was solved by the graphical method in Example 5.2.2 and by the tabular method in Example 5.3.1.)

6.2.1 Example

Maximize $P = 30x + 20y$ when

$$x + y \leq 3000 \qquad (x + y + s_1 = 3000)$$
$$2x + y \leq 4000 \qquad (2x + y + s_2 = 4000)$$

Solution:

Each boundary line of the feasible region (Figure 6.2–1) is determined by one of the four variables being equal to zero. (We include the slack variables.) Thus the intersection of any two boundary lines (in particular, a vertex of the feasible region) is determined by two of the variables being equal to zero. In standard LP problems we begin at the origin; that is, we set the original independent variables all equal

First tableau

	x	y	s_1	s_2	Solution
s_1	1	1	1	0	3000
s_2	②	1	0	1	4000
P	30	20	0	0	0

to zero. To say that $x = 0$ means in this problem that we are selling no Premium bags and to say $y = 0$ means that we are selling no Regular bags. Notice that in this problem s_1 designates the number of unused pounds of wheat and s_2 designates the number of unused pounds of barley. When $x = y = 0$, $s_1 = 3000$ and $s_2 = 4000$.

Figure 6.2–1

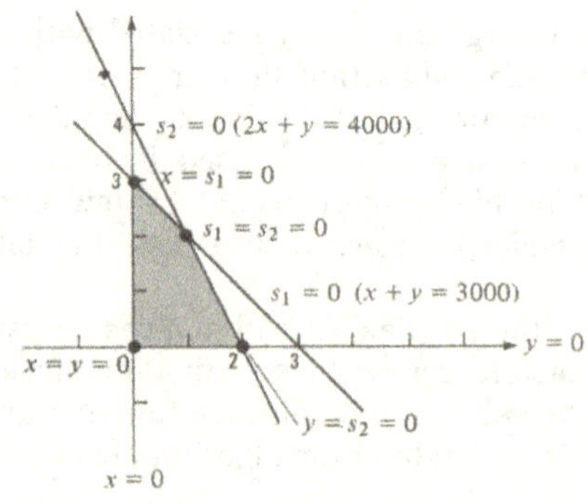

Finding the Pivot

Since the coefficient of x in the objective equation is greater than the coefficient of y, we guess that we can "gain" most by increasing x in the next tableau. In other words, x will become our new basic (possibly nonzero) variable in the second tableau.

How much should x increase from 0? Clearly, x should grow only enough to go to the adjoining vertex; any more would take us outside the feasible region. The quotients help us decide where that next vertex lies. Since x designates the number of Premium bags, we remember that each Premium bag requires 1 lb of wheat and 2 lb of barley. There are 3000 lb of wheat available, so there is enough for $3000/1 = 3000$ Premium bags; the 4000 lb of barley available is enough for $4000/2 = 2000$ bags. Thus we run out of barley before wheat, and $s_2 = 0$ in the next tableau.

Writing the Second Tableau

To change the second constraint so that it reveals the new expression for x, we divide the second row of the inner rectangle by 2; the second constraint now reads $x + \frac{1}{2}y + \frac{1}{2}s_2 = 2000$. To eliminate x from the first constraint, we subtract the new second row from the old first, obtaining $\frac{1}{2}y + s_1 - \frac{1}{2}s_2 = 1000$.

Now we want to eliminate x from the objective function. To do this we subtract the appropriate multiple of the pivot row from the former expression for the objective function:

$$\begin{aligned} P &= 30x + 20y \\ (15)4000 &= 15(2x + y + s_2) \qquad \text{(subtracting)} \\ P - 60{,}000 &= 5y - 15s_2 \end{aligned}$$

Now if we set $y = 0$ and $s_2 = 0$ in this last expression (which is what is implied by y and s_2 being the nonbasic variables), we easily see that $P = 60{,}000$. It appears as $-60{,}000$ in the second tableau because the other numbers in the far right column are positive, and when we subtract, we get a negative number.

Second tableau

	x	y	s_1	s_2	Solution
s_1	0	($\frac{1}{2}$)	1	$-\frac{1}{2}$	1000
x	1	$\frac{1}{2}$	0	$\frac{1}{2}$	2000
P	0	5	0	-15	$-60{,}000$

Now the second tableau is complete. It says that $x = 2000$, $s_1 = 1000$, and the other independent variables are zero. We can check P in the original objective equation: $P = 30(2000) + 20(0) = 60{,}000$.

Finding the Pivot in the Second Tableau

By looking at the expression $P = 5y - 15s_2 + 60{,}000$, we can see that increasing the value of y (which is zero because it is a nonbasic variable) will increase the value of the objective function. Thus we decide to introduce y to the basis—that is, the y-column will be our next pivot column.

If y joins the basis, we are selling a nonzero number of Regular bags. The two $\frac{1}{2}$'s in the y-column tell us that each extra Regular bag would be made from a $\frac{1}{2}$ lb of "extra" wheat (symbolized by s_1) and by using $\frac{1}{2}$ of a Premium bag—from which we would get the other $\frac{1}{2}$ lb of wheat needed and the whole pound of barley that is needed. (*Remember:* Each bag of Premium contains 1 lb of wheat and 2 lb of barley.)

To find whether s_1 or x should leave the basis, we take $1000 \div \frac{1}{2} = 2000$ and $2000 \div \frac{1}{2} = 4000$ to see whether we run out of wheat or Premium bags first. Since we will exhaust the wheat after 2000 Regular bags are made, but have enough Premium bags to make 4000 Regular bags, the wheat is gone first and $s_1 = 0$ in the third tableau.

Thus y will replace s_1 in the left column of the next tableau. We fill in the numbers by pivoting around the $\frac{1}{2}$ in the y-column and s_1-row.

Third tableau

	x	y	s_1	s_2	Solution
y	0	1	2	−1	2000
x	1	0	−1	1	1000
P	0	0	−10	−10	−70,000

This tableau describes the vertex where $x = 1000$, $y = 2000$, $s_1 = 0$, and $s_2 = 0$. At this point $P = 30(1000) + 20(3000) = 70{,}000$, as obtained in Exercises 5.2.2 and 5.3.1. The last row of the tableau expresses P in the form

$$P = -10s_1 - 10s_2 + 70{,}000$$

which, of course, corroborates the value of P above. But the negative coefficients of s_1 and s_2 in this expression also assure us that we cannot make P any bigger by introducing s_1 and s_2 into the basis; if either were made greater than zero, the value of P would decrease. Thus we have achieved the largest possible value of P, and we conclude that $P_{max} = 70{,}000$.

Other Questions

1. What if the largest positive number in the last row appears twice in that row? Which column should I choose?
   Answer: Either choice will work. Choose one and follow the algorithm. (See Examples 6.2.2 and 6.2.3.)
2. What if, when I divide the positive numbers of the pivot column into the solution column, two of the quotients are equal and smaller than the others? Which row do I choose?
   Answer: This means that two of the variables will become zero in the next tableau; one of the basic variables will show a 0 in the solutions column. It will not matter which of the two variables you choose to leave the basis. (Actually, in theory such situations may lead to problems, but in practical problems they do not.) (See Example 6.2.4.)
3. What if none of the numbers in the pivot column in the inner rectangle is positive?
   Answer: Then the solution is unbounded. This indicates that you will not have to make any other of the basic variables smaller as you make the new basic variable larger. In well-formulated practical problems, this usually does not happen.

6.2.2 Example

(See Question 1 above.) Maximize $P = x + y$ when $x + 2y \leq 3$ and $2x + y \leq 3$.

Solution:

First tableau

	x	y	s_1	s_2	Solution
s_1	1	(2)	1	0	3
s_2	(2)	1	0	1	3
P	1	1	0	0	0

In the first solution we use the x-column as the pivot column; in the second solution, we use the y-column.

Second tableau (of first solution)

	x	y	s_1	s_2	Solution
s_1	0	($\frac{3}{2}$)	1	$-\frac{1}{2}$	$\frac{3}{2}$
x	1	$\frac{1}{2}$	0	$\frac{1}{2}$	$\frac{3}{2}$
P	0	$\frac{1}{2}$	0	$-\frac{1}{2}$	$-\frac{3}{2}$

Third tableau (of first solution)

	x	y	s_1	s_2	Solution
y	0	1	$\frac{2}{3}$	$-\frac{1}{3}$	1
x	1	0	$-\frac{1}{3}$	$\frac{2}{3}$	1
P	0	0	$-\frac{1}{3}$	$-\frac{1}{3}$	-2

Second tableau (of second solution)

	x	y	s_1	s_2	Solution
y	$\frac{1}{2}$	1	$\frac{1}{2}$	0	$\frac{3}{2}$
s_2	($\frac{3}{2}$)	0	$-\frac{1}{2}$	1	$\frac{3}{2}$
P	$\frac{1}{2}$	0	$-\frac{1}{2}$	0	$-\frac{3}{2}$

Third tableau (of second solution)

	x	y	s_1	s_2	Solution
y	0	1	$\frac{2}{3}$	$-\frac{1}{3}$	1
x	1	0	$-\frac{1}{3}$	$\frac{2}{3}$	1
P	0	0	$-\frac{1}{3}$	$-\frac{1}{3}$	-2

In the example above, the two choices lead to exactly the same final tableau. In the next example there is a different final point, but the same maximum value of the objective function.

6.2.3 Example

(See Question 1 above.) Maximize $P = x + y$ when $x + y \leq 2$ and $2x + y \leq 3$.

Solution:

First tableau

	x	y	s_1	s_2	Solution
s_1	1	1	1	0	2
s_2	②	1	0	1	3
P	1	1	0	0	0

Again we let the x-column be the pivot column in the first solution and the y-column be the pivot column in the second.

Second tableau (of first solution)

	x	y	s_1	s_2	Solution
s_1	0	$\textcircled{\frac{1}{2}}$	1	$-\frac{1}{2}$	$\frac{1}{2}$
x	1	$\frac{1}{2}$	0	$\frac{1}{2}$	$\frac{3}{2}$
P	0	$\frac{1}{2}$	0	$-\frac{1}{2}$	$-\frac{3}{2}$

Third tableau (of first solution)

	x	y	s_1	s_2	Solution
y	0	1	2	−1	1
x	1	0	−1	1	1
P	0	0	−1	0	−2

Second tableau (of second solution)

	x	y	s_1	s_2	Solution
y	1	1	1	0	2
s_2	1	0	−1	1	1
P	0	0	−1	0	−2

6.2.4 Example

(See Question 2 above.) Maximize $P = 3x + 2y$ when $x + y \leq 2$ and $2x + 3y \leq 4$.

First tableau

	x	y	s_1	s_2	Solution
s_1	①	1	1	0	2
s_2	②	3	0	1	4
P	3	2	0	0	0

This time the x-column must be the pivot column but both test quotients are 2. In the first solution we let the s_1-row be the pivot row, and in the second we let the s_2-row be the pivot row.

Second tableau (first solution)

	x	y	s_1	s_2	Solution
x	1	1	1	0	2
s_2	0	1	-2	1	0
P	0	-1	-3	0	-6

Second tableau (second solution)

	x	y	s_1	s_2	Solution
s_1	0	$-\frac{1}{2}$	1	$-\frac{1}{2}$	0
x	1	$\frac{3}{2}$	0	$\frac{1}{2}$	2
P	0	$-\frac{5}{2}$	0	$-\frac{3}{2}$	-6

SUMMARY

This section shows how the tableaux used in the simplex algorithm describe successive vertices of the feasible region and presents the reasons that the tableaux are constructed as they are. It is *not* necessary to understand all these reasons to correctly solve the problems in Exercises 6.2.A and 6.2.B. These exercises provide more practice in using the simplex algorithm; as you do them, you may want to refer back to this section and see why what you are doing works.

EXERCISES 6.2. A

Use the simplex algorithm to solve the following problems. In the first two exercises, also draw a graph and show how the tableaux march around the feasible region. The last three problems are intended to provide you with more practice using the simplex method for three variables—practice that is probably still needed.

1. Maximize $P = 4x + 3y$ subject to $3x + 2y \leq 12$ and $x + 2y \leq 8$.
2. Maximize $P = 2x + 5y$ subject to $x + 2y \leq 8$ and $x + 3y \leq 9$.

3. Maximize $P = 2x + 3y + z$ when $x + y + 2z \leq 12$, $2x + 2y + z \leq 20$, and $4x + y \leq 6$.
4. Maximize $P = x + y + 3z$ when $x + y \leq 8$, $x + y + 2z \leq 18$, and $y + 4z \leq 16$.
5. Maximize $P = x + y + 2z$ subject to $2x + 2z \leq 10$, $x + 2y + z \leq 12$, and $x + y \leq 6$.

EXERCISES 6.2. B

Use the simplex algorithm to solve the following problems. In the first two exercises, also draw a graph and show how the tableaux trot around the feasible region. The last three problems are intended to provide you with more practice using the simplex method for three variables—practice that is probably still needed.

1. Maximize $P = 2x + 3y$ when $x + y \leq 12$ and $x + 2y \leq 16$.
2. Maximize $P = 4x + 3y$ when $2x + y \leq 16$ and $y \leq 6$.
3. Maximize $P = x + 8y + 3z$ when $2x + 6y + 4z \leq 21$, $3x + 2y \leq 16$, and $9y + 4z \leq 18$.
4. Maximize $P = 7x + y + z$ given that $x + y + 3z \leq 12$. $2x + 2z \leq 16$, and $3y + z \leq 6$.
5. Maximize $P = 2x + 2y + 3z$ subject to $2x + 3y + 4z \leq 10$, $3x + 2y + z \leq 15$, and $3x + y \leq 20$.

EXERCISES 6.2. C

1. What does it mean to have a slack variable nonbasic?
2. Tell in words the meaning of the numbers $\begin{matrix} 2 & -1 \\ -1 & 1 \end{matrix}$ at the right of the third tableau in Example 6.2.1.
3. Solve the problem in Example 5.3.2 using the simplex algorithm and put arrows in Figure 5.3–2 showing how the tableaux progress around that (three-dimensional) feasible region.
4. Maximize $P = 2x + 3y$ when $-x + 4y \leq 22$, $x + 2y \leq 14$, $x + y \leq 9$, and $2x + y \leq 16$. Graph the feasible region for this and solve it using the simplex algorithm. Trace the tableaux going around the feasible region, each time increasing P.

ANSWERS 6.2. A

1.

P_{max} = 17 at (2, 3)
(2, 3); $s_1 = s_2 = 0$
Third tableau
(0, 0); $x = y = 0$
(4, 0); $s_1 = y = 0$
First tableau Second tableau

2.

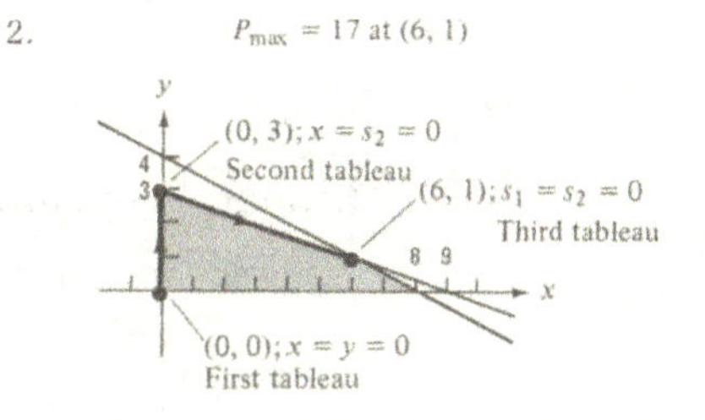

3. $P_{max} = 21$ when $x = s_1 = s_3 = 0$, $y = 6$, $z = 3$, $s_2 = 5$
4. $P_{max} = 20$ when $x = 8$, $y = 0$, $z = 4$
5. $P_{max} = 13.5$ when $x = 0$, $y = 3.5$, $z = 5$

ANSWERS 6.2. C

1. If a slack variable is nonbasic, the ingredient (or machine, or department) from which its constraint is derived is being completely used. There are no "leftovers."
2. If s_1 were increased from 0 to 1 (that is, extra wheat is generated), it would be done by using 2 bags of Regular and making 1 extra bag of Premium; what remained would be an extra pound of wheat. If s_2 were increased from 0 to 1 (that is, extra barley is generated), it would be done by using 1 bag of Premium and making 1 bag of Regular; what remained would be an extra pound of barley.
3.

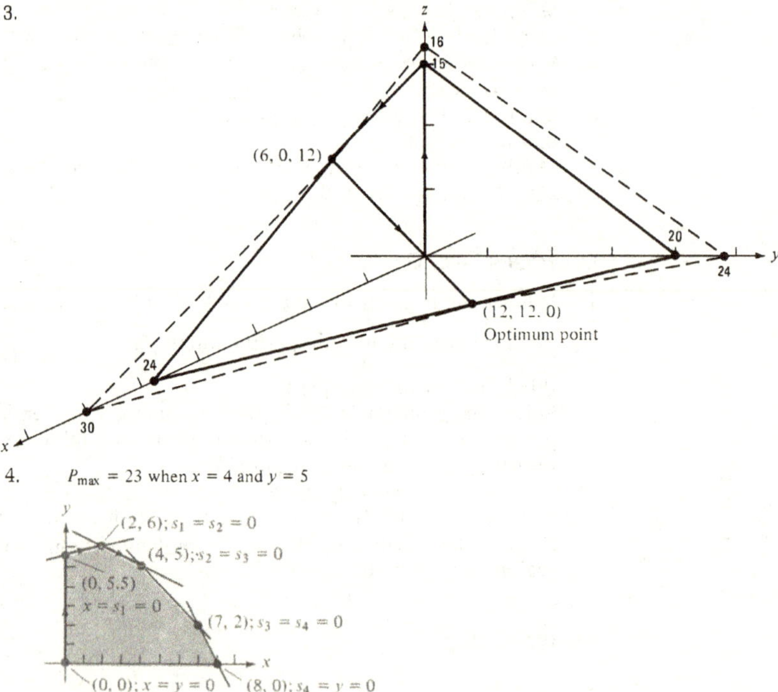

4. $P_{max} = 23$ when $x = 4$ and $y = 5$

★6.3 Linear Programming Problems That Are Not Standard

A linear programming problem is not standard if its objective function is to be minimized, or if some (or all) of its constraints are equalities or "greater than" inequalities, or both. This section adapts the simplex algorithm to such problems.

Minimization Problems

To minimize a function is the same as to maximize its negative:

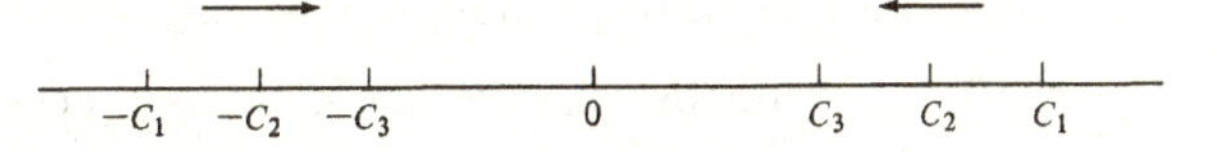

Thus if we want to minimize a function, C, we can multiply it through by -1 and maximize $-C$. Then, after $-C$ has been maximized using the technique of the previous sections, its negative will be the minimum of the original C.

Equalities and "Greater Than" Inequalities

Example 6.3.3 includes the inequality $x + y + z \geq 3$. When this inequality is converted to an equality using a *positive* slack variable on the same side of the equal sign as the other variables, a minus sign must precede the slack variable: $x + y + z - s_1 = 3$. This is because $(x + y + z)$ is bigger than 3, so a positive quantity must be subtracted from it to obtain 3.

But if we set all the original variables to zero for our first tableau, we obtain $s_1 = -3$. This is not allowed; all variables in a linear programming problem must be positive. Therefore, we introduce another variable, called an artificial variable:

$$x + y + z - s_1 + a_1 = 3$$

Now we can set $x = y = z = s_1 = 0$ and have $a_1 = 3$, which is permissible. But the artificial variable has no "real-life" interpretation whatsoever; it is merely a mathematical gimmick used to find an initial feasible solution of an LP problem. Thus the first step in solving such a problem will be to get rid of it! This is done by minimizing a new objective function $A = a_1 + a_2 + \cdots + a_n$—the sum of all the artificial variables, a_i, which it was necessary to introduce. If the original problem was solvable, the minimum must be zero, and when we have obtained it, we have also obtained a feasible solution to the original problem. Then the a_i columns are dropped from the tableau, since the a_i must all have 0 value. At this point we say we have completed "phase 1." "Phase 2" consists of optimizing the original objective function.

6.3.1 Example

An advertising agency is deciding how much of its clients' \$3600 should be spent on advertising in each of three magazines. A formula has been developed for evaluating the importance of the age, income level, sex, and education of each magazine's readers (similar to the idea used in Exercises 1.3.C). This "effectiveness coefficient" for each magazine is then multiplied by the number of readers of that magazine, to obtain the coefficient in the objective equation of a linear programming problem corresponding to that magazine. Thus, if

x = number of advertising pages in magazine A

y = number of advertising pages in magazine B

z = number of advertising pages in magazine C

the objective equation of this linear programming problem might be written

$$P = 0.5(200{,}000)x + 0.4(500{,}000)y + 0.3(1{,}000{,}000)z$$

where 0.5, 0.4, and 0.3 are the effectiveness coefficients for magazines A, B, and C, respectively, and 200,000, 500,000, and 1,000,000 are the number of readers of the three magazines, respectively. Multiplying out the coefficients, we obtain

$$P = 100{,}000x + 200{,}000y + 300{,}000z$$

This will clearly take its maximum or minimum whenever

$$P = x + 2y + 3z$$

does, so we use the simple objective equation $P = x + 2y + 3z$. (In real-world problems the numbers will not divide that evenly, but simplifications like this may be useful after approximations are made.)

If each page of advertising in magazines A, B, and C costs, respectively, \$500, \$1000, and \$1200 and the client has budgeted only \$3600 for advertising, one of the constraints will be

$$500x + 1000y + 1200z \leq 3600$$

which can be written

$$5x + 10y + 12z \leq 36$$

If, in addition, the client insists that at least 6 pages of advertising must be included in magazine A because the basic philosophy of that magazine encourages the use of its product, there is also the constraint $x \geq 6$.

Thus the whole problem can be summarized: Maximize $P = x + 2y + 3z$ subject to $5x + 10y + 12z \leq 36$ and $x \geq 6$.

Solution:

The second constraint is reversed from those in the standard problems of the previous two sections and merits different treatment. First, its slack variable must have a minus sign:

$$x - s_1 = 6$$

And then, since $x = 0$ would imply that $s_1 = -6$ and negative numbers are not permitted, we set

$$x - s_1 + a_1 = 6$$

where a_1 is an artificial variable used to find an initial feasible solution. In the first tableau, $x = 0$ and $s_1 = 0$, so $a_1 = 6$, which is acceptable.

The first constraint is easy to change to an equality:

$$5x + 10y + 12z + s_2 = 36$$

Phase 1: The first project is to rid the problem of a_1 since it has no real-world interpretation. We do this by finding the minimum of $A = a_1$ (which minimum had better be zero), or, equivalently, we find the maximum of $-A = -a$. We might write

First tableau (of Phase 1)

	x	y	z	s_1	s_2	a_1	Solution
a_1	①	0	0	−1	0	1	6
s_2	5	10	12	0	1	0	36
$-A$	0	0	0	0	0	−1	0
$-A$	1	0	0	−1	0	0	6

The row above the dashed line is a summary of the equation $-A = -a$. Alas, it has a nonzero number in the last row of the a_1-column. This is not allowed, because a_1 is a basic variable, and each basic variable must have one 1 in its column and the rest of the numbers 0. To obtain a 0 in the last row in the a_1-column, we add the a_1-row to the last row and write the sum below; then we cross out the original row. Since the only positive number in the last row is 1, it determines the pivot column. Considering $6 \div 1 = 6$ and $36 \div 5 > 7$, we see that 1 in the upper left is the pivot.

Second tableau (of Phase 1)

	x	y	z	s_1	s_2	a_1	Solution
x	1	0	0	−1	0	1	6
s_2	0	10	12	5	1	−5	6
$-A$	0	0	0	0	0	−1	0

Since there is no positive number in the last row of this tableau and the objective function has the value 0, we have found an initial feasible solution to the original problem:

$$x = 6 \qquad s_2 = 6 \qquad y = 0 \qquad z = 0 \qquad s_1 = 0$$

Phase 2: To write the first tableau of the original problem, we must replace the last line of the last tableau in phase 1 with the objective function of the original problem written in terms of the nonbasic variables in the last tableau of phase 1. To do this, we must eliminate the basic variables (in the last tableau of phase 1) from the objective function as originally written. We do this by subtracting (or adding) appropriate multiples of the needed row(s) (of the last tableau of phase 1) to the original objective function. Perhaps the following helps:

	x	y	z	s_1	s_2	Solution	
x	1	0	0	−1	0	6	
s_2	0	10	12	5	1	6	
P	1	2	3	0	0	0	(original form)
P	0	2	3	1	0	−6	(useful form for first tableau in phase 2—obtained by subtracting the x-row from the original form)

Notice that in rewriting the inner rectangle of the last tableau of phase 1, the a_1 column was omitted, since a_1 has become 0 and will not be needed again. Also, the last row was replaced by the original objective function. To obtain a 0 as the coefficient of x in this function (since x is a basic variable and the objective function must be written in terms of the nonbasic variables), the x-row of the inner rectangle is subtracted from the original objective function. You need not write all this on your paper. In the last tableau of phase 1 you would cross out the column of the artificial variable(s) and the last row. Then you could write the coefficients of the original objective function below the tableau and subtract the necessary row(s) to obtain the useful objective function for phase 2. Then if you cross out the old objective function, this is what remains:

First tableau (in Phase 2)

	x	y	z	s_1	s_2	Solution
x	1	0	0	−1	0	6
s_2	0	10	(12)	5	1	6
P	0	2	3	1	0	−6

Since there are positive numbers in the last row, and 3 is the largest, z will now join the basis. Since 12 is the only positive number in its column in the inner rectangle, it must be the pivot.

Second tableau (in Phase 2)

	x	y	z	s_1	s_2	Solution
x	1	0	0	-1	0	6
z	0	$\frac{5}{6}$	1	$\frac{5}{12}$	$\frac{1}{12}$	$\frac{1}{2}$
P	0	$-\frac{1}{2}$	0	$-\frac{1}{4}$	$-\frac{1}{4}$	$-\frac{15}{2}$

Since none of the numbers in the last row are positive, we cannot increase the objective function any more by making some other nonbasic variable positive, so we have found the optimal solution. It is $P_{max} = \frac{15}{2}$ when $x = 6$, $z = \frac{1}{2}$, $y = 0$.

Artificial variables are also used in equality constraints without any slack variables.

6.3.2 Example

A mutual fund is deciding how to divide its investments among bonds, preferred stock, and speculative stock. It does not want to exceed a combined risk rate of 3 when the bonds have been assigned a risk rate of 1, the preferred stock, 3, and the speculative stock, 5. However, the fund does want a total annual yield of at least 10 percent. If the interest rate of the bonds is 8 percent, of the preferred stock is 12 percent, and of the speculative stock is 20 percent, how should the assets be distributed for the greatest annual yield?

Solution:

The total amount to be invested does not matter. We let x, y, and z be the fractions of the whole that will be invested in bonds, preferred stock, and speculative stock, respectively. Clearly these fractions must total the whole portfolio:

$$x + y + z = 1$$

The other two constraints indicate the restrictions on the risk rate and total yield:

$$x + 3y + 5z \leq 3 \qquad \text{(risk rate constraint)}$$
$$8x + 12y + 20z \geq 10 \qquad \text{(annual yield constraint)}$$

The annual yield constraint can be omitted, since whether it is there or not, we will make the left side of the inequality as large as possible. If it must be less than 10, there is no feasible solution to the problem as stated, and the fund must reconsider its goals. If it can equal or exceed 10 subject to the other constraints, that constraint is automatically satisfied. Thus we can summarize the problem: Maximize $P = 8x + 12y + 20z$ subject to $x + y + z = 1$ and $x + 3y + 5z \leq 3$. The nonnegativity constraints, as usual, are assumed.

The first constraint needs no slack variable, but it does need an artificial variable. The second is given a slack variable as usual:

$$x + y + z + a = 1$$
$$x + 3y + 5z + s = 3$$

(Since there is only one artificial variable and one slack variable, we leave off the subscripts.)

Phase 1: To get rid of a, we want to minimize $A = a$, which is the same as maximizing $-A = -a$. Following the same ideas used in Example 6.3.1, we write

First tableau (in Phase 1)

	x	y	z	s	a	Solution
a	(1)	1	1	0	1	1
s	1	3	5	1	0	3
$-A$	0	0	0	0	-1	0
$-A$	1	1	1	0	0	1

Since all three positive numbers in the last row are equal, any one of the first three columns may become the pivot column. We choose arbitrarily to have x join the basis; then a must leave.

Second tableau (in Phase 1)

	x	y	z	s	a	Solution
x	1	1	1	0	1	1
s	0	2	4	1	-1	2
$-A$	0	0	0	0	-1	0

Thus we have found an initial feasible solution to the original problem: $x = 1$, $y = 0$, and $z = 0$. We know that phase 1 is complete because the objective function is 0.

Phase 2: The objective function in the original problem was $P = 8x + 12y + 20z$. To write this without an x-term (because x is in the basis in the initial feasible solution), it is convenient to write the coefficients of this objective function below the inner rectangle of the last tableau in phase 1 and then subtract 8 times the x-row in the inner rectangle from the coefficients of the original objective function:

		x	y	z	s	Solution	
8	x	1	1	1	0	1	
	s	0	2	4	1	2	
	P	8	12	20	0	0	(original form)
	P	0	4	12	0	−8	(useful form for phase 2)

An 8 was written to the left of the x-row to remind us that every element of that row must be multiplied by 8 before it is subtracted from the original objective function.

First tableau (in Phase 2)

	x	y	z	s	Solution
x	1	1	1	0	1
s	0	2	(4)	1	2
P	0	4	12	0	−8

Second tableau (in Phase 2)

	x	y	z	s	Solution
x	1	$\frac{1}{2}$	0	$-\frac{1}{4}$	$\frac{1}{2}$
z	0	$\frac{1}{2}$	1	$\frac{1}{4}$	$\frac{1}{2}$
P	0	−2	0	−3	−14

Since there are no positive numbers in the bottom row, we have obtained the optimum solution. It is $P_{max} = 14$ for $x = \frac{1}{2}$, $y = 0$, and $z = \frac{1}{2}$. Thus the maximum possible yield under these conditions is 14 percent, well above the 10 percent desired minimum.

The next example demonstrates how to handle more than one artificial variable. The statement of the problem is reminiscent of Sections 5.4 and 5.5, but this would be a difficult problem to solve using the graphical or tabular methods.

6.3.3 Example

Suppose that we want to make a fruit drink containing 750 mg of vitamin C and 3000 I.U. of vitamin A. Each can of the red punch contains 250 mg of vitamin C and 500 I.U. of vitamin A and costs 50 cents. Each can of the purple punch costs 60 cents and contains 250 mg of vitamin C and 1000 I.U. of vitamin A. Each can of yellow punch costs 70 cents and contains 250 mg of vitamin C and 2000 I.U. of vitamin A. How much of each punch should be used to provide the required nutrients at least cost?

Solution:

The problem can be summarized: Minimize $C = 50x + 60y + 70z$ subject to $250x + 250y + 250z \geq 750$ and $500x + 1000y + 2000z \geq 3000$.

The objective function will take its minimum at the same time as $C = 5x + 6y + 7z$, which is easier to use. Also, we can divide the first constraint by 250 and the second by 500 to obtain easier numbers: $x + y + z \geq 3$ and $x + 2y + 4z \geq 6$. Each constraint requires both a slack variable and an artificial variable:

$$x + y + z - s_1 + a_1 = 3$$
$$x + 2y + 4z - s_2 + a_2 = 6$$

Phase 1: We first want to minimize $A = a_1 + a_2$, which is the same as maximizing $-A = -a_1 - a_2$. Since both a_1 and a_2 are basic variables in the first tableau, we must obtain zeros in *both* their columns in the last row. To do this, we add both the a_1-row and the a_2-row to the expression summarizing $-A = -a_1 - a_2$.

First tableau (in Phase 1)

	x	y	z	s_1	s_2	a_1	a_2	Solution
a_1	1	1	1	−1	0	1	0	3
a_2	1	2	(4)	0	−1	0	1	6
$-A$	0	0	0	0	0	−1	−1	0
$-A$	2	3	5	−1	−1	0	0	9

Second tableau (in Phase 1)

	x	y	z	s_1	s_2	a_1	a_2	Solution
a_1	($\frac{3}{4}$)	$\frac{1}{2}$	0	−1	$\frac{1}{4}$	1	$-\frac{1}{4}$	$\frac{3}{2}$
z	$\frac{1}{4}$	$\frac{1}{2}$	1	0	$-\frac{1}{4}$	0	$\frac{1}{4}$	$\frac{3}{2}$
$-A$	$\frac{3}{4}$	$\frac{1}{2}$	0	−1	$\frac{1}{4}$	0	$-\frac{5}{4}$	$\frac{3}{2}$

Third tableau (in Phase 1)

	x	y	z	s_1	s_2	a_1	a_2	Solution
x	1	$\frac{2}{3}$	0	$-\frac{4}{3}$	$\frac{1}{3}$	$\frac{4}{3}$	$-\frac{1}{3}$	2
z	0	$\frac{1}{3}$	1	$\frac{1}{3}$	$-\frac{1}{3}$	$-\frac{1}{3}$	$\frac{1}{3}$	1
$-A$	0	0	0	0	0	−1	−1	0

Since the objective function has reached zero, we have obtained an initial feasible solution for the original problem. It occurs when $x = 2$ and $z = 1$, so x and z are in the basis.

Phase 2: The goal of this problem is to minimize $C = 5x + 6y + 7z$, which is the same as maximizing $-C = -5x - 6y - 7z$. Since x and z will be in the basis of the first tableau, we must eliminate them from the objective function. This is accomplished by *adding* multiples of the x-row and z-row to $(-5x - 6y - 7z)$; we add because the negative signs were introduced to convert the minimization problem to a maximization problem. On your paper you could do this by merely crossing out the a_1 and a_2 columns and the last row of the tableau above and then writing in the coefficients in $-C = -5x - 6y - 7z$ below it. But we copy here the needed parts of the last tableau in phase 1 for your convenience.

		x	y	z	s_1	s_2	Solution	
5	x	1	$\frac{2}{3}$	0	$-\frac{4}{3}$	$\frac{1}{3}$	2	
7	z	0	$\frac{1}{3}$	1	$\frac{1}{3}$	$-\frac{1}{3}$	1	
	$-C$	-5	-6	-7	0	0	0	(original objective function)
	$-C$	0	$-\frac{1}{3}$	0	$-\frac{13}{3}$	$-\frac{2}{3}$	17	(original objective function plus 5 times the x-row plus 7 times the z-row)

Thus we can write:

First tableau (in Phase 2)

	x	y	z	s_1	s_2	Solution
x	1	$\frac{2}{3}$	0	$-\frac{4}{3}$	$\frac{1}{3}$	2
z	0	$\frac{1}{3}$	1	$\frac{1}{3}$	$-\frac{1}{3}$	1
$-C$	0	$-\frac{1}{3}$	0	$-\frac{13}{3}$	$-\frac{2}{3}$	17

Since none of the numbers in the last row are positive, we cannot increase $-C$ by putting another variable into the basis. Thus the first tableau in phase 2 happens to give the optimal solution in this example. If there had been a positive number in the last row, we would have pivoted as usual.

The fruit drink in the original problem should be made from 2 cans of red punch and 1 can of yellow punch. It will cost \$1.70. (Remember that we divided the coefficients in the original objective function by 10.)

Understanding the remainder of this section is not necessary in order to do the exercises, but the material will contribute to your general appreciation of linear programming.

Other Questions

1. What if the objective function does not reduce to zero in phase 1?
Answer: Then there is no feasible solution to the original problem; it is impossible to solve. This does not usually happen in carefully formulated practical problems. (See Example 6.3.4.)
2. What if there are still a_i's in the left column when the objective function has become zero?
Answer: Then they must all have the value 0 in the "Solutions" column, since all the variables in an LP problem are nonnegative and the function $A = a_1 + a_2 + \cdots + a_n$ has value 0. The tableau is said to be "degenerate"; there are more variables with value 0 than there are nonbasic variables. There are two possibilities when this occurs:
(i) The only nonzero numbers in the row with an a_i on the left are entries in artificial columns. This means there was a redundancy in the original problem. This situation is handled by eliminating (that is, crossing out) the whole row and the a_i column.
(ii) There is at least one nonzero number in the row with an a_i on the left which is not in an artificial column. Choose one such nonzero entry and use it as a pivot entry. (Note that this pivot entry may be positive or negative.) Since the number in the Solutions column is zero, you will be merely rewriting the same vertex with a different tableau. When you have removed the a_i's from the left column by thus pivoting, drop all of the a_i-columns. (See Example 6.3.5.)

6.3.4 Example

Optimize something when $x + y \leq 1$ and $3x + y = 5$.

Solution:

First tableau (in Phase 1)

	x	y	s	a	Solution
s	①	1	1	0	1
a	3	1	0	1	5
$-A$	3	1	0	0	5

Second tableau (in Phase 1)

	x	y	s	a	Solution
x	1	1	1	0	1
a	0	−2	−3	1	2
$-A$	0	−2	−3	0	2

Since the number in the lower right is not 0 but there are no (other) positive numbers in the last row, we cannot proceed further. The original constraints are inconsistent and there is no feasible region. (See Question 1 above.)

6.3.5 Example

Optimize something when $x + y \leq 2$ and $2x + y = 4$.

Solution:

First tableau (in Phase 1)

	x	y	s	a	Solution
s	(1)	1	1	0	2
a	2	1	0	1	4
$-A$	2	1	0	0	4

Second tableau (in Phase 1)

	x	y	s	a	Solution
x	1	1	1	0	2
a	0	(−1)	−2	1	0
$-A$	0	−1	−2	0	0

The objective function, $-A$, has now been reduced to zero, but to get usable coefficients for the inner rectangle, we must write this vertex without a appearing as a basic variable. Thus we must pivot around some number in the a-row. Since the number on the right of the row is zero, it will remain zero when divided by any number, and we can use a negative number as a pivot. We choose (arbitrarily) to use -1.

Third tableau (in Phase 1)

	x	y	s	a	Solution
x	1	0	−1	1	2
y	0	1	2	−1	0
$-A$	0	0	0	−1	0

The simplex method is sufficiently new that the method for writing it down is not yet standardized. The form used in this text should enable you to read most books about the simplex method. An exception would be those that use "Tucker tableaux," a more succinct, but less intuitive, way of writing simplex algorithm problems. If you are going to do many simplex linear programming problems by hand, you should learn the Tucker tableaux approach; it can be found in such books as *Finite Mathematics* by N. A. Weiss and M. L. Yoseloff (published in 1975 by Worth Publishers, Inc., New York, the publisher of this text). But if you are fated to do many linear programming problems, you will probably be using a preprogrammed computer.

Then it might be good for you to know there is another way to adapt the simplex algorithm to minimization problems. This involves remembering that the last row in each tableau indicates how much can be gained or lost by introducing the variable above it into the basis. If we want to minimize (instead of maximize) the objective function, we need only pivot the inner rectangle of the tableau as usual until all the numbers in the last row are either positive or zero. This would indicate that we cannot decrease the objective function any further.

There is a slightly different method of solving nonstandard linear programming problems called the "Big M method," but it uses ideas similar to those presented here. Some books refer to the slack variables in constraints that require artificial variables as "surplus variables," but the interpretation is the same; it designates the amount of that quantity that is left over.

SUMMARY

This section adapts the simplex method to nonstandard problems. An objective function can be minimized by finding the maximum of its negative. Reversed inequalities subtract their slack variables when they are converted into equations.

Both reverse inequalities and equalities use "artificial variables" in their solution. Artificial variables must eventually disappear since they have no real-world interpretation. This is accomplished by minimizing the sum of all the artificial variables in a problem; if the original problem has a feasible solution, the minimum will be zero, and after it is found, the artificial variables are dropped from the problem. (By contrast, the slack variables can disappear and then reappear as basic variables in a later tableau.)

The process of minimizing the sum of the artificial variables is called phase 1 of a linear programming problem. It is followed by phase 2, which is maximizing (or minimizing) the original objective function; to do this we must write the original objective function in terms of the nonbasic variables in the last tableau of phase 1.

EXERCISES 6.3. A

1. Maximize $P = 5x + 8y + 16z$ subject to $x + y + z = 1$ and $x + 2y + 6z \leq 3$. (See Example 6.3.2 for an economic interpretation and hints on the solution.)

2. Maximize $P = 4x + 5y + 3z$ subject to $x + y \geq 6$ and $500x + 800y + 1000z \leq 4000$. (Patterned on Example 6.3.1.) Remember that if the largest number in the last row appears twice, it is permissible to use either column as the pivot column.
3. Minimize $C = 5x + 6y + 9z$ subject to $300x + 300y + 300z \geq 900$ and $2000x + 4000y + 8000z \geq 12{,}000$. (*Hint*: You may want to refer to Example 6.2.3 during phase 2.)

EXERCISES 6.3. B

1. Maximize $P = 5x + 4y + 3z$ when $400x + 500y + 1000z \leq 3000$ and $x + y \geq 5$. (See Example 6.3.1 for an economic interpretation and hints on the solution.)
2. Maximize $P = x + 2y + 3z$ subject to $x + y + z = 1$ and $x + 2y + 9z \leq 5$. (Patterned on Example 6.3.2.)
3. Minimize $C = 5x + 6y + 4z$ when $200x + 200y + 200z \geq 600$ and $1000x + 2000y + 4000z \geq 6000$.

ANSWERS 6.3. A

1. $P_{max} = 10$ when $x = 0$, $y = \frac{3}{4}$, and $z = \frac{1}{4}$
2. $P_{max} = 32$ when $x = 8$, $y = 0$, $z = 0$
3. $C_{min} = 18$ when $x = 0$, $y = 3$, and $z = 0$

*6.4 A Model of Cleaning a River at Minimum Cost

Linear programming can be used to solve a variety of problems in the real world. This section presents a realistic model of minimizing the costs of cleaning a stream into which a town and a pulp mill are discharging wastes. The numbers are based on those empirically found at the Willamette River in Oregon. This simplified model was devised by the author and her husband, Frederick Chichester, who presented the results as part of a paper at the Joint Automatic Control Conference in July, 1976. As you can see, it might easily be regarded as an example in ecology or business.

Suppose that both a town and a nearby pulp mill are discharging wastes into the given stream and both have a treatment plant at the place where the discharging is done. The goal is to minimize the total cost of running both treatment plants while maintaining a specified level of cleanliness at a certain location (called the critical reach) downstream. Since both the positions of the plants and their costs of operation are different, even this two-part problem has several complexities. If more towns and factories are introduced into the model, it quickly becomes more complicated, but the ideas would be similar.

Since the goal is to minimize total cost, C, we observe that

$$C = (\text{cost for town}) + (\text{cost for pulp mill})$$

There are three constraints in this LP problem, one resulting from physical considerations and the other two arising directly out of the mathematical expression of the costs.

The first constraint describes how much waste must be removed from the stream. Let w_1 and w_2 designate, respectively, the amounts of waste (measured in units of 1000 pounds) that are generated each day by the town and the pulp mill (Figure 6.4–1). Then there are

Figure 6.4–1

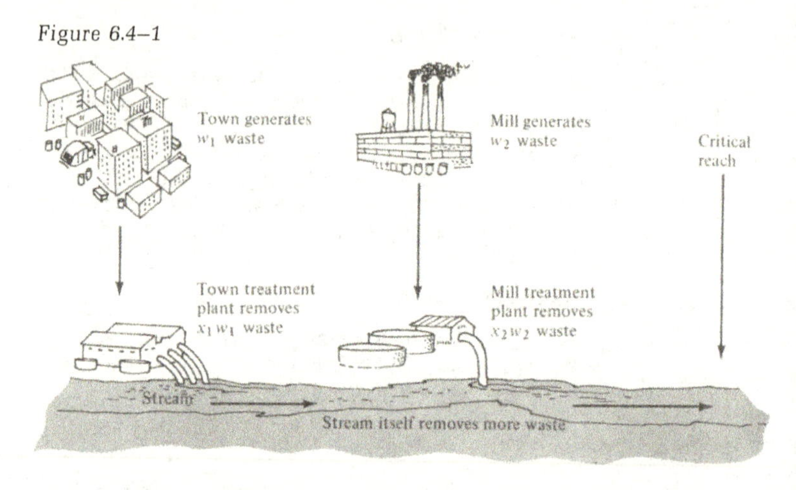

constants n_1 and n_2 determined by nature (and measured by engineers) such that n_1w_1 and n_2w_2 are the quantities of waste that will arrive at the critical reach downstream if there is no treatment (except, of course, the natural processes of the stream). Clearly, $0 \leq n_1 \leq 1$ and $0 \leq n_2 \leq 1$. If, in addition, treatment plants are installed at both the town and the mill, and x_1 is the fraction of waste that is removed by the town's plant and x_2 is the fraction of pulp mill waste that is removed by the mill's treatment plant, then the total amount removed by both plants and the natural processes of the stream will be

$$n_1x_1w_1 + n_2x_2w_2 = R$$

In this expression n_1 and n_2 are constants fixed by nature, w_1 and w_2 are constants fixed by needs of the townspeople and the pulp mill, respectively, and R is a constant determined by a subjective evaluation of how much waste is permissible downstream. The values of x_1 and x_2 are to be determined in such a way as to minimize the cost.

The other two constraints involve the differing costs of removing town and pulp mill waste. The rising cost in both cases is not precisely proportional to the rising percentages of waste removed. This means that x_1 and x_2 cannot be the variables appearing in an LP problem, because the cost is not linear in them.

Figure 6.4–2

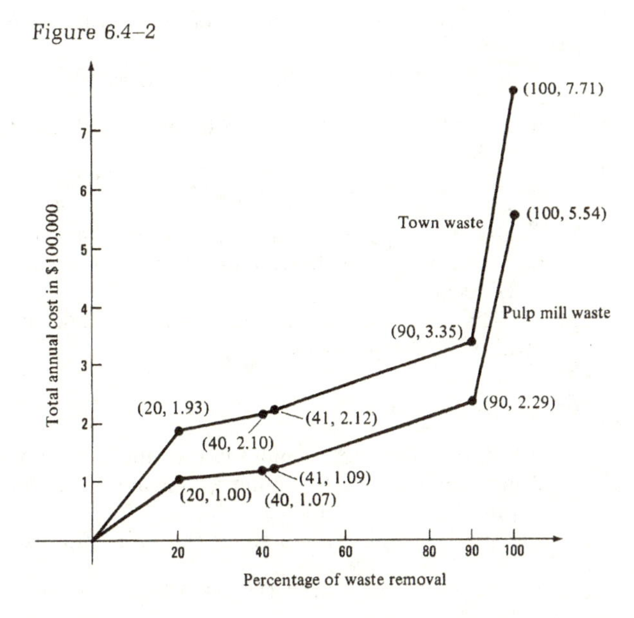

Although the specific numbers differ for the town and the pulp mill (and for other types of industry), there will generally be four breakpoints on the graph—at 20, 40, 41, and 90 percent, where the rate of costs changes (Figure 6.4–2). These breakpoints determine the actual independent variables that will be used in the computations; the intuitive independent variables—the fractions to be removed by each treatment plant—will be written in terms of the breakpoints.

In this presentation we shall assume that each treatment plant will be expected to remove between 40 and 90 percent of the waste that is given to it. This simplifies the computations, but it is also a realistic assumption from an engineering standpoint. With this assumption, we write

$$^{\star}x_1 = 0.4y_{11} + 0.41y_{12} + 0.9y_{13}$$

The three y's are the three new variables that together determine x_1. They are the proportions of each of the fractions 0.4, 0.41, and 0.9 that appear in x_1. For example,

If $x_1 = 0.4$: $y_{11} = 1$, $y_{12} = 0$, and $y_{13} = 0$

If $x_1 = 0.41$: $y_{11} = 0$, $y_{12} = 1$, and $y_{13} = 0$

If $x_1 = 0.9$: $y_{11} = 0$, $y_{12} = 0$, and $y_{13} = 1$

If $x_1 = 0.405$ (that is, halfway between 0.4 and 0.41), then

$$y_{11} = 0.5, \quad y_{12} = 0.5, \quad \text{and} \quad y_{13} = 0$$

If $x_1 = 0.655$ (that is, halfway between 0.41 and 0.9), then

$$y_{11} = 0, \quad y_{12} = 0.5, \quad \text{and} \quad y_{13} = 0.5$$

If x_1 is 30 percent of the way between 0.41 and 0.9, then

$$y_{11} = 0, \quad y_{12} = 0.7, \quad \text{and} \quad y_{13} = 0.3$$

Similarly, if x_1 is any number between 0.4 and 0.41, then $y_{13} = 0$ and if x_1 is any number between 0.41 and 0.9, then $y_{11} = 0$. (See problem 1 in Exercises 6.4.C.)

The sum of these proportions must be 1, so we have another constraint for the problem:

$$y_{11} + y_{12} + y_{13} = 1$$

It will turn out that at most two of these three y's can be nonzero, but this is implicit in the other constraints and need not be expressed explicitly.

Similar observations about the pulp mill lead to setting

$$^\star x_2 = 0.4y_{21} + 0.41y_{22} + 0.9y_{23}$$

where the third constraint is

$$y_{21} + y_{22} + y_{23} = 1$$

Substituting the starred values for x_1 and x_2 back in the first constraint (that is, $n_1 x_1 w_1 + n_2 x_2 w_2 = R$, derived on page 226), our three constraints can be written:

$$\begin{aligned}
n_1(0.4y_{11} + 0.41y_{12} + 0.9y_{13})w_1 + n_2(0.4y_{21} + 0.41y_{22} + 0.9y_{23})w_2 &= R \\
y_{11} + \quad y_{12} + \quad y_{13} \qquad\qquad\qquad\qquad\qquad\qquad &= 1 \\
y_{21} + \quad y_{22} + \quad y_{23} &= 1
\end{aligned}$$

Now we are ready to set up the first tableau in phase 1 of our simplex algorithm solution. Since all three constraints are equalities, we need three artificial variables a_1, a_2, and a_3; the first step in solving the problem will be to minimize $(a_1 + a_2 + a_3)$ in phase 1. The tableau below results when we substitute $n_1 = 0.416$, $n_2 = 0.98$, $w_1 = 40$, and $w_2 = 6.9$ into the first constraint given above. These numbers are taken from the paper "An Economic Approach to Water Quality Control" by Kenneth D. Kerri, which appeared in the *Journal of the Water Pollution Control Federation*, Vol. 38, No. 12 (Dec., 1966), pp. 1883–1897.

We let $R = 11.7$. This means that we want to remove a total of $11.7(1000) = 11{,}700$ lb of waste.

First tableau (in Phase 1)

	y_{11}	y_{12}	y_{13}	y_{21}	y_{22}	y_{23}	a_1	a_2	a_3	Solution
a_1	6.66	6.82	(15)	2.70	2.77	6.09	1	0	0	11.7
a_2	1	1	1	0	0	0	0	1	0	1
a_3	0	0	0	1	1	1	0	0	1	1
$-A$	0	0	0	0	0	0	−1	−1	−1	0
$-A$	7.66	7.82	16	3.70	3.77	7.09	0	0	0	13.7

Second tableau (in Phase 1)

	y_{11}	y_{12}	y_{13}	y_{21}	y_{22}	y_{23}	a_1	a_2	a_3	Solution
y_{13}	0.444	0.455	1	0.18	0.185	0.406	0.067	0	0	0.78
a_2	0.556	0.545	0	−0.18	−0.185	−0.406	−0.067	1	0	0.22
a_3	0	0	0	(1)	1	1	0	0	1	1
$-A$	0.556	0.545	0	0.82	0.815	0.594	−1.067	0	0	1.22

Third tableau (in Phase 1)

	y_{11}	y_{12}	y_{13}	y_{21}	y_{22}	y_{23}	a_1	a_2	a_3	Solution
y_{13}	0.444	0.455	1	0	0.005	0.226	0.067	0	−0.18	0.60
a_2	(0.556)	0.545	0	0	−0.005	−0.226	−0.067	1	0.18	0.40
y_{21}	0	0	0	1	1	1	0	0	1	1
$-A$	0.556	0.545	0	0	−0.005	−0.226	−1.067	0	−0.82	0.40

Fourth tableau (in Phase 1)

	y_{11}	y_{12}	y_{13}	y_{21}	y_{22}	y_{23}	a_1	a_2	a_3	Solution
y_{13}	0	0.020	1	0	0.009	0.406				0.281
y_{11}	1	0.980	0	0	−0.009	−0.406				0.719
y_{21}	0	0	0	1	1	1				1
$-A$	0	0	0	0	0	0	−1	−1	−1	0

The a_i columns in the inner rectangle have been omitted because they will now be dropped, and their computation is purposeless tedium. We now enter phase 2, by introducing the objective equation of real interest. We use the costs graphed in Figure 6.4–2.

$$C = \underbrace{2.10y_{11} + 2.12y_{12} + 3.35y_{13}}_{\text{(cost for town)}} + \underbrace{1.07y_{21} + 1.09y_{22} + 2.29y_{23}}_{\text{(cost for pulp mill)}}$$

Since there were three artificial variables, there are now three basic variables (y_{13}, y_{11}, and y_{21}) that must be eliminated from the objective function. Since the objective function is to be minimized, we write its negative—which is to be maximized—below the inner rectangle of the final tableau of phase 1 and *add* to it the appropriate multiples of all three rows. We write the numbers by which we are multiplying each row to the left of the inner rectangle. The coefficients in the objective equation written in terms of y_{12}, y_{22}, and y_{23} appear in the bottom row; the resulting array can now be considered the first tableau of phase 2 (although it has an extra column on the left and an extra row second from the bottom, which we shall ignore).

First tableau (in Phase 2)

	y_{11}	y_{12}	y_{13}	y_{21}	y_{22}	y_{23}	Solution
3.35 y_{13}	0	0.020	1	0	0.009	0.406	0.281
2.10 y_{11}	1	(0.980)	0	0	−0.009	−0.406	0.719
1.07 y_{21}	0	0	0	1	1	1	1
Old −C	−2.10	−2.12	−3.35	−1.07	−1.09	−2.29	
Useful −C	0	0.01	0	0	−0.08	−0.71	3.52

Second tableau (in Phase 2)

	y_{11}	y_{12}	y_{13}	y_{21}	y_{22}	y_{23}	Solution
y_{13}	−0.02	0	1	0	0.009	0.414	0.266
y_{12}	1.02	1	0	0	−0.009	−0.414	0.734
y_{21}	0	0	0	1	1	1	1
−C	−0.01	0	0	0	−0.08	−0.71	3.51

Thus the minimum cost of 3.51(100,000) = \$351,000 occurs when $y_{12} = 0.735$, $y_{13} = 0.265$, $y_{21} = 1$, and $y_{11} = y_{22} = y_{23} = 0$. This means that

$$x_1 = 0.735(0.41) + 0.265(0.90) = 0.54 \qquad \text{and} \qquad x_2 = 1(0.40) = 0.4$$

For a minimum total cost, therefore, the town's treatment plant should remove 54 percent of the town's waste and the pulp mill's treatment plant should remove 40 percent of the mill's waste.

SUMMARY

This section applies the simplex algorithm to the practical problem of cleaning up a river at minimum total cost. Actual numbers are taken from a study in Oregon, although the problem is simplified by assuming that only one small town and one pulp mill lie in the region. Another simplification is made by the assumption that both treatment plants will remove between 40 and 90 percent of the waste dumped into the stream at that point. The independent variables used to solve the problem arise from the nonlinearities in the cost function for each treatment plant; there are breakpoints on the graph at 20, 40, 41, and

90 percent. Since we assume that the treatment plants will remove between 40 and 90 pèrcent, we need consider only those costs at 40, 41, and 90 percent.

EXERCISES 6.4. A

1. Use the notation of this section.
   (a) If $x_1 = 0.403$, find y_{11}, y_{12}, and y_{13}.
   (b) If $x_1 = 0.408$, find y_{11}, y_{12}, and y_{13}.
   (c) If $y_{11} = 0$, $y_{12} = 0.2$, and $y_{13} = 0.8$, find x_1.
2. Suppose that we had decided to assume that each treatment plant would be permitted to remove between 20 and 90 percent of the waste at its source, depending on what results in a minimum total cost. (a) How many y's will there be now for each x? (b) Write x_1 in terms of y's. (c) How many constraints will the problem have now?
3. Suppose that there are two towns and a pulp mill on the river, and each of the three treatment plants will be permitted to remove between 40 and 90 percent of its waste. (a) How many y's will there be now? (b) How many constraints?
4. Suppose that each treatment plant might remove any amount of waste, from 0 to 100 percent. (a) How many y's will there be for each x? (b) Write x_1 in terms of y's.
5. Suppose that $R = 10.5$; that is, we want to remove 10,500 lb of waste from the stream. Suppose also that the conditions of this section apply except that each treatment plant must remove between 41 and 90 percent of the waste it receives. (a) Write the three constraints arising from these conditions. (b) Write the objective equation for this problem if the goal is to minimize the total cost. (c) Write the first tableau for phase 1 of the solution to this problem.
6. Solve the problem that was set up in problem 5. (*Warning*: This exercise is only for masochists and those in possession of a calculator.)

EXERCISES 6.4. B

1. Use the notation of this section.
   (a) If $x_1 = 0.404$, find y_{11}, y_{12}, and y_{13}.
   (b) If $x_1 = 0.409$, find y_{11}, y_{12}, and y_{13}.
   (c) If $y_{11} = 0$, $y_{12} = 0.3$, and $y_{13} = 0.7$, find x_1.
2. Suppose that we had decided to assume that each treatment plant would be permitted to remove from 40 to 100 percent of the waste discharge at its source, depending on what results in a minimum total cost. (a) How many y's will be needed for each x? (b) Write x_1 in terms of y's. (c) How many constraints will the problem have now?
3. Suppose that there are two towns and two pulp mills on the river, and each of the four treatment plants will be permitted to remove between 40 and 90 percent of its waste. (a) How many y's all told will there be now? (b) How many constraints?
4. Suppose that each treatment plant might remove any amount of waste from 20 to 100 percent. (a) How many y's for each x? (b) Write x_1 in terms of y's.

5. Suppose that the conditions of this section continue except that each treatment plant must remove between 41 and 90 percent of the waste it receives. Suppose also that 12,600 lb of waste is to be removed at the critical reach; that is, $R = 12.6$.
   (a) Write the three constraints arising from these conditions.
   (b) Write the objective equation for this problem if the goal is to minimize the total cost.
   (c) Write the first tableau for phase 1 of the solution to this problem.
6. Solve the problem that was set up in problem 5. (*Hint:* Although the rest of this text opts for easy arithmetic in preference to realistic numbers, this section takes its numbers from the real world. Today's real world involves the use of hand calculators.)

EXERCISES 6.4. C

Show that for any number x_1 between 0.41 and 0.90 there is indeed a y_{12} and a y_{13} such that $x_1 = 0.41y_{12} + 0.9y_{13}$.

ANSWERS 6.4. A

1. (a) $y_{11} = 0.7;\ y_{12} = 0.3;\ y_{13} = 0$ (b) $y_{11} = 0.2;\ y_{12} = 0.8;\ y_{13} = 0$
   (c) $x_1 = 0.802$
2. (a) 4 (b) $x_1 = 0.2y_{11} + 0.4y_{12} + 0.41y_{13} + 0.9y_{14}$ (c) 3
3. (a) 9 (b) 4
4. (a) 6 (b) $x_1 = 0y_{11} + 0.2y_{12} + 0.4y_{13} + 0.41y_{14} + 0.9y_{15} + y_{16}$
5. (a)
$$\begin{aligned} 6.82y_{12} + 15y_{13} + 2.77y_{22} + 6.09y_{23} &= 10.5 \\ y_{12} + y_{13} &= 1 \\ y_{22} + y_{23} &= 1 \end{aligned}$$
   (b) $C = 2.12y_{12} + 3.35y_{13} + 1.09y_{22} + 2.29y_{23}$
   (c)

	y_{12}	y_{13}	y_{22}	y_{23}	a_1	a_2	a_3	Solution
a_1	6.82	15	2.77	6.09	1	0	0	10.5
a_2	1	1	0	0	0	1	0	1
a_3	0	0	1	1	0	0	1	1
$-A$	7.82	16	3.77	7.09	0	0	0	12.5

6. 3.34(100,000) or $334,000

ANSWER 6.4. C

Since for any x_1 between 41 and 90, $x_1 = 41\left(\frac{90 - x_1}{49}\right) + 90\left(\frac{x_1 - 41}{49}\right)$, let $y_{12} = \frac{90 - x_1}{49}$ and $y_{13} = \frac{x_1 - 41}{49}$.

VOCABULARY

algorithm, simplex algorithm, significant constraints, standard linear programming problem, slack variables, tableau, basic variables, basis, pivot, pivot column, artificial variables, surplus variables

SAMPLE TEST

Chapter 6

1. (55 pts) Use the simplex method to maximize $P = 2x + 2y + 3z$ when $x + 4y + 2z \leq 12$, $4.5x + 6y + 3z \leq 24$ and $x + 2y + 2z \leq 16$.
2. (5 pts) Suppose that the simplex algorithm is used to maximize $P = 2x + y$ when $x \leq 6$ and $x + y \leq 9$. Will the tableaux march around the feasible region clockwise or counterclockwise?
3. (5 pts) How can the simplex algorithm as presented in Sections 6.1 and 6.2 be adapted to solve a minimization problem?
4. Suppose that a linear programming problem has the following constraints: $4x + 2y + z \leq 12$, $2x + 3y = 8$, $y + 4z \geq 6$, and $x + z = 3$. (a) (5 pts) How many slack variables are needed? (Include surplus variables.) (b) (5 pts) How many artificial variables are needed? (c) (15 pts) Set up the first tableau for phase 1 of the solution to this problem. Do *not* solve.
5. (10 pts) Suppose that the following is the last tableau of phase 1 of a linear programming problem and the goal is to minimize $C = 2x_1 + 3x_2 + x_3 + 2x_4$. Set up the first tableau in phase 2. Do *not* solve.

	x_1	x_2	x_3	x_4	a_1	a_2	a_3	Solution
x_2	0	1	0	0.4	0.1	1.2	−0.1	0.2
x_1	1	0	0	−0.4	0.1	1.1	0.3	0.8
x_3	0	0	1	1	0	0	1	1
$-A$	0	0	0	0	−1	−1	−1	0

The answers are at the back of the book.

*Chapter 7

DUAL PROBLEMS

7.1 Definition of Dual Problems and Economic Interpretation

The answers to both Examples 5.3.3 and 5.4.2 were 864. Curious! This is a result of their being "dual problems" to each other. We restate them here for your convenience, written with subscript notation.

Example 5.3.3	*Example 5.4.2*
Maximize	Minimize
$P = 48x_1 + 70x_2 + 96x_3$	$C = 200u_1 + 300u_2 + 360u_3$
when $x_1 \geq 0,\ x_2 \geq 0,\ x_3 \geq 0$:	when $u_1 \geq 0,\ u_2 \geq 0,\ u_3 \geq 0$:
$10x_1 + 20x_2 + 20x_3 \leq 200$	$10u_1 + 20u_2 + 30u_3 \geq 48$
$20x_1 + 20x_2 + 30x_3 \leq 300$	$20u_1 + 20u_2 + 30u_3 \geq 70$
$30x_1 + 30x_2 + 40x_3 \leq 360$	$20u_1 + 30u_2 + 40u_3 \geq 96$

If you look at these examples closely, you will see that they are related. The coefficients in the objective function of one are the constants in the significant inequalities of the other. (The "significant inequalities" are those constraints which are not nonnegativity constraints.) The coefficients of one significant inequality (across) in one problem are the coefficients of one variable (down) in the other problem. Furthermore, the inequalities go in reverse directions in the two problems. If a maximization and a minimization linear programming problem are thus related to each other, we say that they are dual problems.

We can write this definition in more mathematical terms.

7.1.1 Definition

If X_1 and X_2 are two matrices (or vectors) of the same dimension, we say that $X_1 \leq X_2$ if there is a matrix (or vector) S of the same dimension such that $X_1 + S = X_2$ and all the elements of S are nonnegative.

7.1.2 Example

Prove that (a) $\begin{bmatrix} 3 \\ 2 \\ 1 \end{bmatrix} \leq \begin{bmatrix} 4 \\ 8 \\ 2 \end{bmatrix}$ and (b) $\begin{bmatrix} 0 \\ -3 \\ 1 \end{bmatrix} \leq \begin{bmatrix} 2 \\ -1 \\ 1 \end{bmatrix}$

Solution:

(a) $\begin{bmatrix} 3 \\ 2 \\ 1 \end{bmatrix} + \begin{bmatrix} 1 \\ 6 \\ 1 \end{bmatrix} = \begin{bmatrix} 4 \\ 8 \\ 2 \end{bmatrix}$ (b) $\begin{bmatrix} 0 \\ -3 \\ 1 \end{bmatrix} + \begin{bmatrix} 2 \\ 2 \\ 0 \end{bmatrix} = \begin{bmatrix} 2 \\ -1 \\ 1 \end{bmatrix}$

Since $\begin{bmatrix} 1 \\ 6 \\ 1 \end{bmatrix}$ and $\begin{bmatrix} 2 \\ 2 \\ 0 \end{bmatrix}$ contain only nonnegative numbers, the inequalities have been proved using Definition 7.1.1.

7.1.3 Definition

Any two problems of the following form are called dual problems:

$$\begin{array}{ll} \text{Maximize } P = BX \text{ when} & \text{Minimize } C = UD \text{ when} \\ AX \le D & UA \ge B \\ \text{and } X \ge 0 & \text{and } U \ge 0 \end{array}$$

(Either of these can be called the "dual" of the other.)

You will notice that the linear programming problem on the left in Definition 7.1.3 is in standard form; Example 5.3.3 was in standard form. You can verify that Examples 5.3.3 and 5.4.2 satisfy Definition 7.1.3 if you substitute in the values:

$$X = \begin{bmatrix} x_1 \\ x_2 \\ x_3 \end{bmatrix} \quad A = \begin{bmatrix} 10 & 20 & 20 \\ 20 & 20 & 30 \\ 30 & 30 & 40 \end{bmatrix} \quad D = \begin{bmatrix} 200 \\ 300 \\ 360 \end{bmatrix}$$

$$B = [48, 70, 96] \qquad U = [u_1, u_2, u_3]$$

The concept of dual linear programming problems is useful because of the following rule.

7.1.4 Rule

If a maximization linear programming problem and a minimization linear programming problem are dual to each other, and either one has an optimal feasible solution, then they both do and $P_{max} = C_{min}$.

7.1.5 Example

Find the dual of the problem given in Example 5.1.5 and check that the optimal values of both objective functions are the same. (Again we use subscript notation because it helps to describe the ideas of this chapter.)

Solution:

This is the problem about the wholesaler who has 3000 lb of wheat and 4000 lb of barley from which she makes Premium and Regular feed bags which bring 30 cents and 20 cents profit, respectively. Thus we want to maximize $P = \mathbf{30x_1} + 20x_2$ when

$$\mathbf{x_1} + x_2 \leq 3000 \qquad x_1 \geq 0$$
$$\mathbf{2x_1} + x_2 \leq 4000 \qquad x_2 \geq 0$$

Each significant inequality in the original problem gives rise to a variable in the dual, so there will be two variables in the dual problem; call them u_1 and u_2. Since the original problem asks for a maximum, the dual problem will ask for a minimum. The coefficients in the objective equation and the constants in the significant inequalities change places. And the coefficients of one variable in the original problem become the coefficients in one constraint in the dual. Thus the dual problem is:

$$\text{Minimize } C = 3000u_1 + 4000u_2 \text{ when}$$
$$\mathbf{u_1} \quad \mathbf{2u_2} \quad \mathbf{30} \qquad u_1 \geq 0$$
$$u_1 + u_2 \geq 20 \qquad u_2 \geq 0$$

Since $P_{max} = 70{,}000$ when Example 5.1.5 is solved in Examples 5.2.2 and 6.2.1, we should get $C_{min} = 70{,}000$ in the dual problem, according to Rule 7.1.4. And we do!

Vertex	$C = 3000u_1 + 4000u_2$
(0, 20)	80,000
(10, 10)	70,000 ←— minimum
(30, 0)	90,000

Figure 7.1–1

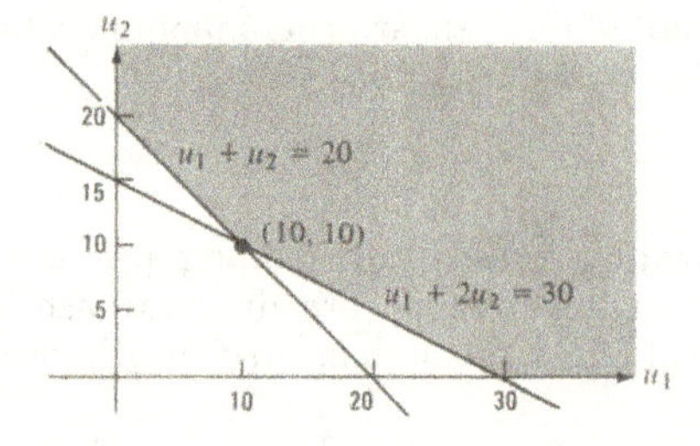

The point (10, 10) (Figure 7.1–1) has economic significance. The 10's tell the maximum amounts that the wholesaler would be willing to pay for extra wheat and barley, respectively. If she could buy each for less than 10 cents, she could make another regular bag, sell it for the customary 20 cents, and make a profit. If she can buy just one for less that 10 cents, she might change her original plans; for example, if only wheat costs less than 10 cents for an extra pound, she could make fewer Premium bags and use the extra barley they supply to make more Regular bags and thus more profit.

The 10's are often called "shadow prices" and are also seen on the last line of the third tableau in Example 6.2.1. They tell how much the wholesaler would lose per unit if s_1 or s_2, respectively, were introduced to the basis—in other words, how much would be lost, per unit, if the available wheat or barley were not completely used. (It is not necessary to understand such an analysis to compute using dual concepts.)

The number of variables is not always the same in an LP problem and its dual. If, for example, Premium and Regular bags in the previous problem also contained oats and there were an "oats constraint" in the original problem, there would be an "oats variable" in the dual. The number of significant constraints in the original problem will be the same as the number of variables in the dual, and vice versa.

7.1.6 Example

State the dual of the problem given in Example 5.3.2 and give the optimum value of C_{min} in the dual.

Solution:

Since there is a "tube-bending machine constraint" and a "webbing machine constraint" in the original problem, there will be two analogous variables in the dual; they will indicate how much extra time on each of the two machines would be worth.

We restate the original problem here for your convenience.

Example 5.3.2

Maximize $P = 2x_1 + 2.5x_2 + 3x_3$

subject to
$$x_1 + x_2 + x_3 \le$$
$$2x_1 + 2x_2 + 3x_3 \le 48$$
$$x_1 \ge 0,\ x_2 \ge 0,\ x_3 \ge 0$$

Dual

Minimize $C = u_1 + 48u_2$

subject to
$$u_1 + 2u_2 \ge 2$$
$$u_1 + 2u_2 \ge 2.5$$
$$u_1 + 3u_2 \ge 3$$
$$u_1 \ge 0,\ u_2 \ge 0$$

The solution of the original problem is $P_{max} = 54$, so the solution of the dual problem is $C_{min} = 54$.

In the next section we will find the values of u_1 and u_2 which yield $C_{min} = 54$, and also develop a general technique for finding the value of the independent variables in a dual problem that yields the optimum value without using the graphical, tabular, or simplex methods. First we look more closely at the economic interpretation of the examples that began this section.

Example 5.3.3 concerned maximizing the profit of processing midget, subcompact, and compact cars if they all required spending time in three departments whose facilities are limited. The solution was to process 9 compact cars and none of the others, so that

$$P_{max} = 48(0) + 70(0) + 96(9) = 864$$

If you substitute (0, 0, 9) into the three constraints (given again on page 234), you will notice that only department III is fully used; the other two have time available that is not used.

Now let us turn to the dual (minimization) problem. The variables can be interpreted as imputed costs of using the departments. These are the "shadow prices" and can be interpreted as the amounts the company would be willing to pay (marginally) to use each department more. Since the first two departments are not fully employed in the given solution, their imputed costs (or shadow prices) are 0; the company will not pay anything for the privilege of using them more. But the third department is in full use for the 360 minutes available, and from this the company profits $864. Thus the imputed cost per minute is 864/360 = $2.40. This coincides with the result of Example 5.4.2 which says that $u_1 = 0$, $u_2 = 0$, and $u_3 = 2.4$.

SUMMARY

The dual of a maximization linear programming problem is a minimization linear programming problem, and vice versa; dual problems come in pairs. The maximum value of the objective equation in a maximization problem is the same as the minimum value of the objective equation in the dual minimum problem.

The coefficients in the objective equation are the constants in the dual problem. The coefficients of all the variables in one significant inequality are the same as the coefficients of one variable in all the significant inequalities of the dual problem. In this manner each independent variable in a linear programming problem corresponds to one of the significant inequalities in the dual problem. In the next section we shall use this correspondence to solve for the values of the independent variables that optimize the objective function in a dual problem where the complete solution of the original problem is known.

EXERCISES 7.1. A

Write the duals to the problems given in Exercises 5.3.A and 5.4.A. Then write the solution to each dual problem. Refer to the hints and answers beginning on page 180 (only) to find the essential mathematical aspects of the problems.

EXERCISES 7.1. B

Follow the instructions in Exercises 7.1.A for Exercises 5.3.B and 5.4.B. Save your answers, which will be used for Exercises 7.2.B.

ANSWERS 7.1. A

Exercises 5.3. A

	Original problems	*Dual problems*
1.	Maximize $P = 15x_1 + 15x_2$	Minimize $C = 20u_1 + 22u_2$
	when $2x_1 + 5x_2 \leq 20$	when $2u_1 + 2u_2 \geq 15$
	$2x_1 + x_2 \leq 22$	$5u_1 + u_2 \geq 15$
	Since $P_{max} = 150$ (when $x_1 = 10$ and $x_2 = 0$), $C_{min} = 150$.	
2.	Maximize $P = 4x_1 + 2x_2 + x_3$	Minimize $C = 8u_1 + 10u_2$
	when $x_1 + x_2 + x_3 \leq 8$	when $u_1 + 2u_2 \geq 4$
	$2x_1 + x_2 \leq 10$	$u_1 + u_2 \geq 2$
		$u_1 \geq 1$
	Since $P_{max} = 23$ (when $x_1 = 5$, $x_2 = 0$, and $x_3 = 3$), $C_{min} = 23$.	
3.	Maximize $P = 60x_1 + 60x_2 + 40x_3$	Minimize $C = 18u_1 + 12u_2 + 12u_3$
	when $3x_1 + 3x_2 \leq 18$	when $3u_1 + 2u_2 \geq 60$
	$2x_1 + x_2 + x_3 \leq 12$	$3u_1 + u_2 + 2u_3 \geq 60$
	$2x_2 + 2x_3 \leq 12$	$u_2 + 2u_3 \geq 40$
	Since $P_{max} = 480$ (when $x_1 = x_2 = x_3 = 3$), $C_{min} = 480$.	
4.	Maximize $P = x_1 + x_2 + x_3 + x_4$	Minimize $C = 6u_1 + 12u_2$
	when $x_1 + x_2 + 2x_4 \leq 6$	when $u_1 + 2u_2 \geq 1$
	$2x_1 + 3x_3 + x_4 \leq 12$	$u_1 \geq 1$
		$3u_2 \geq 1$
		$2u_1 + u_2 \geq 1$
	Since $P_{max} = 10$ (when $x_1 = x_4 = 0$, $x_2 = 6$, $x_3 = 4$), $C_{min} = 10$.	

Exercises 5.4. A

	Original problems	*Dual problems*
1.	Minimize $C = 40x_1 + 18x_2$	Maximize $P = 80u_1 + 2u_2$
	when $32x_1 + 16x_2 \geq 80$	when $32u_1 + 0.5u_2 \leq 40$
	$0.5x_1 + x_2 \geq 2$	$16u_1 + u_2 \leq 18$
	Since $C_{min} = 90$ (when $x_1 = 0$, $x_2 = 5$), $P_{max} = 90$.	
2.	Minimize $C = 5x_1 + 10x_2$	Maximize $P = 15u_1 + 20u_2$
	when $x_1 + 5x_2 \geq 15$	when $u_1 + 2u_2 \leq 5$
	$2x_1 + 5x_2 \geq 20$	$5u_1 + 5u_2 \leq 10$
	Since $C_{min} = 40$ (when $x_1 = 0$, $x_2 = 4$), $P_{max} = 40$.	
3.	Minimize $C = 5x_1 + 3x_2 + 2x_3$	Maximize $P = 12u_1 + 48u_2$
	when $x_1 + 2x_2 + 3x_3 \geq 12$	when $u_1 + 8u_2 \leq 5$
	$8x_1 + 4x_2 \geq 48$	$2u_1 + 4u_2 \leq 3$
		$3u_1 \leq 2$
	Since $C_{min} = 32$ (when $x_1 = x_2 = 4$, $x_3 = 0$), $P_{max} = 32$.	
4.	Minimize $C = 4x_1 + 3x_2 + 3x_3$	Maximize $P = 60u_1 + 30u_2$
	when $x_1 + 2x_2 + 3x_3 \geq 60$	when $u_1 + 2u_2 \leq 4$
	$2x_1 + x_2 + x_3 \geq 30$	$2u_1 + u_2 \leq 3$
		$3u_1 + u_2 \leq 3$
	Since $C_{min} = 78$ (when $x_1 = 6$, $x_2 = 0$, $x_3 = 18$), $P_{max} = 78$.	

7.2 Solving for the Independent Variables in a Dual Problem

Finding C_{min} for the dual problems of Section 7.1 is relatively easy as long as you are willing to accept Rule 7.1.4 without proof. But a bare minimum of mathematical curiosity prompts the question of which values for the independent variables in the dual problem give this minimum value for C. This section answers this question and the next section includes the proof of why the technique given here works. Proofs, admittedly, are icing on the cake of this course, and if you prefer plain cake, you can omit the icing without losing continuity.

We refer again to Example 7.1.6, which we restate here with one modification as Example 7.2.1. The modification involves converting inequalities to equations by using slack variables. As in Chapter 6, a "slack variable" takes up the "slack" in an inequality and makes it an equation. For example, $2x_1 + 3x_2 + 4x_3 \leq 60$ becomes

$$2x_1 + 3x_2 + 4x_3 + r_1 = 60 \qquad \text{for} \qquad r_1 \geq 0$$

where r_1 is the slack variable. The two expressions are equivalent, and since all variables in linear programming problems are assumed to be nonnegative, we need not even take the extra effort to write $r_1 \geq 0$.

If there is a "$\geq$" inequality, a minus sign is used. For example, $2u_1 + 2u_2 \geq 2$ becomes $2u_1 + 2u_2 - s_1 = 2$ for $s_1 \geq 0$. Again it is not necessary to mention that $s_1 \geq 0$, because all variables in a linear programming problem are assumed to be nonnegative. Thus Example 7.1.6 can be written as follows.

7.2.1 Example

Original problem

Maximize $P = 2x_1 + 2.5x_2 + 3x_3$

when $2x_1 + 3x_2 + 4x_3 + r_1 = 60$

$2x_1 + 2x_2 + 3x_3 + r_2 = 48$

Dual problem

Minimize $C = 60u_1 + 48u_2$

when $2u_1 + 2u_2 - s_1 = 2$

$3u_1 + 2u_2 - s_2 = 2.5$

$4u_1 + 3u_2 - s_3 = 3$

Notice that each significant equality has exactly one slack variable. Since each significant constraint in the original problem corresponds to one independent variable in the dual problem, and vice versa, it follows that each slack variable in the original problem corresponds to one independent variable in the dual problem, and vice versa.

In Example 7.2.1 the first constraint in the original problem, containing r_1, corresponds to the independent variable u_1 in the dual problem, so r_1 corresponds to u_1. Similarly, r_2 corresponds to u_2.

Conversely, the independent variable x_1 in the original problem gives rise to the first constraint, containing s_1, in the dual problem, so x_1 corresponds to s_1. Similarly, x_2 corresponds to s_2 and x_3 corresponds to s_3. These correspondences are used in the following rule, which enables us to compute the values of the independent variables in the dual problem that give the optimal value of the objective equation.

7.2.2 Rule

If a linear programming problem and its dual have both been optimized, the product of each independent variable in the original problem and its corresponding slack variable in the dual problem is zero. Similarly, the product of each slack variable in the original problem and its corresponding independent variable in the dual problem is zero.

In Example 7.2.1, Rule 7.2.2 means that the values of the independent and slack variables that give $P_{max} = C_{min} = 54$ must be such that

$$r_1 u_1 = 0$$
$$r_2 u_2 = 0$$
$$x_1 s_1 = 0$$
$$x_2 s_2 = 0$$
$$x_3 s_3 = 0$$

These equations can be used to find the values of u_1 and u_2 (and therefore s_1, s_2, and s_3) which yield $C_{min} = 54$. Referring back to Example 5.3.2, we see that $P_{max} = 54$ when $x_1 = 12$, $x_2 = 12$, and $x_3 = 0$.

Since $x_1 s_1 = 0$ and $x_1 = 12$, we conclude that $s_1 = 0$.

Since $x_2 s_2 = 0$ and $x_2 = 12$, we conclude that $s_2 = 0$.

The expressions $s_1 = 0$ and $s_2 = 0$, respectively, mean that

(1) $2u_1 + 2u_2 = 2$

(2) $3u_1 + 2u_2 = 2.5$

These equations can be solved simultaneously to obtain values for u_1 and u_2. Subtracting (1) from (2), we get

$$u_1 = 0.5$$

Substituting this back in (1), we get

$$u_2 = 0.5$$

Checking in the third significant inequality, we get that

$$4(0.5) + 3(0.5) = 2 + 1.5 = 3.5 \geq 3$$

and the objective equation yields

$$C = 60(0.5) + 48(0.5) = 30 + 24 = 54$$

Thus the values $u_1 = 0.5$ and $u_2 = 0.5$ do satisfy the constraints and give C the value of 54.

By a fundamental rule of dual problem analysis, this checks the problem and shows that we have indeed found the optimal value of the objective equations in both the original problem and the dual. The proof of this rule is postponed until the next section, but we state it here so you can use it now.

7.2.3 Rule

If a maximization linear programming problem and a minimization linear programming problem are dual to each other and if feasible values of the independent variables in both problems have been found so that the two objective equations are equal, then both problems have been optimized—that is, the common value is the minimum in the minimization problem and the maximum in the maximization problem.

This means that if by any means you can find feasible values for the independent variables in both a linear programming problem and its dual that yield the same value for both objective equations, both problems have been solved. Thus dual considerations provide a quick method for checking linear programming problems.

7.2.4 Example

Verify that the solution obtained in Example 5.3.3 was correct by solving its dual problem, Example 5.4.2.

Solution:

We restate both problems here again with slack variables.

Example 5.3.3	*Example 5.4.2*
Maximize $P = 48x_1 + 70x_2 + 96x_3$	Minimize $C = 200u_1 + 300u_2 + 360u_3$
when $10x_1 + 20x_2 + 20x_3 + r_1 = 200$	when $10u_1 + 20u_2 + 30u_3 - s_1 = 48$
$20x_1 + 20x_2 + 30x_3 + r_2 = 300$	$20u_1 + 20u_2 + 30u_3 - s_2 = 70$
$30x_1 + 30x_2 + 40x_3 + r_3 = 360$	$20u_1 + 30u_2 + 40u_3 - s_3 = 96$

$P_{max} = 864$ when $x_1 = 0$, $x_2 = 0$, and $x_3 = 9$.

The products promised in Rule 7.2.2 are

$$r_1u_1 = 0 \qquad x_1s_1 = 0$$
$$r_2u_2 = 0 \qquad x_2s_2 = 0$$
$$r_3u_3 = 0 \qquad x_3s_3 = 0$$

Since $x_3 = 9$ and $x_3 s_3 = 0$, we know that $s_3 = 0$, so

$$\star 20u_1 + 30u_2 + 40u_3 = 96$$

But this is not enough to solve for all three variables.

To get more information, we substitute the values $x_1 = 0$, $x_2 = 0$, and $x_3 = 9$ back into the constraining equations of the original problem to find which slack variables are nonzero.

$$10(0) + 20(0) + 20(9) = 180 \neq 200 \quad \text{so} \quad r_1 \neq 0$$
$$20(0) + 20(0) + 30(9) = 270 \neq 300 \quad \text{so} \quad r_2 \neq 0$$
$$30(0) + 30(0) + 40(9) = 360 \quad \text{so} \quad r_3 = 0$$

Thus $r_1 \neq 0$ and $r_2 \neq 0$. Since $r_1 u_1 = 0$ and $r_2 u_2 = 0$, we conclude that $u_1 = 0$ and $u_2 = 0$. Now it is easy to solve for u_3 using the starred equation above:

$$20(0) + 30(0) + 40u_3 = 96 \quad \text{implies that} \quad u_3 = 96/40 = 2.4$$

Substituting back into the first two constraints, we have

$$10(0) + 20(0) + 30(2.4) \geq 48$$
$$20(0) + 20(0) + 30(2.4) \geq 70$$

And the objective function also checks:

$$C_{min} = 200(0) + 300(0) + 360(2.4) = 864$$

Since $P_{max} = C_{min}$, Rule 7.2.3 assures us that both problems have been correctly solved.

With practice, you can do many of the steps in Example 7.2.4 in your head. Clearly, this method of solving Example 5.4.2 is much less time-consuming than the detailed computations summarized in the table given in Chapter 5.

The economic interpretation of Rule 7.2.2 is not difficult. If we have leftover wheat, barley, tube-bending machine time, or department II time—that is, if r_i does *not* equal zero for some constraint—then we are *not* willing to pay anything for *more* of it. That is, if the slack variable is not zero, the corresponding independent variable in the dual problem is 0; the shadow price (or imputed cost) is 0.

There is an even faster way of finding the values of the independent variables that give the optimum value to the dual problem if you have solved the original problem using the simplex method. The absolute values of the numbers in the last row of the last tableau below the slack variables indicate the values of the corresponding independent variables in the optimal solution of the dual problem.

7.2.5 Example

Find the solution of the dual problem to Example 6.1.1 using the method of the previous paragraph.

Solution:

Consulting the final tableau on page 198 that was used to solve this problem, we see that the numbers in the last row under the slack variables are, respectively, 0, $-\frac{5}{6}$, and $-\frac{1}{3}$. This indicates that the values of the three independent variables in the optimal solution to the dual problem will be

$$u_1 = 0 \qquad u_2 = \tfrac{5}{6} \qquad u_3 = \tfrac{1}{3}$$

If we refer to the original statement of the problem given on page 195, we see that the objective function in the dual problem must be $C = 4u_1 + 6u_2 + 12u_3$. Substituting in the given values, we get $C_{min} = 4(0) + 6 \cdot \frac{5}{6} + 12 \cdot \frac{1}{3} = 5 + 4 = 9$, which is the same as P_{max} in the original problem. We leave it to the reader to check that $(0, \frac{5}{6}, \frac{1}{3})$ satisfies the constraints in the dual problem. Once this is done, Rule 7.2.3 shows that both problems have been correctly solved.

This example indicates a quick way of checking every linear programming problem that has been solved using the simplex method.

SUMMARY

The values of the independent variables that give the optimum value for the objective equation of a dual problem can be computed using the solution of the original problem and the fact that the product of each independent variable in the original problem and the corresponding slack variable in the dual problem is zero, and vice versa. First list the products that equal zero; if there are m independent variables and n slack variables in the original problem, there will be $m + n$ such products. Some of the original variables will have a nonzero value, so their corresponding slack variables will be zero. If this does not yield enough information to solve for the independent variables in the dual problem, substitute the values of the independent variables into the constraining inequalities and find out which slack variables in the original problem are nonzero; the corresponding independent variables in the dual problem will be zero. Substituting these zeros into the constraining equalities that were discovered because their slack variables were zero will often enable you to solve for the remaining independent variables. (The answer may not be unique, in which case the situation is called "degenerate.") This sequence of ideas is sometimes called complementary slackness.

The conclusion of the section mentions a faster way of finding the values of the independent variables for the optimal solution of the dual problem if the original problem was solved using the simplex algorithm. It uses the idea that the amount lost per unit if a slack variable in the original problem reenters the basis will be the maximum amount per unit that would be paid for more of the quantity which gives rise to that constraint.

EXERCISES 7.2. A

Use Rule 7.2.3 to check the "original" problems given in Exercises 7.1.A. That is, use Rule 7.2.2 to find the values of the independent variables in the dual problems that give C_{min}, and verify that $C_{min} = P_{max}$. Work directly from the answers on page 239.

EXERCISES 7.2. B

Use Rule 7.2.3 to check the "original" problems given in Exercises 7.1.B. That is, use Rule 7.2.2 to find the values of the independent variables in the dual problems that give C_{min}, and verify that $C_{min} = P_{max}$. Work directly from your answers to Exercises 7.1.B.

ANSWERS 7.2. A

Exercises 5.3. A

1. $u_1 = 7.5, u_2 = 0$ 2. $u_1 = 1, u_2 = 1.5$ 3. $u_1 = \frac{20}{3}, u_2 = 20, u_3 = 10$
4. $u_1 = 1, u_2 = \frac{1}{3}$

Exercises 5.4. A

1. $u_1 = \frac{9}{8}, u_2 = 0$ 2. $u_1 = 0, u_2 = 2$ 3. $u_1 = \frac{1}{3}, u_2 = \frac{7}{12}$ 4. $u_1 = \frac{2}{5}, u_2 = \frac{9}{5}$

7.3 Dual Problem Proofs

This section presents proofs of the facts about dual problems that were presented in the previous two sections. A theorem is a mathematical fact that can be proved using previously known facts. Advanced math books are usually developed around a series of theorems and their proofs; the rest of the book is discussion and examples involving these theorems. We use this format for this section since it is devoted to "pure" math. A lemma is a preliminary theorem that is used to prove a more important theorem.

Matrix notation is used to simplify the presentation. We remind the reader that any two dual linear programming problems can be expressed in matrix form as

Maximize $P = BX$ when	and	Minimize $C = UD$ when
$AX \leq D$		$UA \geq B$
and $X \geq 0$		and $U \geq 0$

All notation in this section refers to dual problems thus expressed. A is a fixed matrix and B and D are fixed vectors (row and column, respectively). X and U are nonnegative vectors (column and row, respectively) regarded as independent variables, and P and C are numbers that are dependent variables. All dimensions are such that the indicated multiplications can be performed.

In Lemma 7.3.1 we use the fact that when a vector inequality is multiplied by a vector whose elements are all positive, the inequality is maintained in the same direction. The proof uses a similar well-known fact for numbers; it can be found as the answers to problem 3 in Exercises 7.3.A and 7.3.B.

7.3.1 Lemma

If $X \geq O$ and $U \geq O$ are such that $AX \leq D$ and $UA \geq B$, then $BX \leq UD$.

Proof:

Since $AX \leq D$ and $U \geq O$, $UAX \leq UD$.
Since $UA \geq B$ and $X \geq O$, $UAX \geq BX$.
Combining the inequalities on the right we have $BX \leq UAX \leq UD$, which implies that $BX \leq UD$. The lemma is proved.

7.3.2 Theorem

If a linear programming problem and its dual have feasible vectors $X \geq O$ and $U \geq O$ such that $BX = UD$, then $P_{max} = C_{min} = BX = UD$.

Proof:

If BX were not the maximum number for all feasible X, there would be another feasible X, call it X_1, such that $BX_1 > BX = UD$. This means that $BX_1 > UD$, contradicting Lemma 7.3.1.

Similarly, if UD were not the minimum number for all feasible U, there would be a feasible U_2 such that $U_2D < UD = BX$. This means that $U_2D < BX$, again contradicting the lemma. Thus the theorem is proved; no such X_1 or U_2 can exist.

This theorem shows that if we can find feasible values of the independent variables in both a linear programming problem and its dual such that the objective functions take equal values, then both objective functions are optimal in their respective senses. Thus the techniques of the previous two sections indeed provide a relatively easy way to check linear programming problems.

7.3.3 Theorem

If $X \geq O$ and $U \geq O$ are feasible solutions of two dual problems with notation as above and $BX = UD = P_{max} = C_{min}$, and if $R \underset{def}{=} D - AX$ and

$S \underset{\text{def}}{=} UA - B$, then $SX = 0$ and $UR = 0$. (Notice that R and S are the slack variable vectors for the inequalities $AX \leq D$ and $UA \geq B$, respectively, when they are converted to equations.)

Proof:

Since $R = D - AX$, $D = R + AX$. Since $S = UA - B$, $B = UA - S$. Substituting these values of D and B into $BX = UD$, we obtain

$$(UA - S)X = U(R + AX)$$

or

$$UAX - SX = UR + UAX$$

Subtracting, we obtain $-SX = UR$.

Adding SX to both sides of this equation, we obtain $0 = UR + SX$. Since both UR and SX are nonnegative, the only situation in which they can add up to zero is when $UR = 0$ and $SX = 0$. The theorem is proved.

The significance of Theorem 7.3.3 can be better understood if we convert the notation back to that of Section 7.2. Remember that X was the vector $\begin{bmatrix} x_1 \\ x_2 \\ x_3 \end{bmatrix}$ consisting of the independent variables in a maximization LP problem. And S was the vector of slack variables $S = [s_1, s_2, s_3]$ in the dual problem: $UA - S = B$. Thus $SX = 0$ means that

$$[s_1, s_2, s_3] \begin{bmatrix} x_1 \\ x_2 \\ x_3 \end{bmatrix} = 0 \qquad \text{or} \qquad s_1x_1 + s_2x_2 + s_3x_3 = 0$$

Since all variables—both independent variables and slack variables—must be nonnegative, each product must be zero.

SUMMARY

This section provides proof of the statements in the previous section and shows how dual concepts can be used to check linear programming problems.

EXERCISES 7.3. A

1. Write out the values of A, B, and D in each of the odd-numbered problems in Exercises 7.1.A when the problems are expressed in compact notation as

   "Maximize $P = BX$ when $AX \leq D$" and "Minimize $C = UD$ when $UA \geq B$"

2. Discuss the significance of $UR = 0$ as proved in Theorem 7.3.3 in terms of the notation of Section 7.2.

3. The proof of Lemma 7.3.1 uses the fact that $AX \leq D$ implies that $UAX \leq UD$ when U is a row vector of compatible length whose entries are all positive. Suppose that

$$AX = \begin{bmatrix} g_1 \\ g_2 \\ \vdots \\ g_n \end{bmatrix} \qquad D = \begin{bmatrix} d_1 \\ d_2 \\ \vdots \\ d_n \end{bmatrix} \qquad \text{and} \qquad U = [u_1, u_2, \ldots, u_n]$$

Use this notation to prove that $AX \leq D$ implies that $UAX \leq UD$.

EXERCISES 7.3. B

1. Write out the values of A, B, and D in each of the odd-numbered problems in Exercises 7.1.B when the problems are expressed in compact notation as

"Maximize $P = BX$ when $AX \leq D$" and "Minimize $C = UD$ when $UA \geq B$"

2. Discuss the significance of $UR = 0$ as proved in Theorem 7.3.3 in terms of the notation of Section 7.2.
3. The proof of Lemma 7.3.1 uses the fact that $UA \geq B$ implies that $UAX \geq BX$ when X is a column vector of compatible length whose entries are all positive. Suppose that

$$UA = [g_1, g_2, \ldots, g_n] \qquad B = [b_1, b_2, \ldots, b_n] \qquad \text{and} \qquad X = \begin{bmatrix} x_1 \\ x_2 \\ \vdots \\ x_n \end{bmatrix}$$

Use this notation to prove that $UA \geq B$ implies that $UAX \geq BX$.

ANSWERS 7.3. A

1. $A = \begin{bmatrix} 2 & 5 \\ 2 & 1 \end{bmatrix} \qquad B = [15, 15] \qquad D = \begin{bmatrix} 20 \\ 22 \end{bmatrix}$

3. $A = \begin{bmatrix} 3 & 3 & 0 \\ 2 & 1 & 1 \\ 0 & 2 & 2 \end{bmatrix} \qquad B = [60, 60, 40] \qquad D = \begin{bmatrix} 18 \\ 12 \\ 12 \end{bmatrix}$

1. $A = \begin{bmatrix} 32 & 0.5 \\ 16 & 1 \end{bmatrix} \qquad B = [80, 2] \qquad D = \begin{bmatrix} 40 \\ 18 \end{bmatrix}$

3. $A = \begin{bmatrix} 1 & 8 \\ 2 & 4 \\ 3 & 0 \end{bmatrix} \qquad B = [12, 48] \qquad D = \begin{bmatrix} 5 \\ 3 \\ 2 \end{bmatrix}$

2. $U = [u_1, u_2, u_3]$ was the vector of independent variables in the dual problem in Section 7.2 and $R = \begin{bmatrix} r_1 \\ r_2 \\ r_3 \end{bmatrix}$ was the vector of slack variables

in the original problem. To say that $UR = 0$ is to say that

$$[u_1, u_2, u_3]\begin{bmatrix} r_1 \\ r_2 \\ r_3 \end{bmatrix} = 0 \qquad \text{or} \qquad u_1 r_1 + u_2 r_2 + u_3 r_3 = 0$$

Since all variables—both the independent variables and the slack variables—are nonnegative in a linear programming problem, all three products must be zero. This implies that either u_1 or r_1 must be zero, and so forth.

3. The vector inequality says that $g_1 \leq d_1$, $g_2 \leq d_2$, ..., $g_n \leq d_n$. Since all the u_i are nonnegative, and multiplying a numerical inequality by a nonnegative number preserves the inequality, it is true that $u_1 g_1 \leq u_1 d_1$, $u_2 g_2 \leq u_2 d_2 \cdots u_n g_n \leq u_n d_n$. Since adding numerical inequalities of the same sense preserves the sense of the inequality, this implies that

$$u_1 g_1 + u_2 g_2 + \cdots + u_n g_n \leq u_1 d_1 + u_2 d_2 + \cdots + u_n d_n$$

But this is just what we mean when we say that $UAX \leq UD$, so the proof is finished.

VOCABULARY

dual problems, shadow prices, slack variables, complementary slackness, lemma, theorem

SAMPLE TEST

Chapter 7

1. (25 pts) The linear programming problem "Maximize $P = 4x_1 + 2x_2 + 6x_3$ when $x_1 + 3x_2 + x_3 \leq 14$, $x_1 + x_3 \leq 10$, and $2x_2 + x_3 \leq 4$" has the answer $P_{max} = 48$ when $x_1 = 6$, $x_2 = 0$, and $x_3 = 4$. Write the dual of this linear programming problem and give its solution, *including* the values of the independent variables which yield the optimum value of the objective equation.
2. (25 pts) The linear programming problem "Minimize $C = 24u_1 + 20u_2 + 12u_3$ when $2u_1 + 2u_3 \geq 6$, $4u_1 + 6u_2 + 4u_3 \geq 8$, and $4u_2 + 2u_3 \geq 2$" has the answer $C_{min} = 36$ when $u_1 = u_2 = 0$, $u_3 = 3$. Write the dual of this linear programming problem and give its solution, *including* the values of the independent variables that yield the optimum value of the objective equation.
3. (10 pts) What is the dual problem to "Minimize $C = UD$ when $UA \geq B$"?
4. (15 pts) Using the notation of problem 3, what are the values in problem 1 of A, B, and D?
5. (15 pts) Write the three *statements* about dual linear programming problems that were proved in Section 7.3.
6. (10 pts) Prove one of the three statements of problem 5.

The answers are at the back of the book.

★Chapter 8

THE TRANSPORTATION PROBLEM

8.1 Northwest Corner Algorithm and Minimum Cell Algorithm

Many important linear programming problems, but not all, are included in a set of problems called the transportation problem. This set of problems is named after a typical problem which involves minimizing the total cost of transporting some product from factories where it is made to warehouses where it is stored.

8.1.1 Example

Suppose that we have 2 factories and 3 warehouses:

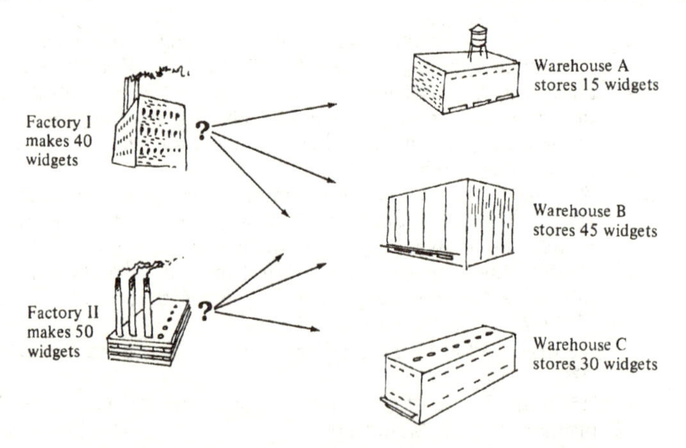

If it costs \$80 to ship one widget from factory I to warehouse A, \$75 to ship one widget from factory I to warehouse B, \$60 to ship one widget from factory I to warehouse C, \$65 per widget to ship from factory II to warehouse A, \$70 per widget to ship from factory II to warehouse B, and \$75 per widget to ship from factory II to warehouse C, what shipping pattern gives the minimum total cost?

Setup: Letting

x_{11} = number of widgets shipped from factory I to warehouse *A*
x_{12} = number of widgets shipped from factory I to warehouse *B*
x_{13} = number of widgets shipped from factory I to warehouse *C*
x_{21} = number of widgets shipped from factory II to warehouse *A*
x_{22} = number of widgets shipped from factory II to warehouse *B*
x_{23} = number of widgets shipped from factory II to warehouse *C*

this problem can be formulated as a linear programming problem: Minimize $C = 80x_{11} + 75x_{12} + 60x_{13} + 65x_{21} + 70x_{22} + 75x_{23}$ subject to $x_{11} + x_{12} + x_{13} \leq 40$, $x_{21} + x_{22} + x_{23} \leq 50$, $x_{11} + x_{21} = 15$, $x_{12} + x_{22} = 45$, and $x_{13} + x_{23} = 30$.

This linear programming problem could then be solved using the tabular method or the simplex method as adapted in Section 6.3, but either would be time-consuming. This chapter presents a faster way.

Northwest Corner Algorithm

8.1.2 Example

Find a feasible (but not necessarily optimal) solution to Example 8.1.1 using the northwest corner algorithm.

Solution:

The facts of the problem can be diagrammed in the following table, where the amounts the factories produce are written on the right, the amounts the warehouses can store are written on the bottom, and the black numbers in the boxes are the costs of shipping from the factory on the left to the warehouse above.

		Warehouse			
		A	*B*	*C*	
Factory	I	80	75	60	**40**
	II	65	70	75	**50**
		15	**45**	**30**	

The northwest corner algorithm first allocates as many widgets as possible to the upper left box (the northwest box). Next, proceed to the nearest box into which something can still be placed, and allocate as much as possible to that one. Then the process continues, each time moving either one box to the right, or one down, or one diagonally down, depending on how the shipments can be made.

Since the 15 at the bottom of the first column is less than the 40 at the right of the first row, factory I can ship only 15 widgets to warehouse *A*, so we write a 15 in the upper left box. Then nothing else can go to warehouse *A*; that is, nothing else will be written in the boxes of the first column.

		Warehouse A	Warehouse B	Warehouse C	
Factory	I	**15** 80	75	60	**40**
	II	65	70	75	**50**
		15	**45**	**30**	

Thus we move right from the 15, making $x_{12} = 40 - 15 = 25$. Now the capacity of factory I has been exhausted, so there can be no more numbers written in the first row. Moving down, we next set $x_{22} = 45 - 25 = 20$ to fill warehouse *B*. Now we must move right and set $x_{23} = 50 - 20 = 30$. The results are as follows:

		Warehouse A	Warehouse B	Warehouse C	
Factory	I	**15** 80	**25** 75	60	**40**
	II	65	**20** 70	**30** 75	**50**
		15	**45**	**30**	

This table tells us we can ship 15 widgets from factory I to warehouse *A*, 25 widgets from factory I to warehouse *B*, 20 widgets from factory II to warehouse *B*, and 30 widgets from factory II to warehouse *C*. It is fairly obvious that this is not the optimum solution (that is, it is not the cheapest), but it is feasible. The total cost is $C = 15 \cdot 80 + 25 \cdot 75 + 20 \cdot 70 + 30 \cdot 75 = 6725$.

8.1.3 Example

Use the northwest corner algorithm to find a feasible solution to the following transportation problem:

		Warehouse A	Warehouse B	Warehouse C	Warehouse D	
Factory	I	7	8	6	5	**30**
	II	9	7	4	6	**15**
	III	6	9	5	9	**20**
	IV	4	5	8	7	**10**
		5	**15**	**20**	**35**	

Solution:

In this example we must go right twice before going down because of the comparatively large capacity of factory I. This causes no problem with the northwest corner algorithm; we just move to the nearest empty box that can be filled at each step.

Warehouse

Factory	A	B	C	D	
I	**5** 7	**15** 8	**10** 6	5	**30**
II	9	7	**10** 4	**5** 6	**15**
III	6	9	5	**20** 9	**20**
IV	4	5	8	**10** 7	**10**
	5	**15**	**20**	**35**	

8.1.4 Example

Use the northwest corner algorithm to find a feasible solution to the following transportation problem:

Warehouse

Factory	A	B	C	D	
I	7	8	6	5	**10**
II	9	7	4	6	**25**
III	6	9	5	9	**15**
IV	4	5	8	7	**20**
	30	**20**	**5**	**15**	

Solution:

Warehouse

Factory	A	B	C	D	
I	**10** 7	8	6	5	**10**
II	**20** 9	**5** 7	4	6	**25**
III	6	**15** 9	5	9	**15**
IV	4	5	**5** 8	**15** 7	**20**
	30	**20**	**5**	**15**	

This example has a "degeneracy" in the sense that the 15 in position x_{32} fills both the needs of warehouse *B* and the capacity of factory III. In Section 8.2 we shall write a zero for either x_{33} or x_{42} so that we have enough "stones" (filled-in boxes) to use the stepping-stone algorithm.

Fictitious Warehouses

The discussion thus far has assumed that the total factory capacity is the same as the total warehouse capacity—that the total across the bottom of each table equals the total down the right side. Clearly, this is not always the case in real life. To handle the situation where the factories manufacture more than the warehouses want, we introduce a fictitious warehouse. All the shipping costs to the fictitious warehouse are zero and its capacity is exactly the difference between the total capacity of the factories and the total capacity of the other warehouses; thus the numbers on the right side will now add to the same sum as the numbers across the bottom. The shipping quantities in the boxes corresponding to the fictitious warehouse can be considered slack variables; the fictitious warehouse assumes the slack. (Similarly, a fictitious factory is used if the total factory capacity is less than the total warehouse capacity.)

8.1.5 Example

Use the northwest corner algorithm to find an initial feasible solution to the following problem. Suppose that a company has two factories, I and II, and two warehouses, *A* and *B*. Factory I has 100 tons of a product on hand and factory II has 125 tons of the product on hand. Warehouse *A* requires 90 tons and warehouse *B* requires 95 tons. If the shipping costs are $4 per ton from I to *A*, $5 per ton from I to *B*, $6 per ton from II to *A*, and $8 per ton from II to *B*, how should the shipments be made to minimize the costs?

Solution:

When writing the information in tabular form, we include a fictitious warehouse because the total factory capacity exceeds the total warehouse needs.

Factory \ Warehouse	A	B	C	
I	4	5	0	**100**
II	6	8	0	**125**
	90	**95**	**40**	

Using the northwest corner algorithm, we obtain the following feasible solution:

Warehouse		A	B	C	
Factory	I	**90** 4	**10** 5	0	**100**
	II	6	**85** 8	**40** 0	**125**
		90	**95**	**40**	

Minimum Cell Algorithm

It is fairly clear that the feasible "solution" found in Example 8.1.2 is not the optimal solution because the shipping costs that are *not* used are both lower than all the shipping costs that *are* used. This can happen because the northwest corner algorithm ignores the costs; Examples 8.1.3 and 8.1.4 have the same costs in their tables, but this is irrelevant to their two (different) initial solutions. The minimum cell algorithm is another method of finding an initial feasible solution; unlike the northwest corner algorithm, it does take the cost into account. Therefore, it usually (not always) gives a better initial solution to a transportation problem.

8.1.6 Example

Use the minimum cell method to find an initial solution to Example 8.1.2.

Solution:
As on page 251, the problem can be summarized in the following table:

Warehouse		A	B	C	
Factory	I	80	75	60	**40**
	II	65	70	75	**50**
		15	**45**	**30**	

Since 60 is the cheapest possible rate, we decide to send as much as possible at that rate. Since warehouse C needs 30 widgets and factory I has 40 widgets, we can send at most 30 widgets from factory I to warehouse C; we do so and write it in the box. At the same time it is useful to circle 60, showing that it has been used. (These circles indicate the "stones" when we use the stepping-stone algorithm in Section 8.2.)

To show that we have accounted for all 30 items of warehouse *C*, we cross off the 30. To show that 30 of the 40 items in factory I have been used, we cross off the 40 and write a 10 beside it to show there are 10 items left in factory I.

		Warehouse A	Warehouse B	Warehouse C	
Factory	I	80	75	30 (60)	~~40~~ 10
	II	65	70	75	50
		15	45	~~30~~	

The next cheapest shipping rate is 65, so we circle it. Since warehouse *A* needs only 15 widgets, we write a 15 in the box with the 65, showing that we will ship 15 widgets from factory II to warehouse *A*. We cross out the 15 below the first column. And we cross out the 50 at the right of the second row and write a 50 − 15 = 35 next to it.

		Warehouse A	Warehouse B	Warehouse C	
Factory	I	80	75	30 (60)	~~40~~ 10
	II	15 (65)	70	75	~~50~~ 35
		~~15~~	45	~~30~~	

The next cheapest shipping rate is 70, so now we circle the 70. We have only 35 of the widgets produced by factory II remaining to send to warehouse *B*, so we write a 35 in the box with the 70. Then we cross out the 35 at the right of the second row, and replace the 45 below the second column with a 45 − 35 = 10.

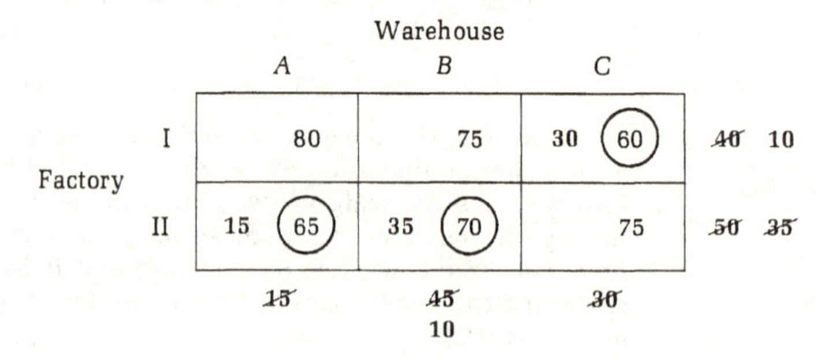

		Warehouse A	Warehouse B	Warehouse C	
Factory	I	80	75	30 (60)	~~40~~ 10
	II	15 (65)	35 (70)	75	~~50~~ ~~35~~
		~~15~~	~~45~~ 10	~~30~~	

Since 75 is the lowest remaining rate, we ship the remaining 10 widgets from factory I to warehouse *B* and indicate the feasible solution we obtain as follows:

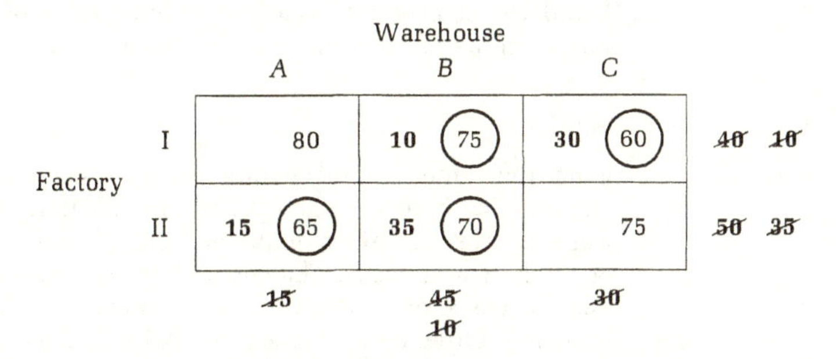

		Warehouse *A*	Warehouse *B*	Warehouse *C*	
Factory	I	80	**10** (75)	**30** (60)	~~40~~ ~~10~~
	II	**15** (65)	**35** (70)	75	~~50~~ ~~35~~
		~~15~~	~~45~~ ~~10~~	~~30~~	

This shipping pattern yields a cost of $15 \cdot 65 + 35 \cdot 70 + 10 \cdot 75 + 30 \cdot 60 = 5975$, which just *happens* to be the minimum cost for this problem (although this fact is not obvious). The minimum cell algorithm *may* yield a larger cost than the northwest corner algorithm, as it clearly does for

6	4	**10**
1000	5	**10**
10	**10**	

Example 8.1.7 shows how an apparently different type of problem can be interpreted as a transportation problem.

8.1.7 Example

Suppose that a company produces three products, *A*, *B*, and *C*. Each of its products can be made on any one of its three machines, but the cost of making the different products on the different machines varies according to the following table. For example, it costs machine I \$4 to produce one of product *A*, but only \$3 to produce one of product *B*.

		Product *A*	Product *B*	Product *C*
Machine	I	4	3	5
	II	6	4	8
	III	7	6	7

Machine I can produce 100 items a day (of any assortment of the products), machine II can produce 110, and machine III can also produce 110. If the company wants to produce 80 of product *A*, 90 of product *B*, and 100 of product *C* each day, how many of each product should each machine make to have a minimum cost?

Solution:

Some reflection should enable you to convince yourself that the mathematics here is exactly the same as that for the transportation problem. The machines play the role of factories and the products "are" the warehouses. The costs of shipping have become, as you can see, the costs of production. We summarize the information in the following table and proceed as before. This time we introduce a "fictitious product" since the real products do not use up the total machine capacity.

		Product				
		A	*B*	*C*	*D*	
Machine	I	4	3	5	0	**100**
	II	6	4	8	0	**110**
	III	7	6	7	0	**110**
		80	**90**	**100**	**50**	

We ignore the fictitious column in choosing our minimum costs; 50 items are fated to be fictitious, and no matter what machine "makes" them, they will cost zero. Since the lowest real production cost is $3, we make 90 of product *B* with machine I and indicate it as follows.

		Product				
		A	*B*	*C*	*D*	
Machine	I	4	**90** (3)	5	0	~~100~~ **10**
	II	6	4	8	0	**110**
	III	7	6	7	0	**110**
		80	~~90~~	**100**	**50**	

Now we ignore the second column. The smallest number in the other two columns is the $4 in the upper left, so we use the remaining capacity of machine I to make 10 of product *A* for $4 each.

Product

		A	B	C	D	
	I	10 (4)	90 (3)	5	0	~~100~~ ~~10~~
Machine	II	6	4	8	0	110
	III	7	6	7	0	110
		~~80~~ 70	~~90~~	100	50	

The first row is now completely used, as well as the second column. The smallest cost elsewhere in the table is \$6, so we next fill the need for product A with machine II.

Product

		A	B	C	D	
	I	10 (4)	90 (3)	5	0	~~100~~ ~~10~~
Machine	II	70 (6)	4	8	0	~~110~~ 40
	III	7	6	7	0	110
		~~80~~ ~~70~~	~~90~~	100	50	

It now remains to use machine III to make the last remaining 100 of product C and to fill in the fictitious column as appropriate.

Product

		A	B	C	D	
	I	10 (4)	90 (3)	5	0	~~100~~ ~~10~~
Machine	II	70 (6)	4	8	40 (0)	~~110~~ 40
	III	7	6	100 (7)	10 (0)	~~110~~ 10
		~~80~~ ~~70~~	~~90~~	~~100~~	50	

The total cost given by the minimum cell algorithm is $C = 10 \cdot 4 + 90 \cdot 3 + 70 \cdot 6 + 100 \cdot 7 + 40 \cdot 0 + 10 \cdot 0 = 1430$. It is easy to check that this is *more* than the cost given by the northwest corner algorithm: $C = 80 \cdot 4 + 20 \cdot 3 + 70 \cdot 4 + 40 \cdot 8 + 60 \cdot 7 = 1400$.

SUMMARY

This section presents two methods for finding an initial feasible solution to a transportation-type linear programming problem. The northwest corner algorithm begins at the upper left of the cost table and "uses up" the factories and warehouses in turn, gradually moving to the lower right. The minimum cell algorithm always uses the least unit cost available at each step. The latter algorithm often yields the smaller total cost of the two, but not always.

The next section will introduce the stepping-stone algorithm, which is a relatively easy adaptation of the simplex algorithm for transportation problems, but one that cannot be used for all linear programming problems. Section 8.3 extends this algorithm for more complicated transportation problems. And Section 8.4 will examine a special type of transportation problem, called the assignment problem, and show an even easier way to solve these problems.

EXERCISES 8.1. A

Find an initial feasible solution for each of the following transportation problems using the northwest corner algorithm and the minimum cell algorithm (or either algorithm as specified by your instructor). Optimal solutions will be found in Exercises 8.2.A and 8.3.A.

1. Suppose that a company has two factories, I and II, both of which have 90 tons of a product on hand. It has two warehouses, *A* and *B*, both of which need 80 tons of the product. If it costs \$4 to ship a ton from I to *A*, \$5 from I to *B*, \$6 from II to *A*, and \$8 from II to *B*, how should the shipments be made to minimize the costs?
2. Suppose that a company has two machines, I and II, either of which can alone make product *A* and product *B*. Each machine can make 80 products (of either type or mixed) per day. If we expect to sell 70 of product *A* per day and 60 of product *B*, how many of each should be made by each machine if it costs machine I \$6 to make product *A* and \$4 to make product *B* and it costs machine II \$7 to make product *A* and \$8 to make product *B*?
3. A company is accepting bids on three jobs; we shall call them jobs *A*, *B*, and *C*. Three workmen are bidding on the jobs. Workman I says he can do 100 jobs per week and will charge \$6 for job *A*, \$5 for job *B*, and \$7 for job *C*. Workman II can do 120 such jobs and charges \$9 for job *A*, \$6 for job *B*, and \$10 for job *C*. Workman III can do 100 jobs in all and would charge \$8 for job *A*, \$10 for job *B*, and \$9 for job *C*. If the company needs to get job *A* done 100 times, job *B* 80 times, and job *C* 90 times, how many of each type of job should it assign to each of the workmen for a minimum total cost?
4. A company has four products, *A*, *B*, *C*, *D*, which it sells anywhere at the same price. It has two plants which make these products and due to various factors (such as availability of machines, skills, and materials) the costs of manufacturing the products vary at the two plants. Plant I can make a total of 15 items; it costs \$900 to make each of product *A*, \$700 for each of product *B*, \$600 for product *C*, and \$800 for product *D*.

Plant II is larger and can make 25 of the products at a cost of \$700 for product *A*, \$900 for product *B*, \$700 for product *C*, and \$700 for product *D*. If the company wants to make 10 of product *A*, 5 of product *B*, 15 of product *C*, and 10 of product *D*, how should the manufacturing be assigned for a minimum total cost?

5.

		Warehouse *A*	*B*	*C*	
Factory	I	5	8	10	10
	II	3	9	2	32
	III	1	4	6	9
		15	15	21	

6.

		Warehouse *A*	*B*	*C*	
Factory	I	4	5	3	25
	II	6	7	8	25
	III	7	8	4	30
		25	20	20	

7.

		Warehouse *A*	*B*	*C*	
Factory	I	4	3	5	90
	II	6	4	8	100
	III	7	6	7	100
		70	80	90	

8.

		Warehouse *A*	*B*	*C*	*D*	
Factory	I	2	5	4	7	4
	II	1	4	4	3	12
	III	6	1	8	5	10
	IV	3	3	7	6	6
		8	11	5	8	

EXERCISES 8.1. B

Find an initial feasible solution for each of the following transportation problems using the northwest corner algorithm and the minimum cell algorithm (or either algorithm as specified by your instructor). Save your answers because optimal solutions will be found in Exercises 8.2.B and 8.3.B.

1. A company has two factories, I and II, with 60 and 80 tons of a product in stock, respectively. It has two warehouses, *A* and *B*, that have each sent in orders for 50 tons of the item. If shipping costs are to be a minimum, how should the shipping be planned assuming that it costs $6 per ton to ship from factory I to warehouse *A*, $4 per ton to ship from factory I to warehouse *B*, $9 to ship from factory II to warehouse *A*, and $8 to ship from factory II to warehouse *B*?
2. Suppose that a company has two machines, I and II, either of which can alone make product *A* and product *B*. Each machine can make 90 products (of either type or mixed) per day. If the company expects to sell 85 of product *A* and 70 of product *B* per day, how many of each should be made by each machine if it costs machine I $4 to make product *A* and $3 to make product *B*, and it costs machine II $7 to make product *A* and $5 to make product *B*? Both products bring the same price.
3. The company is accepting bids on three jobs—*A*, *B*, and *C*. Only three workmen, I, II, and III, are bidding on the jobs. Workman I offers to do 10 jobs per day; he will charge $1 for each time he does job *A*, $2 for job *B*, and $4 for job *C*. Workman II can do 20 jobs per day, but will charge $6 for each time he does job *A*, $8 for job *B*, and $9 for job *C*. Workman III can do 15 jobs total, at the rates of $3 for job *A*, $4 for job *B*, and $5 for job *C*. If the company needs to get job *A* done 5 times a day, job *B* 10 times a day, and job *C* 30 times per day, how many of each type of job should it give to each workman for a minimum total cost?
4. A company has four products, *A*, *B*, *C*, and *D*, which it sells anywhere at the same price. It has two plants which make these four products and due to various factors (such as availability of machines, skills, and materials), the costs of manufacturing the products vary at the two plants. Plant I can make a total of 20 products and plant II a total of 40 products. If it costs $6 to make product *A* at plant I but $10 at plant II; it costs $6 to make product *B* at plant I but $9 at plant II; it costs $3 to make product *C* at plant I and $4 at plant II; and product *D* costs $5 to make at plant I and $9 at plant II; and if the company wants to make a total of 12 of product *A*, 8 of product *B*, 10 of product *C*, and 30 of product *D*, where should the products be manufactured for a minimum total cost?
5.

		Warehouse *A*	*B*	*C*	
Factory	I	3	4	2	20
	II	5	11	7	10
	III	10	9	6	15
		5	15	25	

6.

		Warehouse A	B	C	
	I	8	9	5	60
Factory	II	5	6	4	50
	III	7	8	9	50
		50	45	40	

7.

		Warehouse A	B	C	D	
	I	30	25	40	20	10
Factory	II	29	26	35	40	25
	III	31	33	37	22	15
		10	15	20	5	

8.

		Warehouse A	B	C	
	I	4	3	5	60
Factory	II	6	4	8	55
	III	7	6	7	45
		20	35	75	

ANSWERS 8.1. A

To make it easy for you to check your answers, we print here the shipping quantities only in their appropriate boxes. Variables not mentioned have the value 0. We abbreviate the northwest corner algorithm and the minimum cell algorithm as NCA and MCA, respectively.

1. Both NCA and MCA

80	10	
	70	20

2. NCA

70	10	
	50	30

MCA

20	60	
50		30

3. NCA

100			
	80	40	
		50	50

MCA

20	80		
		70	50
80		20	

4. NCA

10	5		
		15	10

MCA

		15	
10	5		10

5. NCA

10		
5	15	12
		9

MCA

	10	
6	5	21
9		

6. NCA

25			
	20	5	
		15	15

MCA

5		20	
20	5		
	15		15

7. NCA

70	20		
	60	40	
		50	50

MCA

10	80		
60			40
		90	10

8. NCA

4			
4	8		
	3	5	2
			6

MCA

		4	
8			4
	10		
	1	1	4

8.2 The Stepping-Stone Algorithm

We turn now to an adaptation of the simplex algorithm that can be used to find the optimum solution to a transportation problem—the *stepping-stone algorithm.*

8.2.1 Example

Find the optimum solution to the transportation problem for which an initial feasible solution was found in Example 8.1.5.

Solution:

The initial feasible solution, given on page 255, was

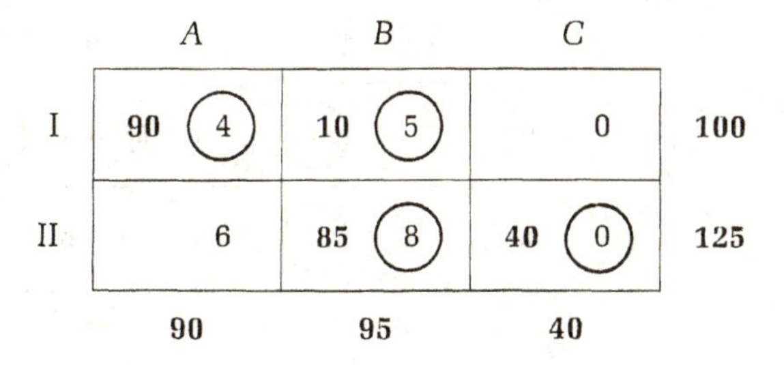

	A	B	C	
I	90 (4)	10 (5)	0	100
II	6	85 (8)	40 (0)	125
	90	95	40	

The costs that are associated with the nonzero variables in this initial feasible solution have been circled so they can be easily identified. Following standard usage, we call the circles stones; the cells that do not contain stones are called empty cells. As in the simplex algorithm as presented in Chapter 6, we shall now try to improve the cost by introducing a single new variable into the given feasible solution.

In this example we notice that the present pattern provides for 0 items to be shipped from II to *A*. Suppose we see what happens to the total cost if this is changed—that is, if we ship a nonzero number from II to *A*. To consider the effect of this, we consider the following path (a term that we shall discuss further later):

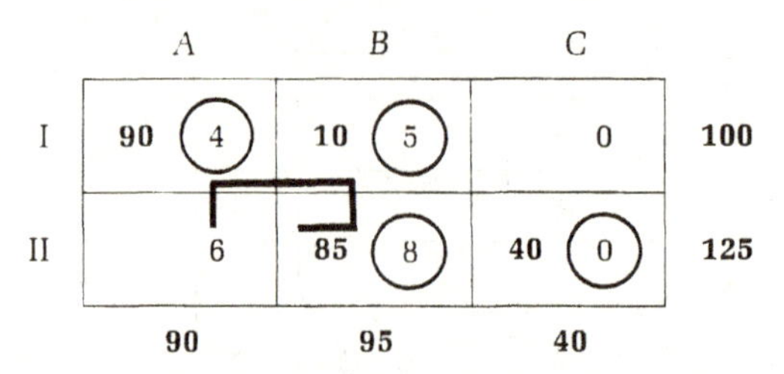

	A	B	C	
I	90 (4)	10 (5)	0	100
II	6	85 (8)	40 (0)	125
	90	95	40	

This path suggests that if we send *more* from II to *A*, we can keep factory II in equilibrium by sending the same amount *less* from II to *B*. Then we can keep *B* in equilibrium by sending an equal amount *more* from I to *B*, and we can keep factory I in equilibrium by sending *less* from I to *A*. This will just balance the extra *A* received from II. How would this affect the total cost? For each unit more that is sent from II to *A*, the total cost will change by $6 - 8 + 5 - 4 = -1$, so it will decrease. Thus it will pay to send *more* from II to *A* and from I to *B* while sending the same amount *less* from I to *A* and from II to *B*.

How many such switches can we make? Since the minimum shipping number in the expensive pair of boxes (x_{11} and x_{22}) is 85, we subtract 85 from both x_{11} and x_{22} and add 85 to x_{12} and x_{21}. Any number larger than 85 would result in a negative number of widgets being shipped from factory II to warehouse *B*, a difficult task indeed. Any smaller number would not take full advantage of the economy we have just discovered. Thus we write the next table as follows:

	A	B	C	
I	90 − 85 = 5 (4)	10 + 85 = 95 (5)	0	100
II	85 (6)	85 − 85 = 0 8	40 (0)	125
	90	95	40	

If you were writing this on your paper, of course you would not write the computations, but merely $\begin{matrix} 5 & 95 \\ 85 & 0 \end{matrix}$ in the appropriate boxes. We have written them out here only for the reader's convenience. Notice that 6 has become encircled by a stone and 8 is now an empty cell.

Would we improve the cost by putting a stone around that 8 again? No, because $8 - 6 + 4 - 5 = 1$, so we would increase the cost by 1 for each unit that we increase x_{22}. We put $+1$ in that box so we can remember this.

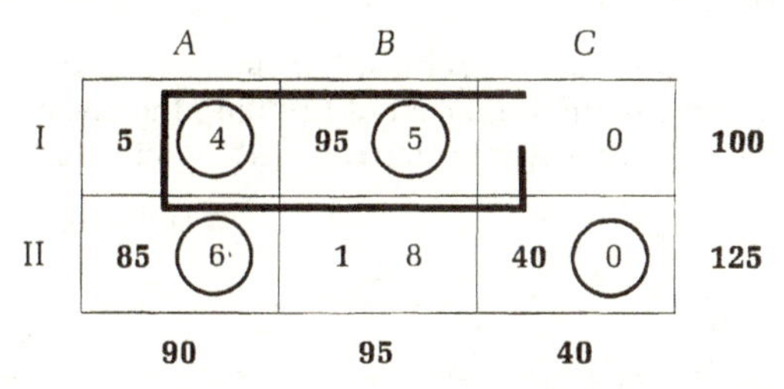

	A	B	C	
I	5 (4)	95 (5)	0	100
II	85 (6)	1 8	40 (0)	125
	90	95	40	

Then we consider whether we should increase the number "shipped" from I to *C*. To do this, we use the path indicated above. Since $0 - 4 + 6 - 0 = 2$, there would be an *increase* of the cost by 2 for each unit that is shipped from I to *C*, so we conclude that the pattern given above yields the minimum possible cost. The minimum cost is $C = 4 \cdot 5 + 5 \cdot 95 + 6 \cdot 85 = 1005$.

If either the northwest corner algorithm or the minimum cell algorithm has been correctly used, the number of stones will not exceed $m + n - 1$ (where m is the number of factories and n the number of warehouses) because at least one row or column was eliminated at each step of the algorithm and the last step eliminated both a row and a column. Thus both algorithms must end after at most $m + n - 1$ steps. The theme of the stepping-stone algorithm is to move the stones in such a way as to decrease the total cost at each step until we can decrease it no more.

The decision as to whether the cost can be decreased by increasing one of the nonzero variables is made by "stepping" along well-defined paths. In each stepping-stone table there will be exactly one path for each empty cell such that the corners of the path are on the stones. These paths are found by inspection in this course; a rule can be given, but it is too complicated to be helpful. The paths assure that the number of items leaving each factory and arriving at each warehouse will not change. In the 2 by n case, the paths will always be rectangles, so there will be two numbers to add and two to subtract. In Section 8.3 we consider more complicated paths.

We use the paths by alternately adding and subtracting the shipping costs on the *corners* of the paths, beginning with adding the cost in the empty cell for which the path was made. If the total change in the cost is negative for some path, we increase the number to be shipped in the corresponding empty cell. The new quantity to be shipped in this cell will equal the smallest amount shipped in the cells from which we are subtracting. That number will become zero, and the shipping amounts in all cells on the path must then be adjusted appropriately.

8.2.2 Example

If a transportation problem is summarized in the following table, what is the minimum total cost of fulfilling the given needs of the warehouse using the indicated resources at the factories?

		Warehouse				
		A	*B*	*C*	*D*	
Factory	I	7	6	4	7	**30**
	II	5	9	5	9	**40**
		20	**10**	**5**	**25**	

Solution:

First we notice that the factories are producing more than the warehouses want, so we introduce a fictitious warehouse with capacity 30 + 40 − 20 − 10 − 5 − 25 = 10.

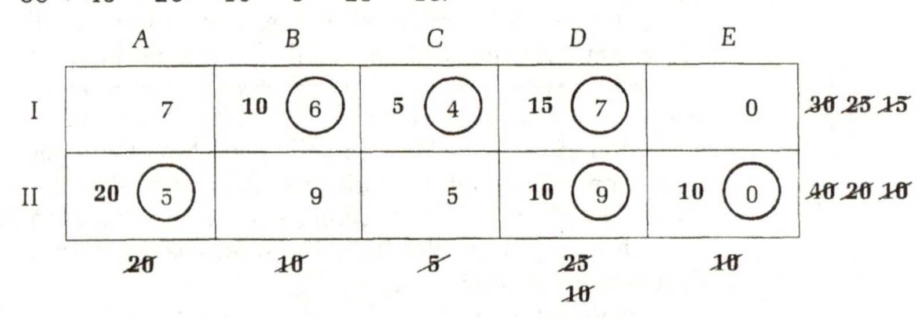

The minimum cell method has been used to get an initial feasible solution, as indicated. Since there are four empty cells, we test each one to see if moving a stone to it would decrease the total cost. To do this we use the following four paths:

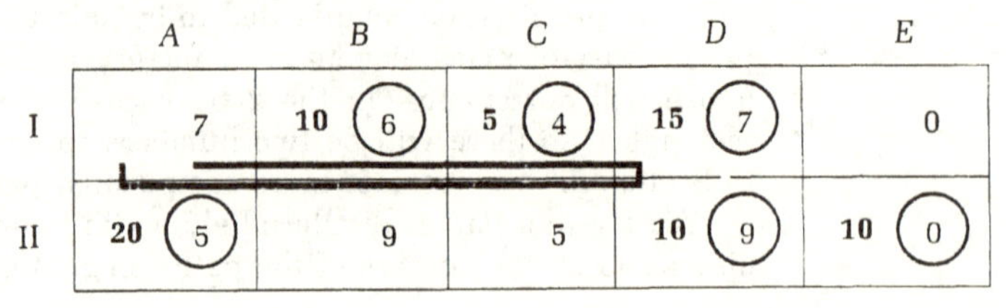

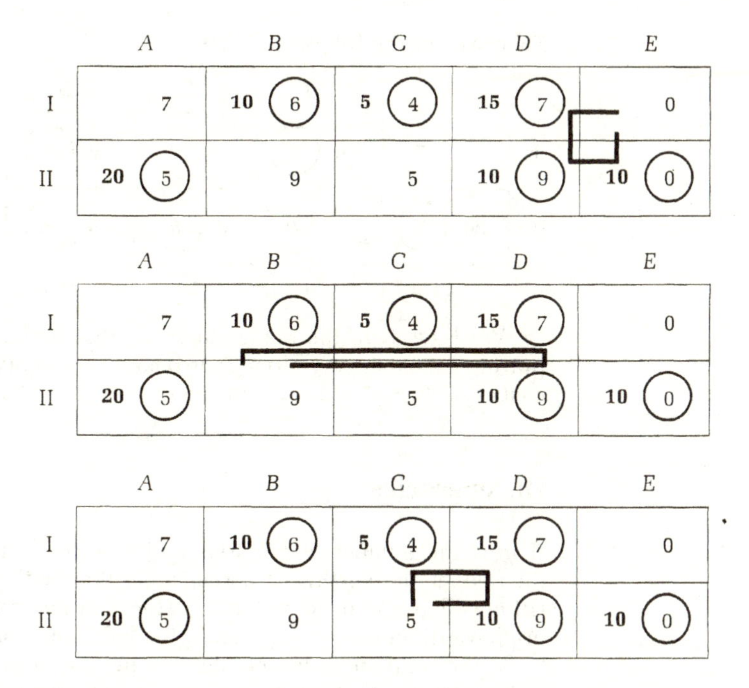

	A	B	C	D	E
I	7	10 (6)	5 (4)	15 (7)	0
II	20 (5)	9	5	10 (9)	10 (0)

	A	B	C	D	E
I	7	10 (6)	5 (4)	15 (7)	0
II	20 (5)	9	5	10 (9)	10 (0)

	A	B	C	D	E
I	7	10 (6)	5 (4)	15 (7)	0
II	20 (5)	9	5	10 (9)	10 (0)

Remember—the corners of the paths must be on stones. With this restriction, there is exactly one path for each empty cell. We make a detailed chart here (one line for each empty cell) for the reader's reference. It is not necessary to do this.

I ⟶ *A*	(use II and *D*)	$7 - 7 + 9 - 5 =$	4	(increases cost)
I ⟶ *E*	(use II and *D*)	$0 - 7 + 9 - 0 =$	2	(increases cost)
II ⟶ *B*	(use I and *D*)	$9 - 6 + 7 - 9 =$	1	(increases cost)
II ⟶ *C*	(use I and *D*)	$5 - 4 + 7 - 9 =$	-1	(decreases cost)

Thus we decide to ship as much as possible from II to *C*, because we can thereby decrease present shipping costs by $1 per item. How much is "as much as possible"? Since $x_{13} = 5$ and $x_{24} = 10$, and the minimum of these is 5, we see that x_{23} can be increased by at most 5:

	C	D
I	**5 − 5 = 0** 4	**15 + 5 = 20** (7)
II	**5** (5)	**10 − 5 = 5** (9)

Then we get the following table:

	A	B	C	D	E	
I	7	10 (6)	4	20 (7)	0	30
II	20 (5)	9	5 (5)	5 (9)	10 (0)	40
	20	10	5	25	10	

We leave it to the reader to verify that nothing can be gained by putting a stone in any of the four boxes now empty. Thus the minimum cost is $6 \cdot 10 + 7 \cdot 20 + 5 \cdot 20 + 5 \cdot 5 + 9 \cdot 5 = 370$.

Degeneracies

If m is the number of factories and n is the number of warehouses, then either the northwest corner algorithm or the minimum cell algorithm will give at most $m + n - 1$ stones in the initial solution. If there are fewer than $m + n - 1$ stones, the solution is said to be degenerate.

If the solution is degenerate, it is necessary to circle enough of the unit costs so that there are $m + n - 1$ stones in order to make all the stepping-stone paths. When the northwest corner algorithm is used, this is done by circling those costs so that a continuous zigzag line of stones is formed from the upper left to the lower right of the table. When the minimum cell algorithm is used, cross out only one row *or* one column at each step, even if filling in an empty cell exhausts both a row and a column. This can and should be done so that neither all the rows nor all the columns are crossed out until the last step, at which point we cross out the last row and the last column and circle the cost at their intersection whether or not a positive quantity is written in its cell.

8.2.3 Example

Find the minimum cost for the transportation problem described in the following table:

	A	B	C	D	
I	11	8	4	8	30
II	7	6	6	5	40
	10	15	20	25	

Solution:

To begin this example using the minimum cell algorithm, we first circle the 4 and then the 5 with the accompanying marks as shown below. The next smallest cost is 6, but we cannot use the 6 below the 4. When we circle the 6 below the 8, a 15 is written in its cell, thereby both exhausting factory II and filling warehouse *B*. But crossing out both the second row and the second column would result in too few stones at the end, so we cross out only one. Arbitrarily, we choose to cross out the second column. We obtain

	A	*B*	*C*	*D*	
I	11	8	**20** (4)	8	~~30~~ **10**
II	7	**15** (6)	6	**25** (5)	~~40~~ **15**
	10	~~15~~	~~20~~	~~25~~	

We must not, of course, try to ship anything more from factory II! Thus the remaining 10 items, destined for warehouse *A*, must come from factory I. Thus we circle the 11 and write a 10 in its box. Which 10 in the margin do we cross out? The 10 at the end of the first row—so that the warehouses are not all crossed out before the factories. Finally, we circle the 7 even though there is no positive quantity to write in its cell, and cross out the numbers at the end of its row and column. Thus the 7 is enclosed in a "fake stone." Completing this, and writing the potential changes in cost by using each one of the empty cells in the shipping pattern, we obtain the following table:

	A	*B*	*C*	*D*	
I	**10** (11)	**−2** 8	**20** (4)	**−1** 8	~~30~~ ~~10~~
II	(7)	**15** (6)	**6** 6	**25** (5)	~~40~~ ~~15~~
	~~10~~	~~15~~	~~20~~	~~25~~	

Yes, those boxes are getting crowded! They contain three types of numbers—shipping costs, amounts shipped, and estimates of how much can be saved (or wasted!) by changing the shipping pattern. Be sure to make your boxes big enough. Three colors of pencils or pens will help keep the three concepts separate.

It is not necessary to continue computing the third set of numbers once you have found one that is negative—you could immediately introduce that cell to the pattern and thereby lower the cost—but computing them all enables you to choose which cell will save the most per item. Since the -2 in the table above will save the most per unit change, we decide to ship from I to *B*.

	A	*B*	*C*	*D*
I	**2** 11	**10** (8)	**20** (4)	**1** 8
II	**10** (7)	**5** (6)	**4** 6	**25** (5)

Now the 7 is encircled by a real stone instead of a fake stone, so the testing of the prospective changes is straightforward. It is impossible to decrease the cost further, so the minimum total cost is $C = 8 \cdot 10 + 4 \cdot 20 + 7 \cdot 10 + 6 \cdot 5 + 5 \cdot 25 = 385$.

If in the course of solving a stepping-stone problem, the shipping quantities in two cells are simultaneously reduced to zero, keep one of the costs circled in a fake stone. It does not matter which one.

When using fake stones, you might encounter a situation such as the following:

5	(6)
10 (7)	**15** (4)

Clearly, costs can be reduced by introducing the box containing the 5 into the shipping pattern, but it is impossible to subtract a positive number from the amount shipped in the cell containing the 6. That's all right—just subtract 0! Then the 6 loses its stone, the 5 acquires a stone, and this is the easiest kind of rearrangement possible.

SUMMARY

After an initial feasible solution has been found for a transportation problem (either by the northwest corner algorithm or by the minimum cell algorithm), the next step is to test each empty cell to see if the total cost could be lowered by using it in the shipping pattern. In the 2 by n case, this involves completing a rectangle for each empty cell, calling the line connecting the corners of that rectangle the path corresponding to the empty cell. If the sum of the cost in the empty cell and the cost in the cell diagonally opposite in its rectangle is less than the total of the costs in the other two corners of the rectangle, it pays to use that empty cell. In this case we put in the empty cell the smaller of the two numbers adjoining it on its path, subtract from those two numbers the smaller of the two (reducing the smaller to zero), and add the smaller of those two numbers to the number in the cell diagonally opposite the (formerly) empty cell. To do this we copy the whole array, putting in the new numbers. Then we test the empty cells in the new array to see if the cost can be further reduced by using one of them.

There can never be more than $(m + n - 1)$ cells with nonzero shipping quantities in any basic feasible solution. If there are fewer than $(m + n - 1)$ stones, we introduce a fake stone in order to be able to draw the needed paths.

EXERCISES 8.2. A

1–2. Complete problems 1 and 2 of Exercises 8.1.A.

For each transportation problem indicated below, find the shipping pattern that gives the least cost.

3.

15	8	7	11	13
10	9	5	6	20
5	6	12	10	

4.

14	11	6	9	35
11	9	5	7	25
10	20	20	10	

5.

11	11	10	11	10
15	14	13	12	30
10	10	10	10	

6. Complete problem 4 of Exercises 8.1.A.

EXERCISES 8.2. B

1–2. Complete problems 1 and 2 of Exercises 8.1.B.

For each transportation problem indicated below, find the shipping pattern that gives the least cost.

3.

15	13	8	12	25
14	10	6	7	30
10	10	15	20	

4.

13	10	8	5	40
10	9	6	4	30
15	20	20	15	

5.

14	16	8	10	15
8	12	6	7	20
10	5	10	10	

6. Complete problem 4 of Exercises 8.1.B.

ANSWERS 8.2. A

Where the answer is not unique, there are many answers, as in any other linear programming problem; we give only those at the vertices of the feasible region.

1. Minimum cost of $860 occurs when 10 tons are sent from I to *A*, 80 tons from I to *B*, and 70 tons from II to *A*.
2. Minimum costs of $710 occur when machine I makes 20 of product *A* and 60 of product *B* and machine II makes 50 of product *A*.
3.

	6	7	
5		5	10

4.

	15	20	
10	5		10

or

	5	20	10
10	15		

5.

10			
	10	10	10

6. Minimum cost of $27,000 when plant I makes 5 of product *B* and 10 of product *C*, and plant II makes 10 of product *A*, 5 of product *C*, and 10 of product *D*.

8.3 Harder Stepping Stones

The problems in Section 8.2 were relatively easy because there were only two factories in each problem. Consequently, finding the stepping-stone paths meant merely completing rectangles. In larger problems we must step on more than three stones in some paths, and those paths are of necessity more complicated than mere rectangles.

The idea can probably be conveyed best through examples. The minimum cell algorithm yielded the following initial feasible solution in Example 8.1.7.

	A	*B*	*C*	*D*	
I	**10** (4)	**90** (3)	5	0	**100**
II	**70** (6)	4	8	**40** (0)	**110**
III	7	6	**100** (7)	**10** (0)	**110**
	80	**90**	**100**	**50**	

The stepping-stone paths for the two middle cells and the two corner cells are easy to find by merely completing appropriate rectangles. The other two paths are longer, as shown below.

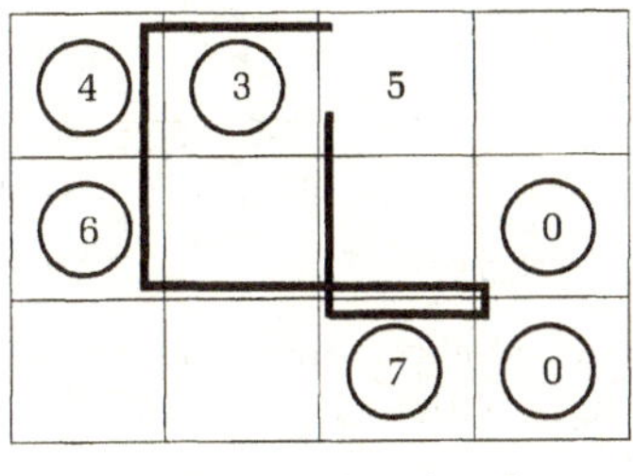

5 − 7 + 0 − 0 + 6 − 4 = 0

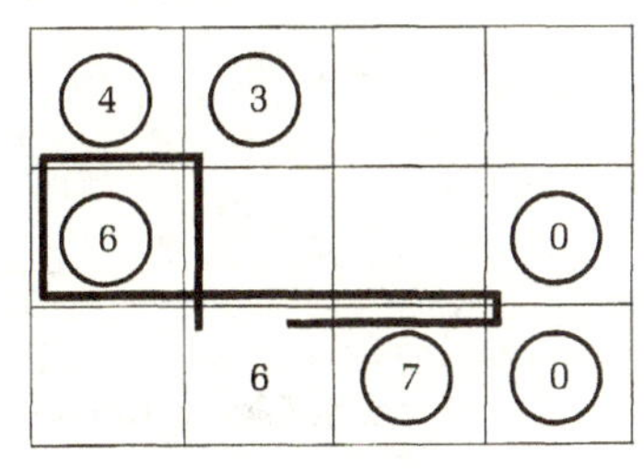

6 − 0 + 0 − 6 + 4 − 3 = 1

The method of measuring the possible gain or loss by introducing a new stone is similar to that used before; we alternately add and subtract the values of the costs on the corners of the path. (Notice that we "jump over" the 3-stone in the left pattern above; factory I will remain in equilibrium if we decrease the number shipped to *A* by the same amount that we increase the number shipped to *C* and do not change those shipped to *B* at all.)

8.3.1 Example

Find the optimum solution to Example 8.1.7 using the stepping-stone algorithm.

Solution:

Using the initial solution found in Example 8.1.7 and the paths suggested above, we write the following table:

10 (4)	**90** (3)	**0** 5	**2** 0
70 (6)	**−1** 4	**1** 8	**40** (0)
1 7	**1** 6	**100** (7)	**10** (0)

Thus we can decrease the cost by making x_{22} positive and decreasing x_{12} and x_{21}. Doing so, we get the following table.

	A	B	C	D
I	80 (4)	20 (3)	5	0
II	6	70 (4)	8	40 (0)
III	7	6	100 (7)	10 (0)

To see which, if any, new stones should be added now, we must consider the following longer paths.

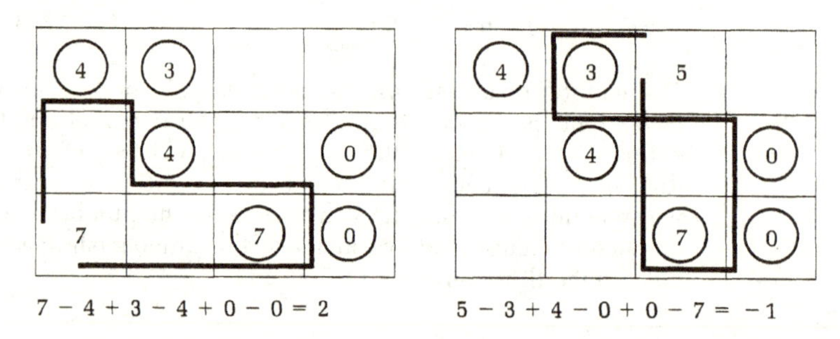

$7 - 4 + 3 - 4 + 0 - 0 = 2$ $\qquad$ $5 - 3 + 4 - 0 + 0 - 7 = -1$

Similarly, we test the other empty boxes, using their rectangular paths. (Be sure to include the empty box corresponding to the fictitious warehouse in the upper right; this is necessary to see if it would be less expensive to allow factory I to be not completely used.) Then the potential increases and decreases of cost can be recorded as follows:

	A	B	C	D
I	80 (4)	20 (3)	**−1** 5	**1** 0
I	**1** 6	70 (4)	**1** 8	40 (0)
III	**2** 7	**2** 6	100 (7)	10 (0)

Thus we now add to x_{13}. How much can we add? We consider the amounts being shipped in the boxes where we will subtract to compensate, and we take the minimum of the numbers now in those boxes. The minimum of 20, 40, and 100 is 20, so we set $x_{13} = 20$. The new table is given:

	A	B	C	D
I	80 (4)	20 − 20 = 0 (3)	20 (5)	0
II	6	70 + 20 = 90 (4)	8	40 − 20 = 20 (0)
III	7	6	100 − 20 = 80 (7)	10 + 20 = 30 (0)

Now we try again! This time it is easy to test all the empty cells except for I $\longrightarrow B$, where we have $3 - 5 + 7 - 0 + 0 - 4 = 1$, and II $\longrightarrow A$, where $6 - 4 + 5 - 7 + 0 - 0 = 0$.

	A	B	C	D
I	**80** (4)	**1** 3	**20** (5)	**2** 0
II	**0** 6	**90** (4)	**1** 8	**20** (0)
III	**1** 7	**2** 6	**80** (7)	**30** (0)

Since none of the estimates is negative, we cannot save any more by moving more stones. Thus we have found the minimum cost, $C = 4 \cdot 80 + 5 \cdot 20 + 4 \cdot 90 + 7 \cdot 80 = 1340$. Notice that we need to write down only three tables; each one corresponds to a tableau when the problem is solved using the simplex method. But we have found the optimal solution without ever writing a fraction!

Sometimes we must use fake stones in solving these more involved stepping-stone problems. The fake stones are used in exactly the same way as they are in the 2 by n case.

There is an easier but less intuitive method for computing the potential gains and losses of using one of the empty cells that is based on duality concepts. To use this method, one must first write down one side and across the bottom of the table under consideration a set of numbers such that each cost inside a stone is the sum of the numbers just written beside it and below it. (This can always be done in many ways.) Then subtract from each other cost in the table the sum of the numbers on its side and below it; this will give the trial costs.

8.3.2 Example

Solve the problem given in Example 8.3.1 using the method described in the previous paragraph.

Solution:

The original table was as follows after the MCA was used.

10 (4)	90 (3)	5	0
70 (6)	4	8	40 (0)
7	6	100 (7)	10 (0)

There is always one free choice of the numbers written around the edges, but it is customary to make the number at the end of the first row a 0. This forces the numbers at the bottom of the first two columns to be a 4 and 3, respectively, so that the costs inside the stones at the upper left will be the sum of the numbers to their right and below in both cases. Then we must have a 2 to the right of the second row, so that the 6 will be the proper sum. This forces a -2 at the bottom of the last column, which forces a 2 at the right of the last row, which forces a 5 at the bottom of the third column. We then have the following:

(4)	(3)	5	0	**0**
(6)	4	8	(0)	**2**
7	6	(7)	(0)	**2**
4	**3**	**5**	**−2**	

We then *subtract* these numbers around the edges from the costs in the corresponding empty cells to get the estimates of the gain or loss from including that cell in the shipping pattern. This will result in the estimates given below—which are exactly the same as the estimates obtained by using the stepping-stone paths at the beginning of Example 8.3.1!

(4)	(3)	0 5	2 0	0
(6)	−1 4	1 8	(0)	2
1 7	1 6	(7)	(0)	2
4	3	5	−2	

SUMMARY

In this section some of the stepping-stone paths involve three factories and three warehouses. Since each path consists of only horizontal and vertical lines and the same number is alternately added and subtracted from the shipping quantities at the corners of the path, each factory and each warehouse gains as much as it loses and thus the quantity shipped from or to it does not change. Since the steps of the algorithm are chosen so that the total cost decreases at each step, the minimum cost pattern has been found when no further decrease is possible.

At the end of the section, a shorter method was presented for estimating the potential gain or loss from using one of the empty cells.

EXERCISES 8.3. A

Complete problems 3, 5, 6, 7, and 8 in Exercises 8.1.A.

EXERCISES 8.3. B

Complete problems 3, 5, 6, 7, and 8 in Exercises 8.1.B.

EXERCISES 8.3. C

Suppose that a product can be produced at 50 per day throughout the year, but because of seasonal fluctuations in the demand, only 20 per day are sold during the first three quarters of the year and 80 per day during the last quarter of the year. Storage costs amount to about $100 per quarter, and the expenses of production are expected to rise at a rate of about $50 per quarter. If the production costs are $2000 during the first quarter, (a) set this up as a transportation problem and (b) find the production schedule that gives a minimum total cost.

ANSWERS 8.3. A

3. Minimum cost of $1890 occurs when workman I does job *A* 100 times, workman II does job *B* 80 times, and workman III does job *C* 90 times.

5.

4	6	
11		21
	9	

or

	10	
11		21
4	5	

6.

25			
	20		5
		20	10

or

5	20		
20			5
		20	10

7.

70		20	
	80		20
		70	30

8.

		4	
3		1	8
	10		
5	1		

ANSWERS 8.3. C

1. (a)

		Quarter sold					
		1st	2nd	3rd	4th	Fictitious	
Quarter made	1st	2000	2100	2200	2300	0	50
	2nd		2050	2150	2250	0	50
	3rd			2100	2200	0	50
	4th				2150	0	50
		20	20	20	80	60	

(b)

20				30
	20			30
		20	30	
			50	

8.4 The Assignment Problem

If each factory makes just 1 unit of a product and each warehouse wants just 1 unit of it, the transportation problem reduces to a simpler type of problem called the assignment problem. In this case it is merely a matter of deciding which factory sends its product to which warehouse—in other words, which factory is "assigned" to which warehouse.

The same mathematical pattern models various other types of practical problems. All such problems can be solved by the assignment algorithm, which is the topic of this section.

8.4.1 Example

Suppose that there are two factories, each of which produces one product, and two warehouses, each of which has ordered one of the products, and the prices for shipping are as in the following table. Which shipping pattern involves a minimum total cost?

		Warehouse	
		A	B
Factory	I	4	5
	II	6	8

Solution:

There are only two feasible solutions to such a problem. The product from factory I can go either to warehouse A or B, and the other product must go to the other warehouse. If factory I sends its product to warehouse A at a cost of \$4 and factory II sends its product to warehouse B at a cost of \$8, the total cost is \$12. If, on the other hand, factory I sends its product to warehouse B at a cost of \$5 and factory II sends its product to warehouse A at a cost of \$6, the total cost is \$11, which is clearly less.

If there were 3 factories and 3 warehouses, there would be $3! = 6$ choices of shipping patterns (instead of $2! = 2$ choices) because the first factory could ship to any of the three warehouses, the second could ship to either of the remaining two, and the third factory would have to ship to the only warehouse remaining.

If there were 4 factories and 4 warehouses, there would be $4! = 24$ choices of shipping patterns, and if there were 10 factories and 10 warehouses, there would be $10! = 3{,}628{,}800$ possible assignments. We could use the simplex algorithm or the transportation algorithm to find which one is best, but the algorithm presented in this section is faster. It is based on the following two rules.

8.4.2 Rule

If all prices are nonnegative, the minimum possible cost is zero.

8.4.3 Rule

If the same number is added or subtracted from all the elements of a row in an assignment problem matrix, the answer does not change.

If the same number is added or subtracted from all the elements of a column in an assignment problem matrix, the answer does not change.

Rule 8.4.2 should be clear; the least that can be charged for any service (if we do not permit negative numbers) is zero. Rule 8.4.3, however, merits more discussion. In Example 8.4.1 adding 2 to each element of the first column corresponds to having all shipping costs to warehouse A rise by \$2; the total costs in the (I $\longrightarrow$ A, II $\longrightarrow$ B) option of that example is \$6 + \$8 = \$14 and for the (I $\longrightarrow$ B, II $\longrightarrow$ A) option is \$5 + \$8 = \$13. The total costs both rise by \$2, but the relationship between them is the same; the first option still costs \$1 more than the second.

Similarly, in a more complicated example a uniform price rise or drop in any one column will affect the total costs, but will not affect the relationship between the various options. Using the same reasoning, convince yourself that adding or subtracting the same number from all the elements of a row (that is, increasing or decreasing the shipping costs from one factory uniformly) will affect the total costs but not the comparative costs of the options.

We use these rules by subtracting (or, on occasion, adding) numbers from whole rows and columns until the assignment can be made in such a way as to have the total cost zero.

8.4.4 Example

Suppose that there are four work teams and four jobs, and we want to assign one job to each team (and thus each team to one job) in such a way as to make the total cost of doing all jobs a minimum. If the following table shows how much each team charges for each job, how should the assignments be made?

		Job			
		1	2	3	4
Work team	1	50	48	55	60
	2	45	51	53	50
	3	47	49	46	51
	4	52	51	52	49

Solution:

Using Rule 8.4.3, we subtract the smallest number in each row from all the numbers in that row:

		Job 1	Job 2	Job 3	Job 4
Work team	1	50–48	48–48	55–48	60–48
	2	45–45	51–45	53–45	50–45
	3	47–46	49–46	46–46	51–46
	4	52–49	51–49	52–49	49–49

or

		Job 1	Job 2	Job 3	Job 4
Work team	1	2	0	7	12
	2	0	6	8	5
	3	1	3	0	5
	4	3	2	3	0

If you look closely at the table just above, you will see that the zeros are so placed that it is possible to pick out four in such a way as to have one in each row and one in each column. There is a total cost of zero in this matrix if the second job is assigned to the first team, the first job is assigned to the second team, the third job is assigned to the third team, and the fourth job is assigned to the fourth team. By Rule 8.4.3 this assignment also gives the best answer to the original problem; the minimum total cost is $C = 48 + 45 + 46 + 49 = 188$.

You would be badly deceived if you concluded that all assignment problems are that easy. We hasten on to another example containing the same numbers—with the important exception of interchanging x_{21} and x_{22}.

8.4.5 Example

Suppose there are 4 stenographers in a stenographic pool and they can do each of 4 needed jobs in the number of minutes given in the following table. How can you assign one job to each stenographer in such a way as to minimize the total time taken for all jobs? (This would be especially important if the jobs are to be done repeatedly.)

		Job 1	Job 2	Job 3	Job 4
Stenographer	1	50	48	55	60
	2	51	45	53	50
	3	47	49	46	51
	4	52	51	52	49

Solution:

As before, we subtract 48 from the first row, 45 from the second, 46 from the third, and 49 from the fourth:

		Job 1	2	3	4
Stenographer	1	2	0	7	12
	2	6	0	8	5
	3	1	3	0	5
	4	3	2	3	0

Since this matrix provides no way to assign the jobs to the teams in a one-to-one fashion, we apply Rule 8.4.3 again, this time subtracting 1 from all the numbers in the first column:

		Job 1	2	3	4
Stenographer	1	1	0	7	12
	2	5	0	8	5
	3	0	3	0	5
	4	2	2	3	0

Alas, although we now have five zeros, there is still no way to assign each stenographer one job in such a way as to have total costs zero, so we use another ingenious trick. We shall add some number to all the rows of the matrix (that is, the whole matrix), and then subtract it from just a few of the rows and columns. If we do this just right, we will then see the answer; if not, we may have to apply the technique again.

To see which rows and columns are "special" and which number should be used, we cover some of the rows and columns with lines, enough to cover all the zeros in the matrix. Try to do this using a minimum number of lines.

		Job 1	2	3	4
Stenographer	1	1		7	1
	2	5		8	
	3	0		0	
	4	2		3	

The smallest number now uncovered by a line is 1, so this is the number we use next. We subtract it from every uncovered number (making the number in the upper left 0) and add it to each number that appears at the intersection of two lines. (Notice that this is equivalent to subtracting 1 from the whole matrix and then adding it to all the covered rows and columns.) We then get the following matrix, where the labels are as above:

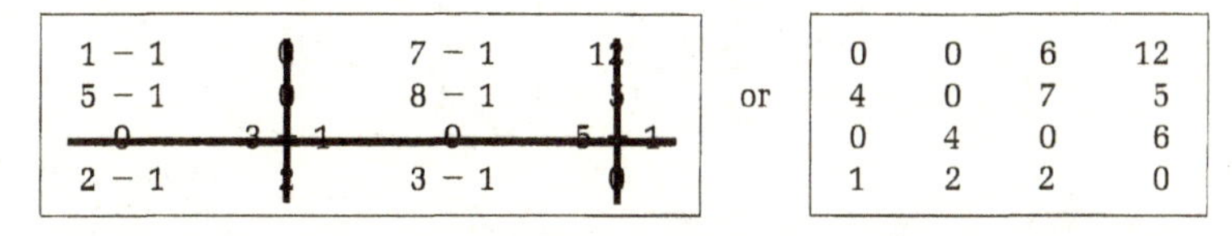

At this point we see that it is possible to assign the jobs so that the total cost is zero: The first stenographer gets the first job, the second stenographer gets the second job, the third stenographer gets the third job, and the fourth stenographer gets the fourth job. This will give a total time of 50 + 45 + 46 + 49 = 190 minutes.

SUMMARY

This section presents the assignment algorithm, which can be summarized as follows.

1. Write the costs in matrix form.
2. Subtract the smallest number in each row from all the numbers in that row.
3. Subtract the smallest number in each column from all the numbers in that column. (If the smallest number is 0, this results in the column remaining unchanged.)
4. If the assignment is not yet obvious, cover all the zeros in the matrix with lines through the rows and columns. Try to use as few lines as possible. Then subtract the smallest uncovered number from all the uncovered numbers, and add it to the numbers at the intersections of the lines. Repeat if necessary.
5. At any point it may become possible to assign the jobs so that the total cost is zero. This assignment pattern is the one that gives a minimum cost in the original problem.

EXERCISES 8.4. A

1. In each of the following assignment matrices, find an assignment such that the total cost is zero.

(a)

		Job 1	2	3	4
Crew	1	1	3	0	4
	2	4	0	2	1
	3	2	3	5	0
	4	0	2	6	1

(b)

Crew \ Job	1	2	3	4
1	2	0	1	3
2	3	4	0	1
3	0	8	7	2
4	1	4	3	0

(c)

Crew \ Job	1	2	3	4
1	2	4	1	0
2	3	0	1	2
3	0	3	1	2
4	2	3	0	1

2. Suppose that three subcontractors have bid on three jobs with prices (expressed in thousands of dollars) as shown in the following table. Because of a need to get the total project completed quickly, the government agency wants to give one job to each of the subcontractors. How can it do so at least cost?

Subcontractor \ Job	1	2	3
1	56	61	54
2	53	57	56
3	58	55	60

3. In each of the following assignment matrices, find a job assignment that makes the total (be it cost or time) a minimum.

(a)

Crew \ Job	1	2	3	4
1	75	80	85	72
2	73	64	76	75
3	71	76	73	77
4	76	80	70	78

(b)

Crew \ Job	1	2	3
1	56	54	61
2	53	57	60
3	58	55	57

(c)

Crew \ Job	1	2	3
1	57	54	61
2	53	57	56
3	58	55	60

(d)

Crew \ Job	1	2	3	4
1	50	60	54	53
2	57	51	58	62
3	52	55	53	54
4	54	55	60	58

(e)

Crew \ Job	1	2	3	4
1	34	40	45	38
2	33	38	35	40
3	39	42	41	36
4	40	37	43	44

(f)

Crew \ Job	1	2	3	4
1	75	80	85	72
2	73	76	76	75
3	71	64	73	75
4	76	70	80	78

(g)

Crew \ Job	1	2	3	4
1	34	33	45	38
2	40	38	35	40
3	39	42	41	36
4	40	37	43	44

(h)

Crew \ Job	1	2	3	4
1	50	60	54	53
2	57	51	58	62
3	55	52	53	54
4	54	55	60	58

4. Suppose that an educational TV station has three programs that it intends to show during prime time. The number of viewers that each program can attract is affected by the other programs being shown at

the same time. Suppose that the following table gives the number of viewers (in thousands) that each program is expected to attract at the given times. Which program should be aired at what time in order to attract the *largest* possible audience?

		Program		
		1	2	3
Hour aired	7 : 00–8 : 00	56	54	61
	8 : 00–9 : 00	53	52	60
	9 : 00–10 : 00	58	55	57

EXERCISES 8.4. B

1. In each of the following assignment matrices, find an assignment such that the total cost is zero.

(a)

		Job			
		1	2	3	4
Crew	1	0	1	4	5
	2	3	4	0	6
	3	2	4	1	0
	4	2	0	1	3

(b)

		Job			
		1	2	3	4
Crew	1	2	0	1	3
	2	1	3	4	0
	3	0	1	5	3
	4	2	4	0	3

(c)

		Job			
		1	2	3	4
Crew	1	1	3	0	5
	2	0	2	1	4
	3	4	0	5	2
	4	3	4	1	0

2. Suppose that a contractor has three employees and three types of jobs that need to be done repeatedly. If the following table tells how long it takes each of the employees to do each job, which person should he assign to which job so as to have the jobs done most quickly?

		Job		
		1	2	3
Employee	1	58	54	56
	2	56	53	55
	3	59	57	55

3. In each of the following assignment matrices, find a job assignment that makes the total (be it cost or time) a minimum.

(a)

Crew \ Job	1	2	3	4
1	34	40	45	38
2	35	38	33	40
3	39	42	41	36
4	40	37	43	44

(b)

Crew \ Job	1	2	3
1	74	76	79
2	85	80	78
3	77	75	80

(c)

Crew \ Job	1	2	3
1	60	58	65
2	57	61	60
3	62	59	64

(d)

Crew \ Job	1	2	3	4
1	48	45	52	51
2	50	52	49	53
3	44	46	51	50
4	50	45	54	46

(e)

Crew \ Job	1	2	3	4
1	50	30	37	42
2	50	53	47	35
3	43	51	50	42
4	46	49	52	50

(f)

Crew \ Job	1	2	3	4
1	55	53	60	62
2	47	51	53	57
3	56	49	55	54
4	60	59	57	55

(g)

		Job 1	2	3	4
Crew	1	54	57	60	55
	2	56	63	62	57
	3	47	49	50	52
	4	50	53	55	56

(h)

		Job 1	2	3	4	5
Crew	1	1	2	6	4	7
	2	9	8	10	12	3
	3	5	11	13	4	3
	4	6	10	11	2	5
	5	7	8	9	12	1

4. Suppose that an educational TV station has three programs that it intends to show during evening prime time. The number of viewers that each program can attract is affected by the other programs being shown at the same time. Suppose that the following table gives the number of viewers (in thousands) that each program is expected to attract at the given times. Which program should be aired at what time in order to attract the *largest* possible audience?

		Program 1	2	3
Hour aired	7 : 00–8 : 00	58	54	56
	8 : 00–9 : 00	56	53	55
	9 : 00–10 : 00	59	57	55

ANSWERS 8.4. A

1. (a) 1st crew gets 3rd job; 2nd crew gets 2nd job; 3rd crew gets 4th job; 4th crew gets 1st job
   (b) 1st crew gets 2nd job; 2nd crew gets 3rd job; 3rd crew gets 1st job; 4th crew gets 4th job
   (c) 1st crew gets 4th job; 2nd crew gets 2nd job; 3rd crew gets 1st job; 4th crew gets 3rd job
2. 1st subcontractor gets 3rd job; 2nd subcontractor gets 1st job, 3rd subcontractor gets 2nd job
3. (a) 1st crew gets 4th job; 2nd crew gets 2nd job; 3rd crew gets 1st job; 4th crew gets 3rd job
   (b) 1st crew gets 2nd job; 2nd crew gets 1st job; 3rd crew gets 3rd job
   (c) 1st crew gets 2nd job; 2nd crew gets 1st job; 3rd crew gets 3rd job
   (d) 1st crew gets 4th job; 2nd crew gets 2nd job; 3rd crew gets 3rd job; 4th crew gets 1st job

(e) 1st crew gets 1st job; 2nd crew gets 3rd job; 3rd crew gets 4th job; 4th crew gets 2nd job
(f) 1st crew gets 4th job; 2nd crew gets 3rd job; 3rd crew gets 2nd job; 4th crew gets 1st job *or* 1st crew gets 4th job; 2nd crew gets 1st job; 3rd crew gets 3rd job; 4th crew gets 2nd job
(g) 1st crew gets 1st job; 2nd crew gets 3rd job; 3rd crew gets 4th job; 4th crew gets 2nd job
(h) 1st crew gets 4th job; 2nd crew gets 2nd job; 3rd crew gets 3rd job; 4th crew gets 1st job

4. Program 3 at 8:00; program 2 at 7:00; program 1 at 9:00

VOCABULARY

transportation problem, northwest corner algorithm, fictitious warehouses, minimum cell algorithm, stones, basic variables, initial solution, degenerate, fake stones, assignment problem

SAMPLE TEST

Chapter 8

1. (20 pts) Tell whether each of the following problems is an assignment problem or a more general transportation problem. Then set up the table needed to apply the appropriate algorithm. Do not solve!
   (a) Suppose that a piano broker has located three very similar pianos which she can sell for the same price (postpaid) to any one of three willing customers. It would cost \$130 to send the first piano to the first customer, \$145 to send it to the second customer, and \$140 to the third. It would cost \$125 to send the second piano to the first customer, \$145 to send it to the second, and \$135 to send it to the third. The third piano would cost \$150 to send to the first customer, \$170 to the second, and \$160 to the third. Which piano should be shipped to which customer for a minimum total cost?
   (b) Suppose that a company has two machines, each of which makes the same three products. It costs machine I \$25 to make product *A*, \$45 to make product *B*, and \$40 to make product *C*. It costs machine II \$30 to make product *A*, \$35 to make product *B*, and \$35 to make product *C*. Machine I can make 35 products a day total and machine II can make 55. If the company wants to manufacture 25 of each product each day, how many of each should each machine make for a minimum total cost?
2. (20 pts) Solve the assignment problem shown in the table.

		Job 1	2	3	4
Team	1	2	4	6	8
	2	3	9	12	15
	3	5	13	18	20
	4	10	1	9	10

3. (40 pts) In the transportation problem shown in the table:
(a) Find the NCA initial solution.
(b) Find the MCA initial solution.
(c) Find the shipping pattern that gives the minimum cost.
(d) Find the minimum cost.

		Warehouse A	Warehouse B	Warehouse C	
Factory	1	14	12	15	55
	2	12	7	16	55
	3	8	6	10	50
		40	45	50	

4. (20 pts) Find a minimum cost pattern for the following transportation problem.

		Warehouse A	Warehouse B	Warehouse C	
Factory	1	4	6	5	100
	2	5	9	7	110
	3	7	8	8	110
		90	100	80	

The answers are at the back of the book.

Chapter 9

PROBABILITY

9.1 Basic Concepts

Probability theory has been developed to model events that are more predictable in the long run than in the short run. For example, if you flip a coin once, it is uncertain whether it will show a head, but if you flip it 1000 times, it will show heads about half the time. Such indefinite observations have been the subject of lengthy philosophical debates over the centuries.

But probability can become a precise mathematical topic by making certain definitions and assumptions and deducing results from these. This section discusses the "sample space" of a probability experiment and presents two methods—intuition and past experience—for assigning probability values to each element of the sample space. The models that result in either case will not predict future events in the real world with complete certainty, but no model does. Probability models, however, are sufficiently accurate to be extremely useful in business, biology, and most of the social sciences.

9.1.1 Example

If a fair coin is tossed, what is the probability that it will come down heads?

Solution:

The answer is $\frac{1}{2}$. The coin has two sides, heads and tails, and even children will generally agree that each is "equally likely" to fall if the coin is tossed by an impartial person. (This is worth remembering in case you are ever caring for two children, and both want the last jelly bean or neither wants to take a bath first.)

The reason that coins provide a peaceful device for a harried parent or sitter making a difficult decision is that almost all people perceive *tossing a coin* as a probability experiment. There are two basic outcomes to this probability experiment—heads and tails—and it is generally accepted that while the outcome cannot be predicted for certain at any particular toss of the coin, in the long run heads will appear $\frac{1}{2}$ the time and tails will appear $\frac{1}{2}$ the time. We say "the probability" of heads is $\frac{1}{2}$ and "the probability" of tails is $\frac{1}{2}$. This conclusion used only intuition—the process of thinking about a problem.

9.1.2 Example

Suppose that a die (singular of dice) is thrown.
(a) What is the probability that a 2 will show?
(b) What is the probability that a number less than 3 will show?
(c) What is the probability that an even number will show?

Solution:

A die is a small cube used in many children's games (and some more adult pastimes) such that each number from 1 to 6 appears on exactly one of its six sides. Thus, when a die is tossed, the top face may show "1," "2," "3," "4," "5," or "6." We say there are six basic outcomes to the probability experiment of *tossing a die*.

(a) Using intuition again, we claim that each of the six basic outcomes seems equally likely, so we say that each has a probability $\frac{1}{6}$. In particular, the probability of rolling a 2 is $\frac{1}{6}$.
(b) There are two numbers less than 3—1 and 2. The probability of rolling either a 1 or a 2 is $\frac{1}{6} + \frac{1}{6} = \frac{1}{3}$.
(c) There are three even numbers among the basic outcomes—2, 4, and 6. Since each has probability $\frac{1}{6}$, the probability of rolling an even number is $\frac{1}{6} + \frac{1}{6} + \frac{1}{6} = \frac{1}{2}$.

With these two examples in mind, we pause to give some definitions that will be useful in our later presentation of probability concepts. Then we shall return to more applications.

9.1.3 Definition

A set of possible outcomes of a probability experiment is said to be a set of basic outcomes if the set is:
(a) Exhaustive—one of the outcomes in the set must occur each time the experiment is performed.
(b) Exclusive—no two outcomes in the set can occur at the same time.

9.1.4 Definition

A set of basic outcomes in a probability experiment is called a sample space of that experiment.

We can diagram the sample space of tossing a coin as in Figure 9.1–1. The sample space of rolling a die can be diagrammed as in Figure 9.1–2.

Figure 9.1–1

Heads	Tails

Figure 9.1–2

1	3	5
2	4	6

The sample space of a probability experiment may not be unique. For example, if we pull a card out of a bridge deck, we may only be interested in its suit (in which case the sample space contains 4 basic outcomes), or we may be interested in its denomination (in which case the sample space has 13 basic outcomes), or we may want to know the exact card (in which case the sample space has 52 basic outcomes).

Once the sample space for a probability experiment has been established, the next step in developing a probability model is to assign probability values to each basic outcome.

9.1.5 Definition

A probability distribution assigns a probability (a number) to each basic outcome of a sample space so that:

(a) The probability of each basic outcome is between zero and 1.
(b) The sum of the probabilities of all the basic outcomes is 1.

If something is sure to happen, we say it has probability 1; if it is impossible, we say it has probability zero. The probability that the sun will shine in Kentucky some time next year may be said to be 1. On the other hand, the probability that a college student will be alive 200 years from now would be said to be zero. (These two probabilities leave room for philosophical debate, but are valid for practical purposes.) Probability values range from 0 to 1—the nearer the event is to being certain, the nearer the probability to 1. The probability that a random college student will be alive 5 years from now is greater than the probability that she or he will be alive 50 years from now.

The usual ways of assigning probabilities to the probability experiments of tossing a coin and rolling a die are given in Figures 9.1–3 and 9.1–4, respectively.

Figure 9.1–3

Heads: $\frac{1}{2}$	Tails: $\frac{1}{2}$

Figure 9.1–4

1: $\frac{1}{6}$	3: $\frac{1}{6}$	5: $\frac{1}{6}$
2: $\frac{1}{6}$	4: $\frac{1}{6}$	6: $\frac{1}{6}$

The numbers associated with each basic outcome in these experiments are those that would measure the likelihood of each outcome, assuming that the coin or die is "fair." If either does not have the probability distribution given above, then we would conclude that it is "weighted" or "unfair."

9.1.6 Definition

An event (or outcome) in a probability experiment consists of one or more basic outcomes of the experiment, and is therefore any nonempty subset of the sample space.

Each basic outcome of a probability experiment is an event, but there may also be other events. If a die is rolled, the following are "events":

(a) "Having a number less than 3 show." (This event consists of basic outcomes 1 and 2.)

(b) "Having an even number show." (This event consists of basic outcomes 2, 4, and 6.)

9.1.7 Definition

The probability of an event E in a probabilistic model is the sum of the probabilities of all the basic outcomes contained in E.

For example, since the event "having an even number show" when rolling a die is the union of basic outcomes 2, 4, and 6, its associated probability is $\frac{1}{6} + \frac{1}{6} + \frac{1}{6} = \frac{1}{2}$.

Not all probability models are as easy to describe as those for a coin or a die. In these two cases our intuition alone enables us to guess the probabilities. But often we use past tests to guess at probabilities. This technique assumes that the future will be somewhat like the past—an assumption open to doubt, but the best we can do.

When discussing the probability of a particular event in a probability model, it is common to refer to that event as "success"; the word "success" is used in a specialized way that need not include the ordinary connotation of "desirable."

9.1.8 Rule

If a probability experiment has been performed n times in the past, where n is large, and m of these times resulted in "success," we assume that the probability of success in the future is m/n.

9.1.9 Example

Suppose that a company employs numerous glassblowers who make a difficult type of bottle. Suppose that someone chooses at random 250 of the bottles made by one blower and discovers that 200 of them are acceptable; the others are defective. What is the probability that a bottle made by this blower will be acceptable?

Solution:

There have been $n = 250$ experiments, of which $m = 200$ resulted in "success." By Rule 9.1.8 we therefore assume that the probability of success is

$$\frac{m}{n} = \frac{200}{250} = \frac{4}{5} = 0.8$$

9.1.10 Example

Suppose that a baby is expected. What is the probability that the baby will be a girl?

Solution:

Let $P(G)$ denote the probability of having a girl. It is common to assume $P(G) = 0.5$ because intuition suggests this, and this assumption is practical for most purposes.

However, the statistics reveal that out of about 3,731,000 babies born in the United States in 1970, only about 1,816,000 were girls. Therefore, if we base our answer to the posed question on national experience for 1970, we conclude there is a probability of $\frac{1816}{3731} = 0.487$ of an American baby being a girl. (Other years yield a similar result. Rounding off to the nearest tenth, of course, we obtain 0.5.)

In the previous example the question might be raised as to the probability $P(B)$ of having a boy. Since a child must be either a boy or a girl, the assumption $P(G) = 0.5$ yields $P(B) = 1 - 0.5 = 0.5$. But if we assume that $P(G) = 0.487$, then $P(B) = 1 - 0.487 = 0.513$. In either model [$P(G) = 0.5$ or $P(G) = 0.487$] we must have $P(G) + P(B) = 1$. This is a special case of the following rule.

9.1.11 Rule

If $P(E)$ denotes the probability of the event E, then P (not E) $= 1 - P(E)$.

9.1.12 Example

If a die is rolled, what is the probability that a number other than 2 will show?

Solution:

Let E be the event that a 2 shows. Then

$$P\ (\text{not } E) = 1 - P(E) = 1 - \frac{1}{6} = \frac{5}{6}$$

When there are more than two basic outcomes, it is often useful to consider the relationship between probabilities of two events.

9.1.13 Example

If a die is rolled, what is the probability that it will show *both* a number less than 3 *and* an even number?

Solution:

Figure 9.1–5

$1: \frac{1}{6}$	$3: \frac{1}{6}$	$5: \frac{1}{6}$
$2: \frac{1}{6}$	$4: \frac{1}{6}$	$6: \frac{1}{6}$

If E_1 denotes the event "a number less than 3" and E_2 denotes the event "an even number," then the probability that *both* will happen is the probability of the intersection of sets E_1 and E_2, as shown in Figure 9.1–5. Consulting this figure, we see that

$$P(E_1 \text{ and } E_2) = P(E_1 \cap E_2) = \tfrac{1}{6}$$

9.1.14 Example

If a die is rolled, what is the probability that it will show *either* a number less than 3 *or* an even number?

Solution:

Again we consult Figure 9.1–5, but this time we are interested in the points that lie in *either* E_1 or E_2 (their union) instead of the points that lie in *both* E_1 and E_2 (their intersection). We are adding the probability of four basic outcomes—1, 2, 4, and 6. We cannot simply add $P(E_1)$ and $P(E_2)$ to get $P(E_1 \text{ or } E_2) = P(E_1 \cup E_2)$ because the event "2" occurs in both E_1 and E_2 and $P(2)$ would be counted twice. Thus we must subtract $P(2) = P(E_1 \text{ and } E_2)$ from $P(E_1) + P(E_2)$ to obtain $P(E_1 \text{ or } E_2) = P(E_1 \cup E_2)$. We get

$$P(E_1 \text{ or } E_2) = P(E_1 \cup E_2) = P(E_1) + P(E_2) - P(E_1 \text{ and } E_2) = \tfrac{1}{3} + \tfrac{1}{2} - \tfrac{1}{6}$$
$$= \tfrac{5}{6} - \tfrac{1}{6} = \tfrac{2}{3}$$

We can generalize the ideas appearing in Example 9.1.14.

9.1.15 Rule

If E_1 and E_2 are two events associated with a probability experiment (that is, they are two subsets of the sample space), then

$$P(E_1 \text{ or } E_2) = P(E_1) + P(E_2) - P(E_1 \text{ and } E_2)$$

If the two events should happen to have no basic outcomes in common (for example, when rolling a die E_1 might be "a number less than 3" and E_2 might be "a number greater than 4"), then $P(E_1 \text{ and } E_2) = 0$, so the last term in Rule 9.1.15 disappears and Rule 9.1.17 is obtained.

9.1.16 Definition

Two events associated with a probability experiment are said to be mutually exclusive if they do not contain any basic outcomes in common—that is, they are subsets of the sample space with a void intersection.

9.1.17 Rule

If E_1 and E_2 are mutually exclusive events in one probability experiment, then

$$P(E_1 \text{ or } E_2) = P(E_1) + P(E_2)$$

In other words, if two events are mutually exclusive, then we may add the probabilities to obtain the probability of the union of the two events.

9.1.18 Example

If a die is rolled, what is the probability that it will show a number less than 3 or a number greater than 4?

Solution:

Since a number cannot be both less than 3 and greater than 4, we have two mutually exclusive events, and by Rule 9.1.17 we can add the probabilities.

$$P(E_1 \text{ or } E_2) = P(E_1) + P(E_2) = \tfrac{1}{3} + \tfrac{1}{3} = \tfrac{2}{3}$$

9.1.19 Example

Suppose that an administrator wonders how lucrative a certain toll booth is on a major parkway. A survey is taken during 100 minutes, selected at random, to see how many vehicles pass through the booth during each of the sampled minutes. The following chart is constructed.

Number of vehicles	0	1	2	3
Number of times this number of vehicles passed through the toll booth during an observed minute	20	40	25	15

Use Rule 9.1.8 and this information to answer the following questions.

(a) What is the probability that no one passes through this toll booth during a random minute?

(b) What is the probability that at least one vehicle passes through the toll booth during a random minute?

(c) What is the probability that two or three vehicles will hurry through during a random minute?

Solution:

(a) The experiment was tried $n = 100$ times of which $m = 20$ are designated as "success." Thus if E_0 is the event of having no car pass through the toll booth, we conclude that

$$P(E_0) = \frac{20}{100} = 0.2$$

(b) We can use Rule 9.1.8 directly and conclude that

$$P\text{ (not } E_0) = \frac{40 + 25 + 15}{100} = \frac{80}{100} = 0.8$$

Alternatively, we could use Rule 9.1.11 and write

$$P\text{ (not } E_0) = 1 - 0.2 = 0.8$$

The answer is the same.

(c) Let E_2 denote the event of having 2 cars pass through the booth. Using Rule 9.1.8, we have

$$P(E_2) = \frac{25}{100} = 0.25$$

Let E_3 denote the event of having 3 cars pass through the booth. Rule 9.1.8 yields

$$P(E_3) = \frac{15}{100} = 0.15$$

Since E_2 and E_3 are mutually exclusive, we can add the probabilities (Rule 9.1.17):

$$P(E_2 \text{ or } E_3) = P(E_2) + P(E_3) = 0.25 + 0.15 = 0.4$$

If there are only a small number of basic outcomes, it can be helpful to write the probabilities in a *probability vector*, each element of which is the probability of one basic outcome. For example, the probability vector describing the tossed coin would be (0.5, 0.5) and the probability vector describing the toll booth in Example 9.1.19 is (0.2, 0.4, 0.25, 0.15).

9.1.20 Definition

A probability vector is a vector $P = [p_1, p_2, \ldots, p_n]$ such that (1) $p_i \geq 0$ for all i and (2) $p_1 + p_2 + \cdots + p_n = 1$.

9.1.21 Example

A supervisor of an encyclopedia sales staff has records showing that of the 200 college students who have sold encyclopedias for a summer job in the past, 120 sold no sets of encyclopedias on their first day working, 60 sold one set the first day, and 20 sold two sets. Assume that there is no drastic change in recruitment, training, or marketing potential.

(a) Write a probability vector describing this situation.
(b) What is the probability that a new staff member this coming summer will sell nothing on the first day?
(c) What is the probability of someone selling exactly one set?
(d) What is the probability that a newcomer will sell at least one set?

Solution:

(a) The basic outcomes are the number of sets that can be sold: 0, 1, and 2. The probability of these is, respectively, $\frac{120}{200} = 0.6$, $\frac{60}{200} = 0.3$, and $\frac{20}{200} = 0.1$, so the probability vector is $[0.6, 0.3, 0.1]$.
(b) 0.6 (c) 0.3 (d) $0.3 + 0.1 = 0.4$

9.1.22 Example in Biology

Probability is indispensable when studying heredity. The simplest mathematical model occurs when only two genes determine an inherited trait and one of the characteristics is dominant over the other. (Tall in pea plants dominates short. Black in guinea pigs dominates white. Brown eyes in humans dominates blue.) The dominant gene might be denoted by an A and the recessive gene by an a. Then individuals possessing either AA or Aa will manifest the dominant trait of the pair; only those with aa will exhibit the recessive characteristic.

Unlike the rest of the body, reproductive cells have only one gene for each trait—either A or a. When the sperm joins the egg, the offspring will then have two genes, one from each parent; these two genes will determine the corresponding trait in the offspring.

(a) Suppose that two brown-eyed Aa-type humans marry. What is the probability that a child of theirs will have blue eyes?
(b) Write the probability vector for the gene type of their children.
(c) Suppose that a black Aa guinea pig mates with a white (and therefore aa) guinea pig. What will be the probability vector describing their offsprings' color?

Solution:

(a) Both the father and the mother can contribute an A or an a with equal probability to the offspring. Thus the possibilities can be charted as follows, with each situation equally likely:

		Father's gene A	a
Mother's gene	A	AA	Aa
	a	Aa	aa

Only if the child has the gene pattern in the lower right box will the eyes be blue. Since this is one out of four equally likely possibilities, the probability of a child of this marriage having blue eyes is $\frac{1}{4}$.

(b) Only if the upper left box describes the inheritance will the offspring have brown eyes of the *AA* type. Half the offspring will inherit the *Aa* gene type: they will have brown eyes, but some of their children may have blue eyes. The probability vector is [0.25, 0.5, 0.25] where the elements describe the probability of *AA*, *Aa*, and *aa*, respectively.

(c) The pattern is

		Black parent A	a
White parent	a	Aa	aa
	a	Aa	aa

so the vector is [0.5, 0.5], where the vectors describe *Aa* and *aa*.

(Biological models, like business models, are only approximations of the truth. In particular, because of other factors not mentioned in this model, it is possible that two blue-eyed parents could have a brown-eyed child.)

SUMMARY

Probability theory models uncertain situations by defining a sample space (an exhaustive and exclusive set of basic outcomes) and assigning a number between 0 and 1 (called its probability) to each basic outcome of the sample space. The sum of the probabilities of all the basic outcomes on one sample space must be 1.

Sometimes the probability numbers are assigned to a sample space by common agreement because they seem obvious (as in the tossing of a coin). Other times we use past experience to establish probabilities; if n repetitions of the same probability experiment have been made resulting in m successes, we say that the probability of success for that probability experiment is m/n.

An event is any nonempty subset of the sample space; an event consists of one or more basic outcomes. The probability of each event is the sum of the probabilities of the basic outcomes included in that event. If E_1 and E_2 denote events and $P(E)$ denotes the probability of the event E, then P (not E) $= 1 - P(E)$ and $P(E_1$ or $E_2) = P(E_1) + P(E_2) - P(E_1$ and $E_2)$. If E_1 and E_2 are mutually exclusive, then $P(E_1$ or $E_2) = P(E_1) + P(E_2)$.

A probability model can be represented by a probability vector, $P = [p_1, p_2, \ldots, p_n]$, where $p_i \geq 0$ for all i and $p_1 + p_2 + \cdots + p_n = 1$. Such a probability vector models a given probability experiment if, when the sample space of that experiment consists of the basic outcomes $b_1, b_2, \ldots, b_n$, the probability that the basic outcome b_i will occur is given by p_i for each i.

EXERCISES 9.1. A

1. Toss a coin 10 times and record the number of heads. Do it again. Make a total of 100 tosses and record how many heads there are in each group of 10. Does the total number of heads get nearer to half the number of tosses as the number of tosses gets larger? You may want to tabulate the results of the whole class.
2. Suppose that there are 15 defective watches in a batch of 200 watches.
   (a) If a watch is bought at random, what is the probability that it will be defective?
   (b) Write a probability vector that describes this situation.
3. Suppose that a 300-page book has exactly one misprint on each of 90 pages, exactly 2 on each of 22 pages, and no misprints on the other pages.
   (a) If I read a page at random, what is the probability that it will have exactly 1 misprint?
   (b) What is the probability that it will have no misprints?
   (c) What is the probability that it will have at least one misprint?
   (d) Write a probability vector describing this situation.
4. Suppose that the following chart describes the number of cars that arrived at a toll barrier during 100 random minutes sampled during rush hour.

Number of cars that arrived	0	1	2	3	4	5	6
Number of times the above number of cars was observed	9	21	26	21	13	7	3

   Assume that this sample can be used to predict the future.
   (a) What is the probability that exactly 2 cars will arrive at the toll barrier in a random minute during rush hour?
   (b) Write a probability vector describing this situation.
   (c) What is the probability that more than 3 cars will arrive at the toll barrier during a random minute?
   (d) What is the probability that fewer than 3 cars will arrive?
5. If a die is thrown, what is the probability of getting:
   (a) The number 3? (b) Either a 5 or a 6? (c) A number less than 5?

(d) An odd number? (e) A number that is both less than 5 and odd? (f) A number that is either less than 5 or odd? (g) A number that is neither less than 5 nor odd?

6. A bridge deck has 52 cards, 13 in each of its four suits, which are clubs, diamonds, hearts, and spades. If a card is picked at random from the deck, what is the probability of getting:
(a) A heart? (b) A card that is not a heart? (c) A 4 in any one of the four suits? (d) Either a 4 or an 8? (e) Either a 4 or a heart? (f) Either a 4 or an 8 or a heart?
7. Suppose that I buy one ticket in a raffle that sells 2000 tickets.
   (a) What are my chances of winning first prize in the raffle?
   (b) If Tricky Tom makes 100 copies of his ticket and throws them all into the raffle, what is the probability of his winning the grand prize?
8. The sex of an offspring is determined by the process described in Example 9.1.22. Every female has XX sex genes; she received one X gene from each parent. Every male has Xy genes; he received the X from his mother and the y from his father. Construct a chart for the offspring as in Example 9.1.22 and write the probability vector for the sex of the offspring.
9. Some X genes—we shall call them X_b—carry hemophilia, "the bleeder's disease." The blood of a victim of this disease does not clot normally, and the victim is in constant danger of dying from a minor scratch. The combinations $X_b y$ and $X_b X_b$ result in a "bleeder" but $X_b X$ does not. A woman with the genes $X_b X$ is called a "carrier" of hemophilia.
   (a) Suppose that an $X_b X$ woman marries a normal Xy man. Make a chart as was done in Example 9.1.22. What is the probability that a son of theirs will be a bleeder? What is the probability that a daughter of theirs will be a carrier?
   (b) Why does hemophilia affect more men than women? (Incidentally, colorblindness is another sex-linked characteristic which, more innocently, is inherited through the same pattern as hemophilia.)

EXERCISES 9.1. B

1. Same as problem 1 of Exercises 9.1.A.
2. Suppose that there are 40 defective pairs of scissors in a batch of 250.
   (a) If you buy one of them at random, what is the probability that it will be defective?
   (b) Write the probability vector for this situation.
3. Suppose that a secretary has typed 200 letters; he made exactly one mistake on 65 of them, exactly two mistakes on 10 of them, and no mistakes on the rest of the letters.
   (a) If the boss picks one up at random to check carefully, what is the probability that he can find two mistakes?
   (b) What is the probability that the one he checks will have exactly one mistake?
   (c) What is the probability that it will have at least one mistake?
   (d) Write the probability vector.

4. Suppose that the following chart describes the number of telephone calls received by one company during 100 random minutes sampled during a variety of working days.

Number of calls received	0	1	2	3	4	5	6
Number of times the above number of calls was received	5	15	22	24	18	10	6

Assume that this sample can be used to predict the future.
(a) What is the probability that this company will receive exactly 3 telephone calls during a randomly chosen minute?
(b) Write a probability vector describing this situation.
(c) What is the probability that more than 2 calls will be received during a random minute?
(d) What is the probability that fewer than 2 calls will be received?

5. If a die is thrown, what is the probability of getting:
(a) The number 4? (b) Either a 4 or a 5? (c) A number other than 4 or 5? (d) A number strictly less than 4? (e) A number that is both strictly less than 4 and odd? (f) A number that is either strictly less than 4 or odd? (g) A number that is neither strictly less than 4 nor odd?

6. A bridge deck has 52 cards, 13 in each of its four suits, which are clubs, diamonds, hearts, and spades. If a card is picked at random from the deck, what is the probability of getting:
(a) A diamond? (b) A card that is not a diamond? (c) A 5 in any one of the four suits? (d) Either a 5 or a 9? (e) Either a 5 or a diamond? (f) Either a 5 or a 9 or a diamond?

7. Suppose that I buy one ticket in a raffle that sells 1000 tickets.
(a) What are my chances of winning first prize in the raffle?
(b) If Deceitful Dora makes 200 copies of her ticket and throws them all into the raffle, what is the probability of her winning the grand prize?

8. There are human pairs of genes such that the recessive trait causes its victim to die in early childhood. Thus possessing *aa* results in death before the reproductive years. A person with an *Aa* gene structure is called a "carrier" of this disease.
(a) If one of your brothers or sisters died as a result of one of these diseases, what do you know about your parents' genes?
(b) What is the probability that you are a carrier?

EXERCISES 9.1. C

1. Suppose that the raffle in problem 7 of Exercises 9.1.B involves two drawings for first and second prize. What is the probability that two of Deceitful Dora's tickets will be drawn so that she will be found out? Do not simplify the arithmetic.

ANSWERS 9.1. A

2. (a) $\frac{3}{40}$ (b) $[\frac{3}{40}, \frac{37}{40}]$
(The order of the elements could be reversed; it does not matter as long as you are consistent throughout any one problem.)
3. (a) $\frac{3}{10}$ (b) $\frac{47}{75}$ (c) $\frac{28}{75}$ (d) $[\frac{47}{75}, \frac{3}{10}, \frac{11}{150}]$
4. (a) $\frac{13}{50}$ (b) [0.09, 0.21, 0.26, 0.21, 0.13, 0.07, 0.03] (c) 0.23 (d) 0.56
5. (a) $\frac{1}{6}$ (b) $\frac{1}{3}$ (c) $\frac{2}{3}$ (d) $\frac{1}{2}$ (e) $\frac{1}{3}$ (f) $\frac{5}{6}$ (g) $\frac{1}{6}$
6. (a) $\frac{1}{4}$ (b) $\frac{3}{4}$ (c) $\frac{1}{13}$ (d) $\frac{2}{13}$ (e) $\frac{4}{13}$ (f) $\frac{19}{52}$
7. (a) $\frac{1}{2000}$ (b) $\frac{101}{2100}$

8.

		Father	
		X	y
Mother	X	XX	Xy
	X	XX	Xy

(0.5, 0.5)

9. (a)

		Father	
		X	y
Mother	X_b	$X_b X$	$X_b y$
	X	XX	Xy

The right column yields a son; the probability is $\frac{1}{2}$ that the son will be a bleeder. The left column yields a daughter; there is a probability of $\frac{1}{2}$ that she will be a carrier.

(b) Male bleeders occur about half the time that a female carrier has a son. Since the mother is healthy, there is no reason why she should not have her share of sons, and a healthy father does not help the sons inherit any less tendency for bleeding. By contrast, a female hemophiliac can occur only when *both* parents contribute an X_b gene. But this implies that her father is a bleeder himself, and many bleeders die before reproductive age or choose not to have children.

ANSWERS 9.1. C

1. $\frac{201}{1200} \cdot \frac{200}{1199}$

★9.2 Counting

Many probability models require special counting techniques to determine the probability of an event happening merely by chance. In this section we consider three of these counting techniques, and then we use them to compute probabilities.

Multiplication Rule

9.2.1 Example

OVER FIFTY TYPES OF PIZZA! says the sign as you drive up. Inside you discover only the choices "onions, peppers, mushrooms, sausages, anchovies, and meatballs." Did the advertiser lie?

Solution:

No, because the cook will be glad to put any one of the six choices on your pizza or leave it off. Thus there are 2 choices for onions—on or off. This is multiplied by the 2 choices for peppers to get 4—you can have both, just onions, just peppers, or neither. This is multiplied by the 2 choices for mushrooms—you can have all three, just onions and peppers, just onions and mushrooms, just peppers and mushrooms, just onions, just peppers, just mushrooms, or none of the three. This is multiplied by the 2 choices for sausages to get 16, which is multiplied by the 2 choices for anchovies to get 32, which is multiplied by the 2 choices for meatballs to get 64. So the advertisement is correct; there are more than 50 choices.

9.2.2 Multiplication Rule

If there are m choices of one item and there are n independent choices of another item, there are mn choices altogether.

This rule can be used repeatedly as it was in Example 9.2.1—six times, once for each added ingredient.

9.2.3 Example

Suppose that a man has 3 pairs of trousers, 2 jackets, 5 shirts, and 8 ties. If he delights in combining them in all possible ways, how many different outfits can he make?

Solution:

$3 \cdot 2 \cdot 5 \cdot 8 = 240$ outfits.

Permutations

9.2.4 Definition

If r items are selected from among n items and are arranged in a specific order, each such ordered selection is called a permutation of n items taken r at a time.

9.2.5 Example

Suppose that there are 5 delegates to be arranged at a prestigious meeting where each of the 5 chairs has political subtleties associated with it. How many choices are there for arranging the 5 people on the 5 chairs?

Solution:

Let us call the people A, B, C, D, and E and let us seat them in turn. Person A has 5 choices of chairs. But once A is placed, there are only 4 choices for B. Thus there are $5 \cdot 4 = 20$ choices for seating A and B. Once this is accomplished there are only 3 chairs left for C. Thus there are $5 \cdot 4 \cdot 3 = 60$ arrangements for A, B, and C. There are only 2 chairs left now, so D has only two choices, yielding $5 \cdot 4 \cdot 3 \cdot 2 = 120$ potential arrangements. Then E has only one remaining choice of a chair, so there are still "only" 120 ways to arrange 5 people on 5 chairs.

We say that there are 120 permutations of 5 items taken 5 at a time.

9.2.6 Example

Suppose that a book company has divided the country into 6 territories and has hired 6 travelers to sell their books. In how many ways can the 6 territories be assigned to the 6 travelers?

Solution:

The first territory can be assigned to any one of the 6 travelers, the second to any one of the remaining 5, the third to any one of the remaining 4, and so forth. Thus there are $6 \cdot 5 \cdot 4 \cdot 3 \cdot 2 \cdot 1 = 720$ ways to assign the territories.

We say that there are 720 permutations of 6 items taken 6 at a time.

9.2.7 Rule

If n distinct objects (or people) are to be arranged in n different positions, there are $n! = n(n-1)(n-2)\cdots 3 \cdot 2 \cdot 1$ different ways to make the arrangement. We say that the number of permutations of n items taken n at a time is $n!$.

The expression $n! = n(n-1)(n-2)\cdots 3 \cdot 2 \cdot 1$ is called "n factorial."

Sometimes we want to choose only some (not all) of the elements from a group for an arrangement. Then we modify Rule 9.2.7.

9.2.8 Example

In how many ways is it possible to pick a president, vice president, secretary, and treasurer out of a group of 10 people?

Solution:

Any one of the 10 can be president, and one of the remaining 9 can be vice president, any one of the remaining 8 can be secretary, and any one of the remaining 7 can be treasurer. Thus there are $10 \cdot 9 \cdot 8 \cdot 7 = \frac{10 \cdot 9 \cdot 8 \cdot 7 \cdot 6 \cdot 5 \cdot 4 \cdot 3 \cdot 2 \cdot 1}{6 \cdot 5 \cdot 4 \cdot 3 \cdot 2 \cdot 1} = \frac{10!}{6!} = 5040$ ways these officers can be chosen.

We say that there are $\frac{10!}{6!} = 5040$ permutations of 10 items taken 4 at a time.

9.2.9 Example

Suppose that a company has 6 sales territories and 14 applicants for the 6 jobs of covering them. In how many ways can the company assign an applicant to each territory?

Solution:

Any one of the 14 can be given the first territory, any one of the remaining 13 can be given the second, any one of the remaining 12, the third, and so forth. Thus there are $14 \cdot 13 \cdot 12 \cdot 11 \cdot 10 \cdot 9 = \frac{14!}{8!}$ possible ways of making the assignment.

We say that there are $\frac{14!}{8!}$ permutations of 14 items taken 6 at a time.

Generalizing from Examples 9.2.8 and 9.2.9, we have the following rule.

9.2.10 Permutations Rule

The number of permutations of n items taken r at a time (when order counts) is denoted by ${}_nP_r$ and equals

$$n(n-1)(n-2)\cdots(n-r+1) = \frac{n!}{(n-r)!}$$

We write

$${}_nP_r = \frac{n!}{(n-r)!}$$

Notice that Rule 9.2.7 merely says that the number of permutations of n things taken n at a time is $n!$ Since $0!$ is conventionally defined to be 1, if $r = n$ in Rule 9.2.10, we obtain Rule 9.2.7, that is,

$${}_nP_n = \frac{n!}{(n-r)!} = \frac{n!}{(n-n)!} = n!$$

Combinations

If we do not care which applicant gets which territory, but are merely choosing 6 out of 14 applicants, then we are considering a combination instead of a permutation. The number of ways of choosing 6 people from among 14 without regard to order is "the number of combinations of 14 people taken 6 at a time." Similarly, the number of ways of choosing 4 delegates from among 10 people without regard to who gets which post is called "the number of combinations of 10 people taken 4 at a time," or ${}_{10}C_4$.

9.2.11 Definition

If r items are to be selected from among n items in such a way that order does not matter, each such selection is called a combination of n items taken r at a time. In other words, a combination of n items taken r at a time is a subset of r items taken from a universe of n items.

The number of combinations of n things taken r at a time is denoted ${}_nC_r$. We now present a method for evaluating ${}_nC_r$.

Each combination of r things can be arranged in r! distinct ways by Rule 9.2.7. Thus there are $r!{}_nC_r$ ways of arranging n things r at a time where order counts. But this just equals ${}_nP_r$. So

$$r!{}_nC_r = {}_nP_r$$

Dividing by r! yields

$${}_nC_r = {}_nP_r\frac{1}{r!} = \frac{n!}{(n-r)!}\frac{1}{r!} = \frac{n!}{(n-r)!r!}$$

9.2.12 Combinations Rule

The number of ways to choose r items out of a set of n items without regard to order is

$${}_nC_r = \frac{n!}{r!(n-r)!}$$

We now motivate Rule 9.2.12 through an example.

9.2.13 Example

Suppose that only 3 people among a group of 5 with similar jobs can be given identical pay raises this year. How many ways are there to choose the lucky 3?

Solution:

Since the pay raises are identical, order does not matter, so this is a combination problem. Again let us call our people A, B, C, D, and E. Their names are to be arranged in the following boxes:

Get a pay raise			Don't get a pay raise	
☐	☐	☐	☐	☐
1	2	3	4	5

As in Example 9.2.5, there are $5! = 120$ distinct ways of putting the 5 people in the 5 places. But this time we need not distinguish among places 1, 2, and 3—each is equally good. Similarly, we need not distinguish between places 4 and 5—each is equally bad. Thus the arrangement

$$A \quad B \quad C \qquad D \quad E$$

is for this purpose the same as any of the following:

$$\begin{array}{ccccc} A & C & B & D & E \\ B & C & A & D & E \\ B & A & C & D & E \\ C & A & B & D & E \\ C & B & A & D & E \end{array}$$

All the permutations of A, B, and C on the left could also be combined with $E\ D$ on the right and yield the same practical result. Thus there are $6 \cdot 2 = 12$ different arrangements that will result in the raises going to A, B, and C. By the same reasoning there will be 12 arrangements that would give any other trio the raises. Thus there are $\frac{120}{12} = 10$ different ways to choose the three for raises.

In the previous example we again saw that ${}_5C_3 = \frac{5!}{3!2!}$. This was because the number of permutations of 5 items had to be divided by the number of ways of rearranging the chosen 3 items (that is, 3!) and also by the number of ways of rearranging the not-chosen 2 items (that is, 2!). Similarly, in Example 5.3.2 we saw there were $10 = \frac{5!}{3!2!}$ ways of choosing trios of equations from a set of 5 equations.

Notice that ${}_nC_r$ must be the same as ${}_nC_{n-r}$ [the number of ways to choose $(n - r)$ items from a set of n] because the items chosen when computing ${}_nC_r$ are exactly the items "not chosen" when computing ${}_nC_{n-r}$.

9.2.14 Example

If a camera shop contains 100 different types of merchandise of which 5 are to be displayed in the window this week, how many different combinations of 5 items can be chosen for display?

Solution:

$$_{100}C_5 = \frac{100!}{5!95!} = \frac{\overset{5}{\cancel{100}} \cdot 99 \cdot 98 \cdot 97 \cdot \overset{16}{\cancel{96}} \cdot 95!}{\cancel{5} \cdot \cancel{4} \cdot \cancel{3} \cdot \cancel{2} \cdot 1 \; 95!} = 75{,}287{,}520$$

different combinations! Thus more than 75 million weeks could pass before the choice must be repeated. (A student without a calculator would be justified in leaving the answer in the form $\frac{100!}{5!95!}$.)

The next examples indicate how counting problems can be useful in computing probabilities.

9.2.15 Example

Suppose that you have 4 chickens and one small feeding dish in the corner of their coop. One day after putting grain in the dish you observe the order in which they eat. The next day you observe the order again and it is the same. What is the probability that this is sheer chance?

Solution:

This time order does matter, so it is a permutations problem. By Rule 9.2.7 there are $4! = 4 \cdot 3 \cdot 2 \cdot 1 = 24$ different orders in which the chickens could eat. If they were all equally likely, there would be a probability of $\frac{1}{24}$ that on the second day the order would be the same as on the first.

9.2.16 Example

Suppose that there are 4 trees in your backyard, 2 maple and 2 oak. Two pairs of robins are nesting in your trees and you assume that they prefer to nest in distinct trees. If each pair chooses a tree in your yard at random, what is the probability that both pairs will choose maple trees?

Solution:

There are $_4C_2 = \frac{4!}{2!2!} = \frac{4 \cdot 3 \cdot 2}{2 \cdot 2} = 6$ ways of choosing 2 trees out of 4, and of these combinations, only one includes both the maple trees. If each of the 6 combinations is equally likely, there is probability $\frac{1}{6}$ that both pairs of robins will choose maple trees.

9.2.17 Example

In the situation described in Example 9.2.16, suppose that the pairs of robins do not care whether they nest in the same tree or not. What is the probability now that both pairs will both end up in the maple trees if their choice is determined solely by chance?

Solution:

This time we use Multiplication Rule 9.2.2. The first pair has 4 choices and so does the second, so there are $4 \cdot 4 = 16$ different arrangements. Among these there are 2 ways the first pair can choose a maple tree and 2 ways the second can, so there are $2 \cdot 2 = 4$ arrangements in which they both nest in maple trees. If all arrangements were equally likely, there is a probability $\frac{4}{16} = \frac{1}{4}$ that both will nest in the maple trees.

9.2.18 Example

Suppose that there are 3 job openings and only 2 of the 7 applicants are women. If the hiring is a matter of sheer chance, what is the probability that at least one of the women will get a job?

Solution:

First, we compute the total number of ways of filling the jobs:

$$_7C_3 = \frac{7!}{3!4!} = \frac{7 \cdot 6 \cdot 5 \cdot 4!}{3 \cdot 2 \cdot 1 \cdot 4!} = 7 \cdot 5 = 35$$

Now we ask: In how many of these 35 trios chosen from the 7 does at least one woman appear? Let us, with our usual display of imagination, call the applicants A, B, C, D, E, F, and G; the women are A and B. There are three possibilities for "success"—both get jobs, A gets a job but not B, and B gets a job but not A.

If both A and B are hired, then exactly one of the five men is too. There are 5 such possibilities, since $_5C_1 = \frac{5!}{1!4!} = 5$.

If A but not B is hired, then 2 of the 5 men are hired. There are $\frac{5!}{2!3!} = 10$ such possibilities.

If B but not A is hired, then again 2 of the 5 men are hired; again there are $_5C_2 = 10$ possibilities.

Thus in 35 distinct, equally likely cases, $5 + 10 + 10 = 25$ of them include at least one woman. Thus, we conclude that there is a probability of $\frac{25}{35} = \frac{5}{7}$ that at least one woman will be hired. This implies that there is less than one third of a chance that they will both be left out even if only 2 of the 7 applicants are women—*if* the decision is based on chance.

SUMMARY

Counting the number of items combined, arranged, or chosen can be important in devising a probability model for an applied problem. This section presents three rules for such counting and gives examples illustrating the use of these rules.

EXERCISES 9.2. A

If the arithmetic is overwhelming, leave the answers in factorial form.

1. Find: (a) ${}_4C_3$ (b) ${}_8C_6$ (c) ${}_6C_2$ (d) ${}_{20}C_{10}$
2. If a style of shoes comes in 5 sizes, 7 widths, and 3 colors, how many pairs of this style must the store keep on hand to supply any customer who might want to buy the shoe?
3. (a) In how many distinct orders can 6 political speakers be scheduled on a program?
   (b) If the scheduling is done at random, what is the probability that your favorite candidate will appear either first or last?
4. How many ways can a set of 4 equations be chosen from a list of 8 equations?
5. How many ways can a sample of 50 light bulbs be chosen from a batch of 1000?
6. If a certain species of pea plant can have long or short stems, smooth or wrinkled pods, and red or white flowers, how many different varieties of this species are thus distinguished?
7. If a given model car can be bought in 10 outside colors; 5 colors of upholstery; 3 different gear types; radial, white-wall, or regular tires; air conditioning or not; radio or not; limited slip differential or not; power steering or not; power brakes or not; power windows or not; and 4 different engine types, how many distinct kinds of this model car are there?
8. If a furniture store has decided to run feature ads once a week for 10 weeks on 10 items, in how many different orders can it feature the products?
9. How many ways can a baseball team of 9 be chosen from 12 players if there is no concern as to who gets which position?
10. If there are 4 job openings and 8 applicants, 2 of whom are from a minority group, what is the probability that at least 1 of the 2 will get a job if the workers are chosen at random?
11. A wine taster claims that he can distinguish good from poor wines; an experiment is designed to test his skills. Suppose that three pairs of glasses of wine are provided, each pair including an expensive and an inexpensive wine. The taster tastes each pair and tells which is the expensive wine. With no skill at all, what is the probability that he will be right for at least two of the three pairs?

EXERCISES 9.2. B

If the arithmetic is overwhelming, do not simplify the answers.

1. Find: (a) ${}_5C_4$ (b) ${}_8C_3$ (c) ${}_6C_3$ (d) ${}_{15}C_{11}$
2. (a) How many distinct abbreviations for organizations (such as IRS, AMA, and AAA) can be formed from 3 letters of our alphabet?

(b) How many potential distinct New Jersey license plates are there if each consists of three letters followed by three digits?
3. How many ways can 8 cub scouts be lined up to get a treat?
4. How many ways can a set of 4 equations be chosen from a list of 9 equations?
5. How many ways can 6 cereals be chosen from a box of 8 distinct cereals?
6. How many ways can a sample of 40 thermometers be chosen from a batch of 500? (Do not simplify!)
7. If a menu has 5 appetizers, 10 main dishes, and 6 desserts, how many different dinners can be chosen?
8. Suppose that 12 patients are examined and are classified as to "healthy," "needing outpatient care," "needing hospitalization." What is the number of basic outcomes in this sample space? (*Hint*: Classify each of the 12 patients in any one of 3 ways.)
9. If 5 executives in a company are to be given two-week summer vacations so that no two overlap, in how many different orders can the executives have their vacations?
10. Suppose that a given style of model house comes in 20 colors, with 3 types of siding, 4 qualities of roofing materials, 4 ranges of air conditioning, 3 types of interior floor finish, 3 types of stairway banisters, 3 types of furnaces, 2 types of stoves, 10 styles of bathroom fixtures, 4 types of fireplaces, and 3 types of front porches. In how many ways can the builder thus vary his homes without making any two exactly identical?
11. If a committee of 3 is to be chosen at random from a group of 8 people, 2 of whom are from some minority group, what is the probability the group will be completely excluded from the committee?
12. Two people claim to use ESP to communicate whether a coin has fallen "heads" or "tails." Suppose that one of them flips a coin three times, looks at it but says nothing, and the other, standing across the room, says "heads" or "tails." With no special powers, what is the probability that the caller will be correct at least 2 of the 3 times?

ANSWERS 9.2. A

1. (a) 4 (b) 28 (c) 15 (d) $\frac{20!}{10!10!} = 184{,}756$ 2. 105
3. (a) $6! = 720$ (b) $\frac{1}{3}$ 4. 70 5. $\frac{1000!}{50!950!}$ 6. $2^3 = 8$
7. $10 \cdot 5 \cdot 4 \cdot 3^2 \cdot 2^6 = 115{,}200$ types 8. $10! = 3{,}628{,}800$ 9. 220
10. $\frac{11}{14}$ 11. $\frac{1}{2}$

9.3 Conditional Probability

Conditional probability is used to model a probability experiment when a certain event is known or assumed to have occurred. For example, we might want to know the probability that a randomly thrown die shows a two on the top face, given that we know the top face is less than four. Let A be the event that the top face is a two and let B be the event that the top face is less than four. We represent the probability that A will occur given that B has occurred by $P(A \mid B)$. In this case $P(A \mid B) = \frac{1}{3}$ even though $P(A) = \frac{1}{6}$, since obtaining a two is one of the three equally likely outcomes that are possible when the number on top of the die is less that four.

Conditional probability is often useful in modeling situations where two or more characteristics are used to describe a sample space. For example, we might consider not only whether a certain bottle is of acceptable quality but also who made it. If so, we might wonder if the probability that a bottle is acceptable "depends on" who made it. The fact that 0.8 of all the bottles are acceptable does *not* imply that 0.8 of those made by each worker is acceptable.

These considerations motivate the following definition.

9.3.1 Definition

Two events A and B are called independent if $P(A) = P(A \mid B)$.

9.3.2 Example

Suppose that a glassblowing center surveys 1000 of its products chosen at random and compiles the following chart.

	Tozzi B_1	Berlinski B_2	
Acceptable A_1	200	600	800
Defective A_2	50	150	200
	250	750	1000

The A's and B's are abbreviations for distinct (but *not* mutually exclusive) events. The four basic outcomes for this probability experiment are (A_1 and B_1), (A_1 and B_2), (A_2 and B_1), and (A_2 and B_2). Probabilities such as $P(A_1 \text{ and } B_1)$ are called joint probabilities. We now consider several types of questions that can be answered using this chart. We assume, as in Rule 9.1.8, that the past can be used to predict the future.

(a) If a bottle is chosen at random from among these 1000 bottles, what is the probability that it is acceptable?

(b) If a bottle is chosen at random from among the bottles made by Tozzi, what is the probability that it is acceptable?

(c) What is the probability that a bottle chosen at random from among the entire 1000 was made by Tozzi?

(d) What is the probability that a bottle chosen at random from among the acceptable bottles was made by Tozzi?

Solution:

(a) $P(A_1) = \frac{800}{1000} = \frac{4}{5} = 0.8$

(b) Out of 250 past experiments (the bottles made by Tozzi), 200 were "successes," so

$$P(A_1|B_1) = \frac{200}{250} = \frac{4}{5} = 0.8$$

(c) $P(B_1) = \frac{250}{1000} = \frac{1}{4} = 0.25$

(d) Out of 800 past experiments (the acceptable bottles), 200 were "successes," so

$$P(B_1|A_1) = \frac{200}{800} = \frac{1}{4} = 0.25$$

We see in Example 9.3.2 that the events "made by Tozzi" and "acceptable" are independent, no matter which one of the two events is assumed. First, let us consider a less direct method for computing conditional probabilities.

After a table such as that in Example 9.3.2 is made, it is often useful to write a probability distribution, which is a similar table, but the numbers are all divided by whatever number appears in the lower right—the total number of trials. This division changes all the numbers in the table to probabilities. The probability distribution displays all the values of the joint probabilities and makes it easy to calculate the conditional probabilities. The probability distribution for the table in Example 9.3.2 is

	B_1	B_2	
A_1	0.2	0.6	0.8
A_2	0.05	0.15	0.2
	0.25	0.75	1.0

Each conditional probability is, by definition, the quotient of some number in the interior of the *original* table divided by the number either at the end of its row or beneath its column. But when the entire original table is divided by the same number (in the lower right) to obtain the corresponding probability distribution, these quotients do not change. Thus we can also find the conditional probabilities by dividing quantities that appear in the *probability distribution* table. It follows that (whether or not the events are independent)

$$P(A_i|B_j) = \frac{P(A_i \text{ and } B_j)}{P(B_j)} \quad \text{and} \quad P(B_j|A_i) = \frac{P(A_i \text{ and } B_j)}{P(A_i)}$$

Using the probability distribution table, parts (b) and (d) of Example 9.3.2 could have been computed as follows:

(b) $P(A_1|B_1) = \frac{0.2}{0.25} = 0.8$ (d) $P(B_1|A_1) = \frac{0.2}{0.8} = 0.25$

Now look again at the joint probability table on page 317. Each one of the joint probabilities (the four numbers in the inner rectangle) can be obtained by multiplying the corresponding probabilities at the end of its row and column. This will always be true if and only if the characteristics events are independent. This follows from the fact that we always have $P(A|B) = \dfrac{P(A \text{ and } B)}{P(B)}$; clearly, if $P(A|B) = P(A)$, it follows that $P(A)P(B) = P(A \text{ and } B)$. Conversely, if $P(A)P(B) = P(A \text{ and } B)$, division yields $P(A) = \dfrac{P(A \text{ and } B)}{P(B)}$ and we have, therefore, that $P(A) = P(A|B)$. These ideas motivate the following definition.

9.3.3 Definition

Two characteristics (such as "acceptability" and "maker" in Example 9.3.2) are independent if and only if $P(A_i \text{ and } B_j) = P(A_i) \cdot P(B_j)$ for every i and j. They are called "dependent" if there is an i and j such that $P(A_i$ and $B_j) \neq P(A_i)P(B_j)$.

9.3.4 Example

Suppose that the same glassblowing center surveys 1000 bottles made by Suzuki, Chang, and Kim and compiles the following table.

	Suzuki B_1	Chang B_2	Kim B_3	
Acceptable A_1	160	410	230	800
Defective A_2	40	90	70	200
	200	500	300	1000

(a) Write the corresponding probability distribution table.
(b) Are "Acceptable" and "Made by Chang" independent here? Why?
(c) Find the probability that a bottle chosen from the 1000 is acceptable.
(d) Find the probability that a bottle made by Suzuki is acceptable.
(e) Find the probability that a bottle made by Chang is acceptable.
(f) Find the probability that a bottle made by Kim is acceptable.
(g) Find the probability that an acceptable bottle was made by Suzuki.
(h) Find the probability that an acceptable bottle was made by Chang.
(i) Find the probability that a bottle is both acceptable and made by Kim.

Solution:

(a)

	B_1	B_2	B_3	
A_1	0.16	0.41	0.23	0.8
A_2	0.04	0.09	0.07	0.2
	0.2	0.5	0.3	1.0

(b) No, because we cannot obtain 0.41 by multiplying 0.8 and 0.5.
(c) $P(A_1) = 0.8$
(d) $P(A_1 | B_1) = \frac{0.16}{0.2} = 0.8$
(e) $P(A_1 | B_2) = \frac{0.41}{0.5} = 0.82$
(f) $P(A_1 | B_3) = \frac{0.23}{0.3} \approx 0.77$ (rounded off)
(g) $P(B_1 | A_1) = \frac{0.16}{0.8} = 0.2$
(h) $P(B_2 | A_1) = \frac{0.41}{0.8} \approx 0.51$ (rounded off)
(i) $P(A_1 \text{ and } B_3) = 0.23$

9.3.5 Example

Suppose that a mail-order house surveys 500 of its credit-card holders selected at random and makes the following table relating the number who used their cards in the month just past to those who used it in the month before that. That is, (A_1 and B_1) indicates the card holders who used their cards in both of the two just-preceding months.

		Card used last month B_1	Card not used last month B_2	
Card used the month before last	A_1	225	75	300
Card not used the month before last	A_2	40	160	200
		265	235	500

(a) Write a probability distribution describing this table. Use the probability distribution to answer the remaining questions.
(b) Are the purchasing trends of the two months independent?
(c) What is the probability that a customer used his or her card both of the two preceding months?
(d) What is the probability that a card was used neither month?
(e) What is the probability that a card was used one month but not both?
(f) What is the probability that a random card was used in the month just past?
(g) If a card is picked at random from those used this past month, what is the probability that it was also used the month before last?
(h) If a card is picked at random from those used the month before last, what is the probability that it was also used last month?

Solution:

(a)

	B_1	B_2	
A_1	0.45	0.15	0.6
A_2	0.08	0.32	0.4
	0.53	0.47	1.0

(b) No, because, for example, $0.45 \neq (0.6)(0.53)$.
(c) $P(A_1 \text{ and } B_1) = 0.45$
(d) $P(A_2 \text{ and } B_2) = 0.32$
(e) $P(A_1 \text{ and } B_2) + P(A_2 \text{ and } B_1) = 0.15 + 0.08 = 0.23$
(f) $P(B_1) = 0.53$
(g) $P(A_1 | B_1) = \frac{0.45}{0.53} \approx 0.85$ (rounded off)
(h) $P(B_1 | A_1) = \frac{0.45}{0.6} = 0.75$

SUMMARY

If two events (such as the number of buyers this month and the number of buyers last month) are involved in one probability experiment, it is often useful to consider conditional probabilities. The conditional probability $P(A | B)$ gives the probability that A will occur if it is already known that B has occurred. It can be computed using the formula

$$P(A | B) = \frac{P(A \text{ and } B)}{P(B)}, \text{ provided } P(B) > 0$$

Two events A and B are said to be "independent" if $P(A) = P(A | B)$.

EXERCISES 9.3. A

1. Suppose that a random sample is taken of 1000 people leaving a store the day after an ad appears in the local paper.
   (a) If 0.5 of those sampled saw the ad and 0.8 of them made a purchase before leaving the store, does it follow that $0.4 = (0.5)(0.8)$ of them both saw the ad and made a purchase?

   Suppose that the following table summarized the survey. Use it and Rule 9.1.8 to answer the remaining questions.

		Made a purchase at the store B_1	Did not make a purchase at the store B_2	
Saw the ad	A_1	450	50	500
Did not see the ad	A_2	350	150	500
		800	200	1000

   (b) What is the probability that a random person sampled had neither seen the ad nor made a purchase?
   (c) What is the probability that a random person saw the ad but did not make a purchase?
   (d) Assuming that a person sampled had made a purchase, what is the probability that she or he saw the ad?
   (e) Assuming that a person sampled had seen the ad, what is the probability that she or he made a purchase?
   (f) Write a probability distribution table from these data.
   (g) $P(A_2 | B_1) =$ (h) $P(B_1 | A_2) =$ (i) $P(A_1 | B_2) =$ (j) $P(B_2 | A_1) =$

2. Suppose that in a large random sample of ball bearings bought by a company, 0.9 were acceptable while 0.1 were defective. Furthermore, 0.4 of them were bought from manufacturer 1 and 0.6 from manufacturer 2.
   (a) Does it follow that 0.1 of those received from manufacturer 1 were defective?
   (b) Suppose that the characteristic "defective" is independent of the manufacturer. Complete the probability distribution table. Let A_1 be the event the ball bearing is acceptable, and A_2 be the event that it is defective. The manufacturers are identified in a natural way.

	B_1	B_2	
A_1			
A_2			

   (c) Suppose now that the acceptability of the ball bearings depends on the manufacturer and suppose the probability of a ball bearing being both acceptable and made by manufacturer 1 is 0.3. Complete the probability distribution table.

	B_1	B_2	
A_1			
A_2			

   Using the table you made in part (c), find:
   (d) $P(A_1|B_1)$ (e) $P(B_1|A_1)$ (f) $P(A_2|B_1)$ (g) $P(B_1|A_2)$
3. A three-year study of about 120,000 Japanese men completed in 1970 revealed that 1225 of them had died of lung cancer during the study. Of the 25,000 nonsmokers in the group, 25 had been victims of lung cancer; and of the 20,000 who smoked more than 2 packs a day, 480 had died of lung cancer. (The numbers have been rounded for your convenience.)
   (a) Complete the following probability density table.

	Non-smokers B_1	Moderate smokers B_2	Those who smoked more than 2 packs B_3	
Died of lung cancer A_1				
Did not die of lung cancer A_2				

(b) Are smoking and death from lung cancer statistically independent?
(c) How do you know?
(d) Write out in words the meaning of the following expressions: $P(A_1)$; $P(A_1 | B_1)$.
(e) Compute each of the probabilities indicated in part (d).

4. In 1970 there were about 64 million housing units in the United States. They were distributed as in the following chart, where the numbers are in millions:

	Owner-occupied B_1	Renter-occupied B_2	
White A_1	37	20	57
Nonwhite A_2	3	4	7
	40	24	64

(a) Write a probability distribution table for this information.
(b) Was home ownership in the United States in 1970 independent of race of the family?
(c) What is the probability that a white U.S. family in 1970 owned its home?
(d) What is the probability that a nonwhite U.S. family in 1970 owned its home?

EXERCISES 9.3. B

1. Suppose that a random sample of 1000 licensed drivers emerging from the cafeteria of a large commuter college reveals that 0.6 of them drove less than 10,000 miles last year and 0.1 of them were involved in a reportable traffic accident during the year.
(a) Does it follow that $0.06 = (0.6)(0.1)$ of them both drove less than 10,000 miles and were involved in a reportable accident?

Suppose that the following table summarized the survey. Use it and Rule 9.1.8 to answer the remaining questions.

		Had a reportable traffic accident B_1	Did not have a reportable accident B_2	
Drove less than 10,000 miles last year	A_1	40	560	600
Drove 10,000 miles or more last year	A_2	60	340	400
		100	900	1000

(b) What is the probability that a random person sampled had driven more than 10,000 miles last year without having an accident?
(c) What is the probability that a random person sampled drove less than 10,000 miles last year without having an accident?

(d) Assuming that a person sampled had an accident, what is the probability that she or he drove less than 10,000 miles last year?
(e) Assuming that a person sampled drove less than 10,000 miles last year, what is the probability that she or he had an accident?
(f) Write a probability distribution from these data.
(g) $P(A_2|B_1) =$ (h) $P(B_1|A_2) =$ (i) $P(A_1|B_2) =$ (j) $P(B_2|A_1) =$

2. Suppose that in a large random sample of mouth thermometers, 0.8 were acceptable and 0.2 defective.
(a) If 0.7 were bought from firm 1 and 0.3 from firm 2, does it follow that 0.2 of those bought from firm 2 were defective?
(b) Suppose that the characteristic "defective" is independent of the supplying firm. Complete the probability distribution table, including the marginal probabilities. Let A_1 be the event the thermometer is acceptable, and A_2 be the event that it is defective. The firms are identified in a natural way.

	B_1	B_2	
A_1			
A_2			

(c) Suppose now that the acceptability of the thermometers depends on the firm supplying them, and the probability of a thermometer being both acceptable and supplied by firm 1 is 0.6. Complete the probability distribution table.

	B_1	B_2	
A_1			
A_2			

Using the table you made in part (c), find:
(d) $P(A_1|B_1)$ (e) $P(B_1|A_1)$ (f) $P(A_2|B_2)$ (g) $P(B_2|A_2)$

3. Complete the accompanying probability distribution table if $P(A_1) = 0.6$, $P(B_1) = 0.3$, $P(B_3) = 0.2$, $P(A_1 \text{ and } B_2) = 0.35$, and $P(A_2 \text{ and } B_3) = 0.05$.

	B_1	B_2	B_3	
A_1				
A_2				

4. There were about 3.6 million babies born in the United States in 1969; they could be classified as in the following chart, where the numbers are in thousands:

	White	Nonwhite	
Babies lived	2947	581	3528
Babies died in the first year	53	19	72
	3000	600	3600

(a) Write a probability distribution table for this information.
(b) Was infant mortality independent of race in the United States in 1969?
(c) What was the probability of a random white baby dying in his or her first year?
(d) What was the probability of a random nonwhite baby dying in his or her first year?

ANSWERS 9.3. A

1. (a) No (b) 0.15 (c) 0.1 (d) 0.5625 (e) 0.9

(f)

	B_1	B_2	
A_1	0.45	0.05	0.5
A_2	0.35	0.15	0.5
	0.8	0.2	1.0

(g) 0.4375 (h) 0.7 (i) 0.25 (j) 0.1

2. (a) No

(b)

	B_1	B_2	
A_1	0.36	0.54	0.9
A_2	0.04	0.06	0.1
	0.4	0.6	1.0

(c)

	B_1	B_2	
A_1	0.3	0.6	0.9
A_2	0.1	0	0.1
	0.4	0.6	1.0

(d) 0.75 (e) $\frac{1}{3}$ (f) 0.25 (g) 1

3. (a)

	B_1	B_2	B_3	
A_1	0.0002	0.0060	0.0040	0.0102
A_2	0.2081	0.6190	0.1627	0.9898
	0.2083	0.6250	0.1667	1.0000

(b) No

(c) $(0.0102)(0.2083) = 0.0021 \neq 0.0002$

(d) $P(A_1)$ indicates the probability that a man selected at random from such a group will die of lung cancer during such a study. $P(A_1 | B_1)$ indicates the probability that a nonsmoker selected at random will die of lung cancer during such a study.

(e) $P(A_1) = 0.0102$; $P(A_1 | B_1) = 0.0010$

4. (a)

	B_1	B_2	
A_1	0.58	0.31	0.89
A_2	0.05	0.06	0.11
	0.63	0.37	1.00

(b) No (c) 0.65 (d) 0.43

★9.4 Regular Markov Matrices

When the outcome of a probability experiment depends on the result of a just-previous experiment but not on earlier trials, we have what is known as a Markov process, named after Andrei Andreevich Markov (1856–1922), who investigated it thoroughly. For example, the probability of a customer using a charge card this month might depend on whether or not the card was used last month. The probability of your car needing a major repair next year might depend on how many major repairs it had last year. And the probability of persuading a college alumnus to donate to the alumni fund this year is related to whether or not he gave last year. Such situations may generate Markov processes and Markov matrices.

Suppose that a mail-order house has a probability distribution similar to that obtained in part (a) of Example 9.3.5 several months in a row. Then it might conclude that the probability of one of last month's customers using his or her card again this month is 0.75 [because $P(B_1 | A_1) = 0.75$] and the probability of his not using his card again is 0.25 [because $P(B_2 | A_1) = 0.25$]. On the other hand, if a customer did not use his card last month, the probability of his using it this month is $0.08/0.4 = 0.2$ and the probability of his not using it this month is $0.32/0.4 = 0.8$.

Thus we have *two* probability vectors, [0.75, 0.25] and [0.2, 0.8], that describe this month's purchases, depending on whether the sampled customer made a purchase last month or not. If we write these vectors together, we obtain a Markov matrix:

		This month	
		Card used	Card not used
Last month	Card used	0.75	0.25
	Card not used	0.2	0.8

9.4.1 Definition

A Markov matrix is a square matrix such that all elements are greater than or equal to zero and the sum of each row is 1.

Using the notation of Chapter 1, a_{ij} in a Markov matrix often represents the probability of the jth event, given that the ith event has happened just previously. In Example 9.3.5 the first event was "card used" and the second was "card not used." The months were the intervals between which the probability experiments were tried.

9.4.2 Example

Using the Markov matrix derived just before Definition 9.4.1 as a summary of the ideas in Example 9.3.5, find the expected number of charge-card customers for May if 300 of the card holders made purchases in April and 600 did not. (In Example 9.3.5, only 500 customers were surveyed, but there are 900 in total.)

Solution:
Three fourths of the 300 April purchasers can be expected to return in May along with 0.2 of the 600 nonpurchasers. Thus the expected number of purchasers in May is 0.75(300) + 0.2(600) = 345. The computations are automatic using a Markov matrix:

$$[300, 600]\begin{bmatrix} 0.75 & 0.25 \\ 0.2 & 0.8 \end{bmatrix} = [345, 555]$$

Notice that the left element of each vector indicates the buyers, and the right element indicates the nonbuyers. Thus the store can expect 345 buyers and 555 nonbuyers in May.

Encouraged by such good news (45 more buyers in May than in April) the market analyst might hasten to examine the prospects for June:

$$[345, 555]\begin{bmatrix} 0.75 & 0.25 \\ 0.2 & 0.8 \end{bmatrix} = [370, 530]$$

Progress is clearly slower—only 25 more purchasers in June than in May. One might wonder if it will continue to July:

$$[370, 530]\begin{bmatrix} 0.75 & 0.25 \\ 0.2 & 0.8 \end{bmatrix} = [384, 516]$$

The question arises as to whether the number of customers per month is approaching a limit and, if so, what it is. That is, we are wondering if the sequence of vectors [300, 600], [345, 555], [370, 530], [384, 516], and so on, is getting closer and closer to some "fixed" vector, which we shall call X. To answer this question it might seem as if we, or our computer friend, would have to keep multiplying matrices for a long time. But if we look at the question from another angle, it turns out there is a faster way to get the answer.

Suppose that there were a limit to the given sequence of vectors, and we have named it X. Then we would expect that there would be no change in next year's number of purchasers once we have "reached" X. This means that $XM = X$. Thus to find the value of X, if it exists, we might solve the equation $XM = X$.

9.4.3 Example

Find X such that $XM = X$ when $M = \begin{bmatrix} 0.75 & 0.25 \\ 0.2 & 0.8 \end{bmatrix}$.

Solution:
We are looking for an $X = [x, y]$ such that

$$[x, y]\begin{bmatrix} 0.75 & 0.25 \\ 0.2 & 0.8 \end{bmatrix} = [x, y]$$

This matrix equation is equivalent to

$$0.75x + 0.2y = x$$
$$0.25x + 0.8y = y$$

Either of these two equations can be changed to $0.25x - 0.2y = 0$ by subtracting. It follows that $0.25x = 0.2y$ and that $x = 4(0.2)y = 0.8y$, so any vector satisfying $x = 0.8y$ will satisfy $XM = X$. Checking, we write

$$[0.8y, y]\begin{bmatrix} 0.75 & 0.25 \\ 0.2 & 0.8 \end{bmatrix} = [0.8y, y]$$

In Example 9.4.2 the total number of customers is $300 + 600 = 900$, so we want $x + y = 900$. Thus we might solve

$$0.8y + y = 900$$

or

$$1.8y = 900$$

Dividing, we obtain $y = 500$. Since $x + y = 900$, $x = 400$. Checking again, we see that

$$[400, 500]\begin{bmatrix} 0.75 & 0.25 \\ 0.2 & 0.8 \end{bmatrix} = [400, 500]$$

so if there are exactly 400 buyers one month, this trend will tend to continue. In Example 9.4.2 you can see that the number of buyers is getting nearer and nearer to 400.

Is this method justified? It turns out that for many Markov matrices, M, it will always work, although we will need a short mathematical discussion to describe precisely when.

9.4.4 Definition

If M is any square matrix and X is a vector such that $XM = X$, then X is called a fixed vector for M. (Some books use the term "constant vector," or "stationary vector," or "characteristic vector," or "eigenvector.")

Example 9.4.3 showed that [400, 500] is a fixed vector for

$$\begin{bmatrix} 0.75 & 0.25 \\ 0.2 & 0.8 \end{bmatrix}$$

9.4.5 Definition

If M is a Markov matrix and P is a probability vector such that $PM = P$, then P is called a fixed probability vector for M.

If a matrix M has a fixed vector X, it is possible to find a fixed probability vector for M by dividing X by the sum of the elements in X. Thus a fixed probability vector for

$$\begin{bmatrix} 0.75 & 0.25 \\ 0.2 & 0.8 \end{bmatrix} \quad \text{is} \quad \left[\frac{400}{900}, \frac{500}{900}\right] = \left[\frac{4}{9}, \frac{5}{9}\right]$$

9.4.6 Definition

A Markov matrix is said to be regular if there is some positive integer n such that the elements of M^n are all positive.

Clearly, $\begin{bmatrix} 0.75 & 0.25 \\ 0.2 & 0.8 \end{bmatrix}$ is regular, because all the elements of $M^1 = M$ are positive. The following theorem is not obvious, nor is a proof given here. [For a proof, see *Finite Markov Chains* by J. Kemeny and J. Snell (New York: Van Nostrand Reinhold, 1960).] But several applications will display the power and usefulness of this theorem.

9.4.7 Theorem

If M is a regular Markov matrix,

(a) There is a *unique* fixed probability vector P_M for M.

(b) If Y is any vector that can be postmultiplied by M, then $\lim_{n \to \infty} YM^n$ is a fixed vector for M and (therefore) the product of P_M times the sum of the entries of Y.

Part (a) says that $[\frac{4}{9}, \frac{5}{9}]$ is the *only* fixed probability vector of $\begin{bmatrix} 0.75 & 0.25 \\ 0.2 & 0.8 \end{bmatrix}$.

And part (b) says that $[300, 600]M^n$ approaches $[400, 500]$ as n gets large since $[400, 500]$ is the fixed vector of M with the same total as $[300, 600]$.

9.4.8 Example

Suppose that an alumni association discovers that of those alumni who did not donate to last year's fund only $\frac{1}{10}$ will donate to this year's campaign, and they will give only small gifts. Of those who gave small gifts last year, $\frac{2}{10}$ will not give at all this year and $\frac{1}{10}$ will increase their donations to the "large gift" category; the others will maintain status quo. Of those who gave large gifts last year, $\frac{1}{10}$ will forget this year, $\frac{2}{10}$ will diminish the gift to a "small" gift, and the others will continue to be major supporters of their alma mater.

(a) Write a Markov matrix describing this situation.

(b) If this trend continues, what fraction of the alumni can eventually be expected each year to be nondonors, small donors, and large donors?

Solution:

(a)

		This year		
		No gift	Small gift	Large gift
Last year	No gift	0.9	0.1	0
	Small gift	0.2	0.7	0.1
	Large gift	0.1	0.2	0.7

It is easy (though somewhat tedious) to show that the indicated matrix A is indeed a regular Markov matrix by computing A^2 and verifying that none of the elements are zero.

(b) The fractions of the alumni that will eventually make up the three groups will form a fixed vector for the Markov matrix given above. To find out what these fractions are, we solve:

$$[x, y, z]\begin{bmatrix} 0.9 & 0.1 & 0 \\ 0.2 & 0.7 & 0.1 \\ 0.1 & 0.2 & 0.7 \end{bmatrix} = [x, y, z]$$

This is equivalent to equations (1), (2), and (3) below:

$$
\begin{array}{lrll}
(1) & 0.9x + 0.2y + 0.1z &= x & \\
(2) & 0.1x + 0.7y + 0.2z &= y & \\
(3) & 0.1y + 0.7z &= z & \\
(4) & 0.1y - 0.3z &= 0 & \text{[subtracting z from (3)]} \\
(5) & y &= 3z & \text{[solving (4)]} \\
(6) & -0.1x + 0.6z + 0.1z &= 0 & \text{[substituting (5) into (1)]} \\
(7) & -0.1x \qquad + 0.7z &= 0 & \text{(adding)} \\
 & x &= 7z &
\end{array}
$$

Checking, we see that, indeed,

$$[7z, 3z, z]\begin{bmatrix} 0.9 & 0.1 & 0 \\ 0.2 & 0.7 & 0.1 \\ 0.1 & 0.2 & 0.7 \end{bmatrix} = [7z, 3z, z]$$

Thus for any positive number z, the vector $[7z, 3z, z]$ is a fixed vector for the Markov matrix in part (a). Since a probability vector is needed, $7z + 3z + z = 1$, so $z = \frac{1}{11}$. Thus $P_M = [\frac{7}{11}, \frac{3}{11}, \frac{1}{11}]$. If the patterns remain as stated, Theorem 9.4.7 assures the alumni fund directors that no matter what this year's participation is, eventually the alumni fund will receive large gifts from about $\frac{1}{11}$ of the alumni, small gifts from about $\frac{3}{11}$ of the alumni, and nothing at all from the other $\frac{7}{11}$.

The same ideas as those used in Example 9.4.8 apply not only to other fund-raising situations, but also to projecting the buying habits of a large group of customers—either how much of a certain commodity they will buy or how much they will spend at a given store. The ideas of Markov matrices can be similarly applied to many other situations, including projecting the number of subscribers that do or do not renew subscriptions, the frequency with which an inherited trait disappears and reappears as animals reproduce, the number of American drivers who will or will not have a major accident next year, and the migrations of animals or people (tracing urban–rural shifts in sociology, for example). Our final detailed example has a slightly different emphasis.

9.4.9 Example

We now trace the fate of a rumor. Suppose it has a "yes–no" aspect, such as whether or not two famous movie stars are about to be divorced or whether or not a new disarmament treaty will soon be signed. Somebody knows the truth but does not want to tell the media yet. One friend confides in another until everyone "knows" about this situation. How reliable is this personally passed information?

Solution:

If everyone always told the truth, then everyone who was told anything about the situation would know the truth. But the reader will probably concede that people tell the truth not with probability 1, but with probability p such that $0 < p < 1$. Then for a yes–no rumor (that is, one that is either true or false), the probability that a speaker will reverse what he or she heard is $(1 - p)$, and we can form the following Markov matrix:

		What Everyperson told his or her friend: Yes	No
What Everyperson was told	Yes	p	$1 - p$
	No	$1 - p$	p

By Theorem 9.4.7 the proportion of people who eventually are told "yes" to those who are eventually told "no" will be given by the fixed probability vector of this matrix.

$$[x, y]\begin{bmatrix} p & 1-p \\ 1-p & p \end{bmatrix} = [x, y]$$

(1) $xp + y(1 - p) = x$

(2) $x(1 - p) + py = y$

(3) $y(1 - p) = x(1 - p)$ [subtracting xp from (1) and factoring]

(4) $y = x$ [dividing by $(1 - p)$ since $p \neq 1$]

Equation (4) suggests that for those far removed from the source of the rumor, the number who are told "yes" is approximately equal to the number of people who are told "no"—no matter what the original statement was and no matter how little people err in reporting the truth!

You might wonder how long it takes for the truth to get distorted. Suppose that people tell the truth 0.8 of the time. Then the powers of the appropriate Markov matrix are:

$$M = \begin{bmatrix} 0.8 & 0.2 \\ 0.2 & 0.8 \end{bmatrix} \qquad M^2 = \begin{bmatrix} 0.68 & 0.32 \\ 0.32 & 0.68 \end{bmatrix}$$

$$M^3 = \begin{bmatrix} 0.608 & 0.392 \\ 0.392 & 0.608 \end{bmatrix} \qquad M^4 = \begin{bmatrix} 0.5648 & 0.4352 \\ 0.4352 & 0.5648 \end{bmatrix}$$

This computation implies that the fourth person to hear the rumor has only a 56% probability of getting the truth. To see this, suppose that the truth is "yes." Then the corresponding vector is [1, 0]. But after it has passed through 4 people it becomes

$$XM^4 = [1, 0]\begin{bmatrix} 0.5648 & 0.4352 \\ 0.4352 & 0.5648 \end{bmatrix} = [0.5648, 0.4352]$$

A similar computation shows that if the original truth was "no" symbolized by [0, 1], it will come to the fourth person with a probability about 56%.

It follows that even the fourth person in line has only a 56% probability of receiving the truth—assuming that people are truthful 0.8 of the time! If you are further removed from the primary source than fourth or if some combination of carelessness, dishonesty, and imperfect memory should render the gossiping public less than 80 percent accurate, your chances of hearing the truth are much less. (See problems 5 in Exercises 9.4.A and 9.4.B.)

Notice that Markov matrices are another kind of input–output matrix with input at the left and output at the top. But, in contrast to Leontief matrices, the fractions are taken of the input quantities (*last* month's purchases, *last* year's donors), and so it is the rows—not the columns—that add up to 1. This is also why we use row vectors on the left instead of column vectors on the right. Confusing, perhaps, but conventional. As was observed at the end of Section 1.2, different disciplines use linear mathematics in different forms; this text has conformed to the conventional notation within each applied area.

SUMMARY

If the results of a previous probability experiment affect the probability of a coming experiment, it is helpful to use Markov matrices, square matrices such that every row is a probability vector. If for a square matrix M, X is such that $XM = X$, then X is called a fixed vector for M. If such an X is a probability vector, it is called a fixed probability vector for M.

A Markov matrix M is called "regular" if there is some n such that M^n has no zero elements. This section stated without proof that a regular Markov matrix M has a unique fixed probability vector X_M. Furthermore, if Y is any vector of dimension compatible with M, $\lim_{n \to \infty} YM^n$ exists and is some fixed vector for M—that fixed vector whose sum of elements is the same as the sum of the elements in Y.

EXERCISES 9.4. A

1. Find a fixed vector and the fixed probability vector for each of the following regular Markov matrices:
   (a) $\begin{bmatrix} \frac{1}{4} & \frac{3}{4} \\ \frac{2}{3} & \frac{1}{3} \end{bmatrix}$ (b) $\begin{bmatrix} \frac{4}{5} & \frac{1}{5} \\ \frac{2}{5} & \frac{3}{5} \end{bmatrix}$ (c) $\begin{bmatrix} \frac{1}{2} & \frac{1}{2} \\ \frac{5}{6} & \frac{1}{6} \end{bmatrix}$
2. Write a Markov matrix that models each of the following situations:
   (a) A government survey revealed that 0.04 of the rural population moved to urban areas one year and 0.07 of the urban population moved to rural areas.
   (b) A company survey in two successive years revealed that $\frac{1}{2}$ of those workers whose work was rated as "unsatisfactory" the first year were similarly rated the second and the other $\frac{1}{2}$ were rated satisfactory. Of those deemed "satisfactory" the first year, 0.1 were unsatisfactory the second, 0.8 remained satisfactory, and 0.1 became

"exceptional." Of those rated "exceptional" the first year, 0.1 were unsatisfactory the second, 0.5 were satisfactory, and 0.4 continued to be exceptional.

3. Write the fixed probability vector for both Markov matrices found in problem 2.
4. (a) If the trends observed in problem 2(a) continue, what proportion of the population will eventually end up in rural areas?
   (b) If the trends observed in problem 2(b) continue, what proportion of the workers will eventually be unsatisfactory, satisfactory, and exceptional, respectively?
5. If rumors are spread with $\frac{2}{3}$ accuracy, what is the probability that the fourth person to hear the rumor will get the right report? (See Example 9.4.9.)

EXERCISES 9.4. B

1. Find a fixed vector and the fixed probability vector for each of the following regular Markov matrices:

   (a) $\begin{bmatrix} \frac{1}{2} & \frac{1}{2} \\ \frac{3}{4} & \frac{1}{4} \end{bmatrix}$ (b) $\begin{bmatrix} \frac{3}{5} & \frac{2}{5} \\ \frac{4}{5} & \frac{1}{5} \end{bmatrix}$ (c) $\begin{bmatrix} \frac{1}{3} & \frac{2}{3} \\ \frac{3}{4} & \frac{1}{4} \end{bmatrix}$

2. Write a Markov matrix that models each of the following situations:
   (a) A small-town newspaper discovers that about 0.95 of its subscribers renew their subscription each year. By aggressive advertising, it can convince 0.15 of the households not having a subscription this year to take one next year.
   (b) A large group of mice are housed in a huge cage divided into three parts, as in the accompanying diagram. The mice in R_1 move within a minute to R_2 and R_3 with probability $\frac{1}{2}$ and $\frac{1}{3}$, respectively. The mice in R_2 move to R_1 and R_3 with probability $\frac{1}{2}$ and $\frac{1}{4}$; and the mice in R_3 move to R_1 and R_2 with probability $\frac{1}{3}$ and $\frac{1}{4}$.

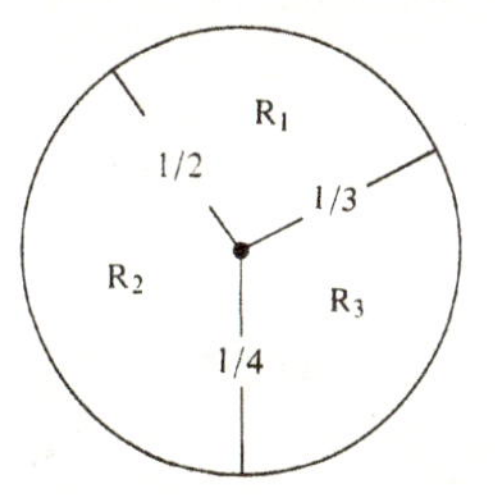

3. Write a fixed vector and the fixed probability vector for both Markov matrices found in problem 2.
4. (a) If the trends observed in problem 2(a) continue, what proportion of the town's households will eventually subscribe to the newspaper?
   (b) If the trends reported in problem 2(b) continue, what proportion of the mice will eventually end up in each room of the cage?
5. If rumors are spread with $\frac{3}{4}$ accuracy, what is the probability that the fourth person to be told the rumor will get the right report? (See Example 9.4.9.)

EXERCISES 9.4. C

1. Find the fixed probability vector of $\begin{bmatrix} 1-a & a \\ b & 1-b \end{bmatrix}$, where $0 < a < 1$ and $0 < b < 1$.
2. Another important and useful fact about Markov matrices is that for every regular Markov matrix $\lim_{n \to \infty} M^n$ exists, and this limit is the Markov matrix such that each row is the unique fixed probability vector P_M. This says, for example, that as n gets large, $\begin{bmatrix} 0.75 & 0.25 \\ 0.2 & 0.8 \end{bmatrix}^n$ gets near to $\begin{bmatrix} \frac{4}{9} & \frac{5}{9} \\ \frac{4}{9} & \frac{5}{9} \end{bmatrix} \approx \begin{bmatrix} 0.44 & 0.56 \\ 0.44 & 0.56 \end{bmatrix}$. For each of the matrices M given in problem 1 of Exercises 9.4.B, find $\lim_{n \to \infty} M^n$. Use the results of problem 1.

ANSWERS 9.4. A

1. (a) A vector is a fixed vector if and only if it is of the form $[\frac{8}{9}y, y]$. The fixed probability vector is $[\frac{8}{17}, \frac{9}{17}]$.
   (b) A vector is a fixed vector if and only if it is of the form $[2y, y]$. The fixed probability vector is $[\frac{2}{3}, \frac{1}{3}]$.
   (c) A vector is a fixed vector if and only if it is of the form $[\frac{5}{3}y, y]$. The fixed probability vector is $[\frac{5}{8}, \frac{3}{8}]$.
2. (a) $\begin{bmatrix} 0.96 & 0.04 \\ 0.07 & 0.93 \end{bmatrix}$ (b) $\begin{bmatrix} 0.5 & 0.5 & 0 \\ 0.1 & 0.8 & 0.1 \\ 0.1 & 0.5 & 0.4 \end{bmatrix}$
3. (a) $[\frac{7}{11}, \frac{4}{11}]$ (b) $[\frac{1}{6}, \frac{5}{7}, \frac{5}{42}]$
4. (a) Eventually $\frac{7}{11}$ will be in rural areas and $\frac{4}{11}$ in urban.
   (b) There will be $\frac{1}{6}$ unsatisfactory, $\frac{5}{7}$ satisfactory, and $\frac{5}{42}$ exceptional.
5. $\frac{41}{81}$

ANSWERS 9.4. C

1. $[b/(b+a), a/(b+a)]$
2. (a) $\begin{bmatrix} \frac{3}{5} & \frac{2}{5} \\ \frac{3}{5} & \frac{2}{5} \end{bmatrix}$ (b) $\begin{bmatrix} \frac{2}{3} & \frac{1}{3} \\ \frac{2}{3} & \frac{1}{3} \end{bmatrix}$ (c) $\begin{bmatrix} \frac{9}{17} & \frac{8}{17} \\ \frac{9}{17} & \frac{8}{17} \end{bmatrix}$

VOCABULARY

sample space, basic outcome, probability distribution, event, mutually exclusive, probability vector, dominant gene, multiplication rule, permutation of n items taken r at a time, combination of n items taken r at a time, n factorial, independent characteristics, dependent characteristics, joint probabilities, conditional probabilities, probability distribution table, Markov matrix, fixed vector, fixed probability vector, regular Markov matrix.

SAMPLE TEST

Chapter 9

1. (10 pts) Suppose an ice cream store has 3 types of cones and 20 flavors of ice cream. How many different double-dip cones can be made?
2. (30 pts) Suppose that a random sample is taken of 100 minutes of activity at a telephone switchboard. The number of calls received in each of these minutes was tabulated as follows:

0	1	2	3	4	5
10	20	30	25	10	5

Assume that the past can be used to predict the future.
(a) What is the probability that this exchange will receive 3 telephone calls in a random minute?
(b) What is the probability that an even number of telephone calls will be received in a random minute?
(c) What is the probability that 3 or more calls will be received?
(d) What is the probability that either 3 or more calls or an even number of calls will be received in a random minute?
(e) Write a probability vector describing this probability model.
3. (10 pts) Suppose that 2 professors are asked independently to rate 5 essay exams in order of quality and they both rate the 5 in exactly the same order. What is the probability that this is due to chance?
4. (10 pts) Suppose that your living room has 2 red chairs and 5 blue chairs. If the first two guests that come to your party both sit on the red chairs, what is the probability that this is due to chance?
5. (25 pts) In 1974 about 2 million Americans died, of which about 46,000 deaths were due to auto accidents. Among the 2 million about 48,000 were Americans between the ages of 15 and 24; about a third of these 48,000 deaths were due to auto accidents.
(a) Make a *probability* distribution table from these data.
(b) Are "age" and "cause of death" statistically independent?
6. (15 pts) Let $M = \begin{bmatrix} 0.6 & 0.4 \\ 0.1 & 0.9 \end{bmatrix}$.
(a) What kind of a matrix is M?
(b) Find a fixed probability vector for M.

The answers are at the back of the book.

Chapter 10

GAME THEORY

10.1 Expected Value

If the sample space of a probability experiment consists of numbers, the variable x that results from performing the experiment is called a random variable. Each random variable takes on the values of the sample space for some probability experiment. Rolling a die, for example, yields a number, so we can speak of the random variable x associated with the probability experiment of rolling a die; in this case x can be 1, 2, 3, 4, 5, or 6.

We can associate with any random variable a number called the expected value (or expectation) in such a way as to give the *average* value of the random variable in the long run—that is, if the experiment is repeated many times.

10.1.1 Rule

The expected value, $E(x)$, of any random variable x that takes on a finite number of values can be computed by postmultiplying the probability vector by a column vector that lists the basic outcomes matching each probability. Thus, if a random variable x takes the values x_1, x_2, and x_3 with probabilities p_1, p_2, and p_3, respectively,

$$E(x) = [p_1, p_2, p_3]\begin{bmatrix} x_1 \\ x_2 \\ x_3 \end{bmatrix} = p_1x_1 + p_2x_2 + p_3x_3$$

10.1.2 Example

Find the expected value of x, where x is the random variable associated with rolling a die.

Solution:

Since each number occurs with probability $\frac{1}{6}$, if the die is tossed many, many times, the average value per throw is

$$\left[\tfrac{1}{6}, \tfrac{1}{6}, \tfrac{1}{6}, \tfrac{1}{6}, \tfrac{1}{6}, \tfrac{1}{6}\right]\begin{bmatrix}1\\2\\3\\4\\5\\6\end{bmatrix} = \tfrac{1}{6}\cdot 1 + \tfrac{1}{6}\cdot 2 + \tfrac{1}{6}\cdot 3 + \tfrac{1}{6}\cdot 4 + \tfrac{1}{6}\cdot 5 + \tfrac{1}{6}\cdot 6 = 3.5$$

Thus if a die is tossed many times, the average value per throw will be approximately 3.5. Notice that the expected value need not be one of the values that the random variable takes; it is merely the average of these values in the long run.

10.1.3 Example

Find the expected value for the number of cars arriving at a toll booth per minute if the probability of 0, 1, 2, and 3 cars arriving, respectively, is 0.2, 0.4, 0.25, and 0.15.

Solution:

The average number of cars per minute in the long run will be

$$[0.2, 0.4, 0.25, 0.15]\begin{bmatrix}0\\1\\2\\3\end{bmatrix} = 0.2(0) + 0.4(1) + 0.25(2) + 0.15(3) = 1.35$$

This means that in the long run there will be about 1.35 cars per minute passing through that toll booth.

10.1.4 Example

The probability that an American man in his early fifties will survive for another year is about 0.99. How much should such a man pay for $20,000 worth of term insurance, exclusive of administrative costs and profit for the company?

Solution:

There are two basic outcomes—getting $0 by living through the year or getting $20,000 by dying in it. The probability vector is [0.99, 0.01], so the expected value is

$$[0.99, 0.01]\begin{bmatrix}0\\20{,}000\end{bmatrix} = 0 + 200 = \$200$$

Thus $200 is the amount to be paid for the insurance itself, exclusive of administrative costs and profits.

10.1.5 Example

Suppose that the probability of finding 0, 1, 2, and 3 people in line ahead of you at the Apex Supermarket is 0.2, 0.4, 0.25, and 0.15, respectively. If the probability of finding 0, 1, 2, and 3 people in line at the B & Q Supermarket across the street is 0.3, 0.3, 0.2, and 0.2, respectively, at which supermarket would you find shorter lines in the long run? (Assume that both markets have the practice of opening up a new line as soon as one exceeds 3 people in length.)

Solution:

Apex Supermarket *B & Q Supermarket*

$$[0.2, 0.4, 0.25, 0.15]\begin{bmatrix} 0 \\ 1 \\ 2 \\ 3 \end{bmatrix} = 1.35 \qquad [0.3, 0.3, 0.2, 0.2]\begin{bmatrix} 0 \\ 1 \\ 2 \\ 3 \end{bmatrix} = 1.3$$

Since the expected length of a line in Apex is 1.35 and that in B & Q is only 1.3, there is slightly less waiting on the average in B & Q than in Apex.

If one random variable y is a multiple of another random variable x, there is a simple relationship between their expected values $E(x)$ and $E(y)$. For example, suppose that the cost of passing through a toll booth is 0.50 per vehicle. Then the total revenue y collected in a random minute will be a multiple of the number of vehicles x that pass through that toll booth per minute. To be specific, $y = 0.50x$. If the expected value of the number of cars passing through the toll booth is 1.35 per minute (as in Example 10.1.3), then the expected value of the gross income at that booth per minute is $(0.50)(1.35) = \$0.675$.

10.1.6 Rule

If $E(x)$ is the expected value of the random variable x, then $E(kx) = kE(x)$ if k is any constant.

10.1.7 Example

What is the expected daily income of each member of the sales force of a car dealership if the expected number of sales $E(x)$ each day per person is 0.75 and there is a commission of \$120 per car?

Solution:

If y is the random variable indicating daily income, then $E(y) = E(120x) = 120E(x) = 120(0.75) = 90$. Thus in the long run the average daily income per person will be \$90.

If two random variables can be added, the rule for finding the expected value of their sum is as straightforward as Rule 10.1.6.

10.1.8 Rule

If $E(x)$ and $E(y)$ are the expected values of the random variables x and y, then $E(x + y) = E(x) + E(y)$.

10.1.9 Example

If two toll booths, side by side, are such that the expected value of their gross receipts per minute is $E(x) = \$0.675$ and $E(y) = \$0.825$, what is their combined expected gross income per *hour*?

Solution:

The expected total gross receipts $E(z)$ per minute for the two booths is $E(z) = E(x + y) = E(x) + E(y) = \$0.675 + \$0.825 = \$1.50$. Thus the expectation of the total gross receipts per hour is $E(w) = E(60z) = 60E(z) = 60(1.50) = \$90$.

10.1.10 Example

Suppose that among a "typical" group of 10 tiny pine trees sold by a nursery, the following number will die in the first year with the given probability.

Number that will die in the first year:	0	1	2	3	4
Probability:	0.1	0.3	0.3	0.2	0.1

(a) What is the expected number that will die in the first year of a group of 10?

(b) If a customer buys 500 trees, what is the expected number that will die?

(c) Suppose that the trees can be purchased under two options. In option *A* the trees cost \$50 for a group of 10 and there is no replacement or compensation for those that die. Under option *B* a group of 10 trees costs somewhat more than \$50 but the customer is given \$5 for each tree that dies during the first year. What would be a fair price for 10 trees under option *B*?

Solution:

(a)

$$[0.1, 0.3, 0.3, 0.2, 0.1]\begin{bmatrix}0\\1\\2\\3\\4\end{bmatrix} = 0 + (0.3)1 + (0.3)2 + (0.2)3 + (0.1)4 = 1.9$$

Thus 1.9 (almost 2) of the 10 trees are "expected" to die the first year.

(b) A customer who buys 500 trees has bought 50 10-tree groups. For $i = 1, 2, \ldots, 50$ we let x_i be the number of trees in the ith group that dies in the first year. Then $y = x_1 + x_2 + \cdots + x_{50}$ is the total number of trees that die. Thus the expected number that will die is $E(y) = E(x_1) + E(x_2) + \cdots + E(x_{50}) = 1.9 + 1.9 + \cdots + 1.9 = (50)(1.9) = 95$.

(c) Since in a very large group of pine trees the average number that will die in the first year out of each 10-tree group is approximately 1.9 and since the price per tree is \$5, the amount that the customer will lose on the average is (\$5)1.9 = \$9.50. Thus a fair cost under option B would be \$9.50 + \$50 = \$59.50.

SUMMARY

A random variable x is a variable whose value is determined by a probability experiment. To each random variable x there corresponds an expected value or expectation $E(x)$ which describes the average value of x in the long run. It can be computed for small sample spaces by forming a vector of the basic outcomes that matches the probability vector and taking the inner product. Several examples of this were given; more appear in the exercises.

EXERCISES 10.1. A

1. If a customer buys 200 trees from the nursery described in Example 10.1.10, what is the expected number that will die in the first year?
2. Suppose that a sales representative makes only 2 presentations a day and that of these, the probability of making exactly 1 sale is 0.5 and of making 2 sales is 0.3.
   (a) What is the probability of his making no sales?
   (b) Write a probability vector describing this situation.
   (c) What is the probability of his making at least 1 sale?
   (d) What is his expected number of sales per day?
   (e) If he makes \$100 per sale, what is his expected income in a 5-day week?
3. Suppose that the company has another sales representative with the same type of job but less experience who has a probability of 0.4 of making 1 sale a day and a probability of 0.2 of making 2 sales a day. Answer parts (a)–(e) of problem 2 for this second salesman.
   (f) What is the expected number of sales that both will make in a day?
   (g) What is the expected number of sales that both will make in a week?
4. Suppose that the company has two representatives with the skills of the one described in problem 2 and 5 others with the skills reported in problem 3.
   (a) How many sales can the company expect in a day?
   (b) If they pay \$100 per sale, what is the expected amount the company pays in commissions in a 5-day week?
5. Recently, the probability of an American white baby surviving for a

month has been about 0.983. Suppose that you are on a volunteer committee studying the medical facilities in a predominantly white community and there are about 2000 babies born each year in the local hospital. How many can reasonably be expected to die within the first month?

6. (a) Suppose that a decal for using a parking lot for a month costs $25 and the probability of getting a ticket for illegal street parking is only 0.2. There is a $3 fine for illegal parking. If there are 22 parking days per month, does it pay financially in the long run to buy the decals?
   (b) Suppose that the probability of getting a ticket rises to 0.5. Does it pay now to buy a parking decal?
   (c) Suppose that the city cannot afford to hire the people necessary to catch parking violators 0.5 of the time, and must maintain the probability of getting caught at 0.2. What is the minimum fine (in whole dollars) that will discourage penny-pinching workers whose only alternative is the $25-a-month parking lot from illegal parking?

EXERCISES 10.1. B

1. Suppose that the glassblower with an 0.8 probability of blowing an acceptable bottle blows 80 bottles this week. About how many of these would be expected to be acceptable?
2. Suppose that a group of day-to-day construction workers can work either 0, 1, or 2 days, depending on what is needed and how much they want to work. (A 2-day workday consists of 12 hours of work with double-time pay for the extra 4 hours.) There is a probability 0.2 of not working at all, 0.7 of working a normal day, and 0.1 of working a 2-day workday in one day.
   (a) Write a probability vector describing this situation.
   (b) What is the probability of working at least some of the time during a day?
   (c) What is the expected amount of work per day?
   (d) If the pay is $4 per hour and there are 8 hours in the standard workday, what is the expected pay per day?
   (e) How much will such a worker expect to make in an average week?
3. Suppose that there is another group who have probability 0.4 of not working at all and probability 0.6 of working a normal 8-hour day.
   (a) What is their probability of working a double day?
   (b) Write a probability vector for this situation.
   (c) What is the expected amount of work per day?
   (d) What is the expected pay per day at $4 per hour?
   (e) What is the expected pay per week?
4. Suppose that the construction company uses 5 workers such as the one described in problem 2 and 10 workers such as the one described in problem 3. What is the expected weekly payroll?
5. The "payoff ratings" in Exercises 1.3.C (page 29) is a kind of expectation. Of the three foundations analyzed in the example and in the two exercises, which has the greatest expectation of granting money to the given project?

6. Suppose a company calculates that it would cost $10 million per month to maintain a treatment plant effective enough to keep the waste products it dumps into a river within legal cleanliness limits.
   (a) If the chances of getting caught at a violation are 0.3 and the annual fine is $100 million, is it in the short-term financial interests of the company to install the treatment plant?
   (b) What is the minimum fine that would be a realistic deterrent?

ANSWERS 10.1. A

1. 38 trees 2. (a) 0.2 (b) [0.2, 0.5, 0.3] (c) 0.8 (d) 1.1 (e) $550
3. (a) 0.4 (b) [0.4, 0.4, 0.2] (c) 0.6 (d) 0.8 (e) $400 (f) 1.9 (g) 9.5
4. (a) 6.2 (b) $3100 5. 34 6. (a) No (b) Yes (c) $6

10.2 Saddle Points and Mixed Strategies

Game theory, like linear programming, is a twentieth-century phenomenon. Although its roots can be traced to earlier ideas, its beginning as a carefully formulated discipline is generally credited to John von Neumann, who lived from 1903 to 1957. Along with being one of the most creative mathematicians of the twentieth century, von Neumann was a friendly, sociable person, much interested in the interactions of human beings, both personally and politically.

Game theory is a mathematical device for analyzing conflict and cooperation between rational beings, either in pairs or among larger groups. It has been used in such varied fields as international relations, business competition, and marital problems. The games themselves are used in psychological studies about people's reactions to the sex or race of his or her opponent in a conflict situation. (For an example, see the *Journal of Conflict Resolution*, Vol. 17; an article there shows how self-proclaimed unprejudiced white college students reacted to nonwhite game opponents.)

Although game theory is only a few decades old, it has many complexities far beyond the scope of this book. We shall confine ourselves to conflict situations (games) involving only two opponents, because then our familiar two-dimensional matrices are used; one person chooses a row and the other a column. Games with more than two players need more than two dimensions, and are beyond the scope of this book.

Some of the most interesting 2-person games—such as checkers, chess, and poker—cannot be analyzed (yet, at least!) by matrix theory. This book will discuss only games that can be expressed as matrix games.

When a matrix describes a game, one player chooses a row and the other simultaneously chooses a column; the element at the intersection of that row and that column is called the payoff of that round of the game. The row player wins by the amount of the payoff and the column player loses by this amount; we might think of this as the column player's "paying" the row player the amount. When the payoff is negative, the row player pays the column player. Thus matrix games are often called zero-sum games; the sum of what one player wins and the other loses is zero.

Saddle Points

10.2.1 Example

Suppose that a group of people, each with a pile of pennies, has gathered for an evening of playing matrix games. At the first round the hostess hands Maria, the row player, and Vincent, the column player, the following matrix game.

$$\text{Maria} \overset{\text{Vincent}}{\begin{bmatrix} -1 & -2 & 4 \\ 3 & -1 & 2 \\ 2 & -2 & 1 \end{bmatrix}}$$

At first Maria is pleased, because there are more positive numbers than negative (indicating that if each chooses at random, she will win more often) and the average of all the numbers is clearly positive. But as she contemplates her choice, she becomes more and more uneasy, until finally she protests to the hostess that this is not a fair game and she will not play. Why?

Solution:

Maria wants the payoff to be as large as possible. Thus she quickly decides she will not play the third row, because, no matter what Vincent does, she can always win more by playing the second. She eyes the "4" in the upper right greedily, but quickly realizes that Vincent (being rational and selfish, as all game players are assumed to be) will never play the third column, because then he is sure to lose more than if he plays the second. Thus she might as well play the second row, since $3 > -1$ and $-1 > -2$. And Vincent, with only a quick glance, immediately decides to play the second column. Thus the payoff will always be $a_{22} = -1$, which means Maria must pay Vincent a penny each time they play the game. No wonder Maria does not want to play!

The hostess could make this into a fair game simply by adding 1 to each element of the matrix.

$$\begin{bmatrix} 0 & -1 & 5 \\ 4 & 0 & 3 \\ 3 & -1 & 2 \end{bmatrix}$$

Then, since the structure of these games is so similar, the same strategy would be used by each player and the payoff for each time the game is played would still be the center element of the matrix. But since the payoff is 0, the game is considered "fair."

When there is one element that will clearly be chosen by both players, assuming they are rational and greedy, it is called a saddle point. In Example 10.2.1 the saddle point was $a_{22} = -1$.

If a game has a saddle point, that saddle point is said to be the value of the game. If the value of the game is 0, it is said to be a fair game.

Not every game has a saddle point, and soon we shall examine some that do not. But first we consider one way to identify saddle points.

Dominated Rows and Columns

If all the elements of one row of a game matrix are less than or equal to the corresponding elements of another row, we say that the row is dominated by the row with larger elements. The row player would never play a row that is dominated, because he or she is sure of a higher payoff by playing the dominating row. Thus dominated rows should always be eliminated from a game matrix before we use other means to analyze the matrix.

Similarly, if the elements of some column are all *greater* than or equal to the corresponding elements of another column, we can eliminate that column from the game matrix. Often after such obviously dominated rows and columns are eliminated, the smaller resulting matrix will have more dominated rows and/or columns which can also be eliminated. This process may continue until none of the rows or columns remaining are dominated.

If only one element remains at the end, that element is the saddle point of the game. (Some saddle points are not revealed by this process; the solution method in Section 10.4 can then be used.) If more than one element remains, other methods of analysis must be applied to the (perhaps smaller) matrix that is left after the dominated rows and columns have been eliminated.

10.2.2 Example

Suppose that two costume companies each make clown, skeleton, and space costumes. They all sell for the same amount and use the same machinery and workmanship. Furthermore, the market for costumes is fixed; a certain given number of total costumes will be sold this

Halloween. But each company has its own individual styles which affect how the costumes sell.

On the basis of past experience, the following matrix has been inferred. It indicates, for example, that if both make clown outfits, then for every 20 that are sold, company I will lose 2 sales to company II. Similarly, if company I makes clown outfits and company II makes space suits, then for each 20 sold, company I will sell 4 more than company II.

		Company II		
		Clown	Skeleton	Space
Company I	Clown	-2	0	4
	Skeleton	0	2	1
	Space	-1	-4	0

How should each company plan its manufacturing?

Solution:

Company I can quickly decide not to make space outfits, because it will do better with skeleton costumes, no matter what company II does. Similarly, company II should not make space outfits, because it can always be better off by making clown outfits, no matter what the choice of company I. Thus we have reduced the matrix as follows:

$$\begin{bmatrix} -2 & 0 & 4 \\ 0 & 2 & 1 \\ -1 & -4 & 0 \end{bmatrix}$$

With the remaining 2 by 2 matrix, it is clear that company I will profit most from playing the second row, and that company II should always play the first column. Thus this game also has a saddle point. It is $a_{21} = 0$. The value of the game is 0; company I should always make skeleton outfits and company II should always make clown costumes.

10.2.3 Example

Find the saddle points of the games described by each of the following matrices, if there are any.

(a) $\begin{bmatrix} 4 & 5 & 4 & 5 \\ -2 & 0 & 3 & 1 \\ 4 & 7 & 5 & 6 \\ 1 & -1 & 1 & 5 \end{bmatrix}$ (b) $\begin{bmatrix} 3 & 0 & 3 \\ 3 & 1 & 1 \\ 0 & 4 & 2 \end{bmatrix}$

Solution:

(a) The procedure is no different if one or both players has more than three choices. In this case both the second and fourth rows are dominated by the third, and both the third and fourth columns are dominated by the first. Thus we obtain

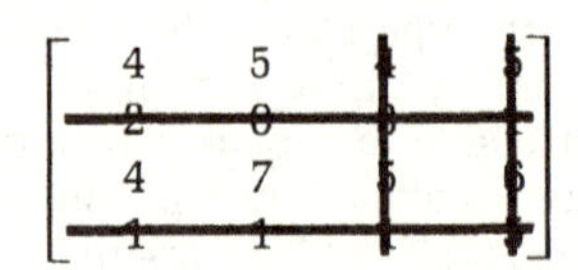

And now we see that the second column is dominated by the first, so we eliminate it. Two numbers—not one—remain, but they are both saddle points. The row player can wisely choose either the first or third row; the column player should choose the first column; and the value of the game is 4.

(b) This time no row or column is dominated by any other row or column, so we cannot reduce the matrix. Playing the game described by such a matrix requires a mixed strategy, that is, making any one choice only some of the time. In Section 10.4 we shall "solve" this game, but in this section we shall confine ourselves to easier, but similar, examples.

Mixed Strategies

10.2.4 Example

Suppose that Peppermint Patty suggests that she and Charlie Brown duke to decide whose baseball team will bat first. She calls, "Odds!" Each of them then puts out either one or two fingers; she wins if the sum is odd and Charlie Brown wins if the sum is even.

(a) Write a matrix for this game.
(b) What strategy should Peppermint Patty use?
(c) Suppose that they play this game over and over and keep score. Then what strategy should she use?

Solution:

(a)

		Charlie Brown (evens)	
		1 finger	2 fingers
Peppermint Patty (odds)	1 finger	-1	1
	2 fingers	1	-1

(b) If they play just once, it does not matter what strategy is used; that is what makes this a good game for fair decisions.

(c) Then she should play each row roughly the same number of times but not in a predictable pattern. She could, for example, toss a coin to decide how many fingers to play.

In the example above it is clear that there are no saddle points. It is also clear (and true) that each player should play each choice half the time in a random pattern. We might summarize Peppermint Patty's strategy by the probability vector $P = [\frac{1}{2}, \frac{1}{2}]$. Such a strategy is called a mixed strategy because the same choice is not made all the time. The value of the game of duking is zero.

Now we turn to a more subtle game. In analyzing the following game we shall make the traditional game theory assumption that each player wants to lose as little as possible.

10.2.5 Example

Suppose that in the course of the merry game between Peppermint Patty and Charlie Brown, Lucy steps in and pursuades Charlie Brown that the game would be better if she would get the total number of points as fingers showing when the number of fingers is odd and he would get the total number of points as fingers showing when the total is even. This seems fair to Charlie Brown because the average number of points available is zero:

		Charlie Brown (evens)	
		1 finger	2 fingers
Lucy (odds)	1 finger	-2	3
	2 fingers	3	-4

But after playing this game 100 times or so, Lucy is so far ahead that Charlie Brown wonders if he has been hoodwinked again. Has he?

Solution:

It is easy to check that no row or column dominates another. Thus we must use another technique to find the best strategy for each player. We shall discover that this time $P = [\frac{1}{2}, \frac{1}{2}]$ is *not* the best strategy for Lucy.

Suppose that Lucy plays one finger with probability p, where p is to be determined. Then she will play two fingers with probability $(1 - p)$, so her strategy can be summarized as $P = [p, 1 - p]$. Let us compute her expected payoff. Actually, we compute two expected payoffs—one for over all the times that Charlie Brown plays one finger, and another for over all the times when he plays two.

When Charlie Brown plays one finger, there is a payoff of either (-2) or 3; this means that Lucy's expected payoff over this time will be $E_1 = p(-2) + (1 - p)3 = 3 - 5p$. Similarly, when Charlie Brown plays two fingers, there is a payoff of either 3 or (-4); over this time the expected payoff will be $E_2 = p(3) + (1 - p)(-4) = 7p - 4$. Let us graph these two expected payoffs as functions of p, remembering that $0 \leq p \leq 1$ because p is a probability (see Figure 10.2–1).

If Lucy plays $P = [\frac{1}{2}, \frac{1}{2}]$, then $E_1 = \frac{1}{2}$ and $E_2 = -\frac{1}{2}$. This is good for her if Charlie Brown plays "one finger" often, but she will loose significantly if he concentrates on "two fingers." Similarly, if Lucy plays any other strategy $P = [p, 1 - p]$, where p is to the left of the intersection of E_1 and E_2 in Figure 10.2–1, her minimum expectation will be $E_2(p)$; if she plays $P = [p, 1 - p]$, where p is to the right of the intersection, her minimum expected gain is $E_1(p)$. To maximize her minimum expected gain, she should play that probability p which occurs at the intersection of the lines $E_1 = 3 - 5p$ and $E_2 = 7p - 4$. If she plays this strategy, her expectation will be exactly the y-coordinate of that intersection, no matter what Charlie Brown does.

Figure 10.2–1

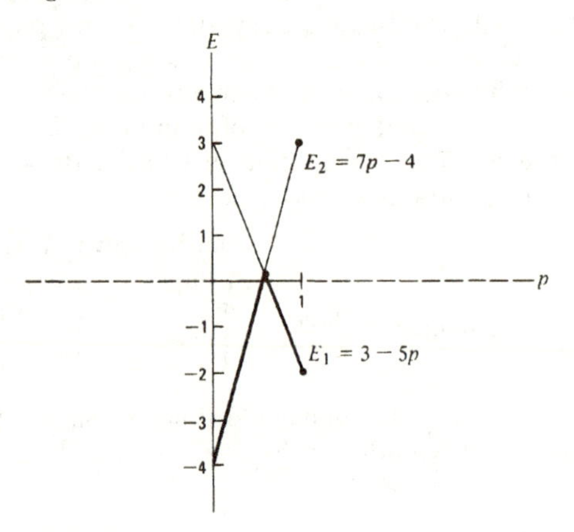

And what is this p? We solve to find the p where E_1 and E_2 are equal:

$$7p - 4 = 3 - 5p$$
$$12p = 7$$
$$p = \tfrac{7}{12}$$

So if Lucy plays $p = \frac{7}{12}$, her expected payoff is $E_1(\frac{7}{12}) = 3 - 5(\frac{7}{12}) = \frac{36}{12} - \frac{35}{12} = \frac{1}{12}$, and in the long run Lucy will indeed win, no matter what her opponent does. Poor Charlie Brown!

You might wonder what the analysis looks like from Charlie Brown's side. Suppose that he plays the first column with probability q. Then when Lucy plays the first row, the expected payoff of the game is $E_3 = -2q + 3(1 - q) = -5q + 3$. If she plays the second row, the expected payoff is $E_4 = 3q - 4(1 - q) = 7q - 4$. Remembering that $0 \leq q \leq 1$ and graphing E_3 and E_4, we get the following picture.

Figure 10.2–2

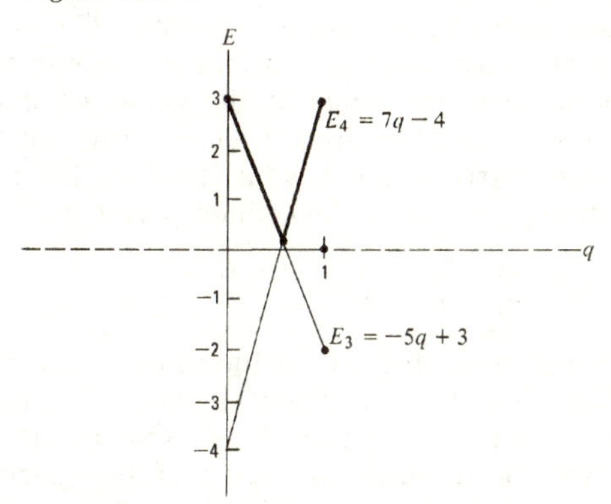

Charlie Brown, of course, would like to make the expected payoff as low as possible. But no matter what he does, he cannot force it to be any lower than $\frac{1}{12}$. He can, however, be sure it gets no higher than $\frac{1}{12}$ by playing one finger exactly $q = \frac{7}{12}$ of the time, because $q = \frac{7}{12}$ determines the intersection of E_3 and E_4 in Figure 10.2–2.

It is essential that each player play with a random pattern not discernible by the other. It is only *in the long run* that each should play one finger $\frac{7}{12}$ of the time. How do they decide which finger to play at each play of game? Perhaps each has a spinner such as the one in Figure 10.2–3 that she or he spins to determine each move. (Or did Lucy do the spinning ahead of the game and then memorize her plays?)

Figure 10.2–3

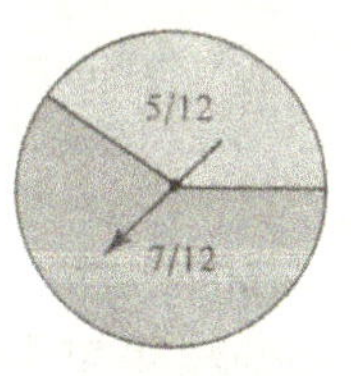

In the previous game, we say that the optimum strategy for Lucy is $P = [\frac{7}{12}, \frac{5}{12}]$ and that the optimum strategy for Charlie Brown is $Q = \begin{bmatrix} \frac{7}{12} \\ \frac{5}{12} \end{bmatrix}$. The fact that they are the same results from the symmetry of the game. The value of the game is $\frac{1}{12}$.

10.2.6 Definition

The value of a matrix game is the expected payoff of that game if both players use their respective optimum strategies.

The optimum strategy for the row player is a probability vector that describes the proportion in which that player should make the choices of the various rows so that in the long run the expected payoff will be at least the value of the game, no matter what the column player does. It is generally written as a row vector so it can premultiply the game matrix and tell what the expected payoff will be for each pure strategy choice of the column player:

$$\left[\tfrac{7}{12}, \tfrac{5}{12}\right]\begin{bmatrix} -2 & 3 \\ 3 & -4 \end{bmatrix} = \left[\tfrac{1}{12}, \tfrac{1}{12}\right]$$

The optimum strategy for the column player is a probability vector that describes the proportion in which that player should make the choices of the various columns so that in the long run the expected value will be at most the value of the game, no matter what the row player does. It is generally written as a column vector so it can postmultiply the game matrix and tell what the expected value will be for each pure strategy choice of the row player:

$$\begin{bmatrix} -2 & 3 \\ 3 & -4 \end{bmatrix}\begin{bmatrix} \tfrac{7}{12} \\ \tfrac{5}{12} \end{bmatrix} = \left[\tfrac{1}{12}, \tfrac{1}{12}\right]$$

A strategy is called a pure strategy if one of the elements in the probability vector is 1 and the others are zero. When a game has a saddle point, the optimum strategy is a pure strategy.

10.2.7 Example

Analyze the following matrix game. That is, find the optimum strategy for each player and find the value of the game.

$$\text{Player I} \quad \overset{\text{Player II}}{\begin{bmatrix} 3 & -1 \\ -4 & 2 \end{bmatrix}}$$

Solution:

A quick check reveals there are no dominated rows or columns.

We assume that player I will play the first row with probability p. Then when player II plays the first column, the expected value is $E_1 = 3p - 4(1 - p) = 7p - 4$; when player II plays the second column, the expected value is $E_2 = -p + 2(1 - p) = -3p + 2$. If we graph these, the lines will cross. (This is true, although we will not prove it, for any 2 by 2 matrix game that does not have a saddle point.) Thus we need only solve the equation $E_1 = E_2$ for p to find the intersection; the actual graphing is not necessary. We set

$$\begin{aligned} 7p - 4 &= -3p + 2 \\ 10p &= 6 \\ p &= 0.6 \end{aligned}$$

When $p = 0.6$, $E_1 = 7(0.6) - 4 = 4.2 - 4 = 0.2$. Thus the optimum strategy for player I is $[0.6, 0.4]$, and the value of the game is 0.2.

Assuming that player II will play the first column with probability q, we have $E_3 = 3q - (1 - q) = 4q - 1$ when player I plays the first row and $E_4 = -4q + 2(1 - q) = -6q + 2$ when player I plays the second row. Setting $E_3 = E_4$, we have

$$\begin{aligned} 4q - 1 &= -6q + 2 \\ 10q &= 3 \\ q &= 0.3 \end{aligned}$$

Thus the optimum strategy for player II is $Q = \begin{bmatrix} 0.3 \\ 0.7 \end{bmatrix}$ and the value of the game for $q = 0.3$ is (still!) $E_3 = 4(0.3) - 1 = 1.2 - 1 = 0.2$.

SUMMARY

Although game theory as a discipline is only a few decades old, the study of two-person zero-sum games is well developed. This section presents the theory of such games that can be represented by 2 by 2 matrices. One player chooses a row, the other chooses a column, and the payoff is the element of the matrix in their intersection; the column player "pays" the row player that number of points. Thus the bigger the payoff, the happier the row player; the smaller the payoff, the happier the column player.

To begin analyzing a game, one might eliminate all dominated rows and columns. This can sometimes be done repeatedly, and if only one element remains, the game has a saddle point. This saddle point is the value of the game, and each player would be wisest to play that pure strategy which results in the saddle point.

If there is no saddle point, there will be a probability vector for each player called that player's optimum strategy which ensures that player's not losing any more than necessary. If both players use their own optimum strategies, the expected value of the game is a certain fixed number called the value of the game. This section showed how to find the value of every 2 by 2 matrix game. In the next section we shall begin to consider larger matrices and, more briefly, two-person games that cannot be expressed by a matrix.

EXERCISES 10.2. A

1. Solve the following games. That is, find the optimum strategy for each player and find the value of the game.

   (a) $\begin{bmatrix} -3 & 1 \\ 1 & 4 \end{bmatrix}$ (b) $\begin{bmatrix} -1 & 2 & -2 \\ 2 & 1 & 1 \\ -3 & 0 & -1 \end{bmatrix}$ (c) $\begin{bmatrix} -1 & -2 & 0 \\ 1 & -5 & -3 \\ 2 & -1 & 2 \end{bmatrix}$

   (d) $\begin{bmatrix} -1 & 3 \\ 2 & 0 \end{bmatrix}$ (e) $\begin{bmatrix} 4 & -2 \\ 1 & 2 \end{bmatrix}$

2. Set up each of the following situations as a matrix game.
   (a) A shopping mall is being planned and two department stores will dominate the mall. Each can choose whether it wants to be in the center of the mall or on the edge. (They are each offered an opposite edge.) If they are both in the center or both on the edge, they will divide the business evenly. But if one chooses the edge and the other chooses the center, the one in the center will get more business than the one on the side.
   (b) There are three possible prices at which each of two competing companies could price their competing products. If they set the prices identically, they will share the market evenly. If either one sets the price one level below the other, it will sell enough more products to make a higher profit than its competitor. But if one company sets the lowest price and the other sets the highest, the one with the highest will get the extra profits.
   (c) Suppose that the "enemy" drives a load of supplies every night from city A to outpost B, either along the main highway or along the dark mountain road. The rebels are trying to decide whether to wait for them along the highway or along the road. If they meet them on the road, they are sure to be able to overpower the truck, with a maximum payoff of 10. If they meet them along the highway, there is a chance that the truck might get through anyway, so the payoff is only 6. If they miss the truck, the payoff is negative, but if they wait on the road, they lose only their time and some fuel for a payoff of -1, whereas the highway risks also include that of being discovered and arrested and are rated at -3.

EXERCISES 10.2. B

1. Solve the following games. That is, find the optimum strategy for each player and find the value of the game.

   (a) $\begin{bmatrix} 0 & -2 \\ 2 & -1 \end{bmatrix}$ (b) $\begin{bmatrix} -1 & 0 & 3 \\ 2 & 3 & 2 \\ 0 & -2 & 1 \end{bmatrix}$ (c) $\begin{bmatrix} 3 & 0 & 3 \\ 0 & -1 & 1 \\ 4 & -3 & -2 \end{bmatrix}$

   (d) $\begin{bmatrix} 4 & 0 \\ -3 & 3 \end{bmatrix}$ (e) $\begin{bmatrix} 3 & 1 \\ -1 & 2 \end{bmatrix}$

2. Set up each of the following situations as a matrix game.
   (a) Two supermarkets are planning their weekend sales. Suppose that the profit from yogurt if neither advertises it will be \$100. If supermarket I has a sale and the other does not, so many customers will buy yogurt at supermarket I that its profit will be \$110; conversely if the other runs a sale but supermarket I does not, its profit will be only \$90. But if both have sales, the number sold at supermarket I will not change and its profit will drop to \$95.
   (b) A gasoline sales company is about to place a new station in a moderate-sized town through which a major highway runs. The management is trying to decide whether the station should be on the highway at the west of town, or on the highway at the east, or in the center business district. There is a rumor that another

company, with higher-priced gasoline and more aggressive advertising, is making a similar decision. The lower-priced company figures that if both stations are located near the same place, it will steal some of the market from the other company. If the stations are located at opposite ends of town, the company with the aggressive advertising will get the majority of the market with roughly the same imbalance. Otherwise, both companies can be expected to divide the market evenly.

(c) The robbers have hurried off with their booty and one witness saw them take a southern route. There is one police car near enough to give chase, and the police know there is a fork in the road not far south of town. If the robbers take the small road over the mountain and the police follow them there, the chances are very good that the robbers can be sighted and then stopped at the Mexican border for a net police gain of 5. If both the police and the robbers take the highway, the robbers may get lost in the crowd and slip over the border unnoticed and away from detection, but on the other hand they may be intercepted; the net advantage to the police is 3. If the police take the wrong road, it is certain they will not arrest the robbers; but if they take the highway, the inconvenience amounts to a gain of -1, whereas the risks on the mountain road decrease the police gain to -2.

EXERCISES 10.2. C

Solve the following matrix games, using the techniques explained in this section.

1. (a) $\begin{bmatrix} 1 & -2 & -1 & 0 \\ 2 & -1 & 4 & -1 \\ 1 & -3 & 0 & -2 \\ 2 & -2 & 3 & -1 \end{bmatrix}$ (b) $\begin{bmatrix} -2 & 2 & -1 & 0 \\ -1 & 1 & -2 & 1 \\ -2 & 1 & -1 & -1 \\ 0 & 3 & 0 & 1 \end{bmatrix}$

   (c) $\begin{bmatrix} -2 & -1 & 2 & 0 \\ 2 & 1 & -1 & 2 \\ -2 & 3 & 0 & 4 \\ 0 & 4 & 3 & 5 \end{bmatrix}$

2. Find a 3 by 3 matrix game that has a saddle point but no dominated rows or columns. That is, there is a specific row and a specific column that the two players will always play if they are smart.

ANSWERS 10.2. A

1. (a) Saddle point at $a_{21} = 1$. Thus the value of the game is 1; the optimum strategy for the row player is $P = [0, 1]$ and the optimum strategy for the column player is $Q = \begin{bmatrix} 1 \\ 0 \end{bmatrix}$

   (b) Saddle point at $a_{23} = 1$. Thus the value of the game is 1; the optimum strategy for the row player is $P = [0, 1, 0]$, and the optimum strategy for the column player is $Q = \begin{bmatrix} 0 \\ 0 \\ 1 \end{bmatrix}$

(c) Saddle point at $a_{32} = -1$. Thus the value of the game is -1; the optimum strategy for the row player is $P = [0, 0, 1]$, and the optimum strategy for the column player is $Q = \begin{bmatrix} 0 \\ 1 \\ 0 \end{bmatrix}$.

(d) There is no saddle point. The value of the game is 1. The row player should use strategy $P = [\frac{1}{3}, \frac{2}{3}]$—that is, the first row should be played about $\frac{1}{3}$ of the time in a random pattern, and the second row should be played $\frac{2}{3}$ of the time. The column player should use strategy $Q = \begin{bmatrix} \frac{1}{2} \\ \frac{1}{2} \end{bmatrix}$—that is, play each column half the time in a random pattern.

(e) The value of the game is $\frac{10}{7}$; $P = [\frac{1}{7}, \frac{6}{7}]$; $Q = \begin{bmatrix} \frac{4}{7} \\ \frac{3}{7} \end{bmatrix}$.

2. (a)

	Center	Edge
Center	0	1
Edge	−1	0

(b)

	Lowest price	Medium price	Highest price
Lowest price	0	1	−1
Medium price	−1	0	1
Highest price	1	−1	0

(c)

	Enemy	
Rebels	Highway	Dark road
Highway	6	−3
Dark road	−1	10

ANSWERS 10.2. C

1. (a) Saddle point at $a_{22} = -1$, so $v = -1$; the row player should play the second row and the column player should play the second column.

(b) Saddle points at a_{41} and a_{43}, so $v = 0$; the row player should play the fourth row and the column player should play either the first or third column.

(c) $v = 1$; $P = [0, \frac{1}{2}, 0, \frac{1}{2}]$; $Q = \begin{bmatrix} \frac{2}{3} \\ 0 \\ \frac{1}{3} \\ 0 \end{bmatrix}$

2. There are many such matrices, but one is

$$\begin{bmatrix} -1 & -2 & 1 \\ -7 & -3 & 4 \\ 6 & -4 & -5 \end{bmatrix}$$

It is easy to check that this has no dominated rows or columns, but not so easy to see that $a_{12} = -2$ is a saddle point. However, if the row player plays the first row, the column player will only do worse by not playing the second column. Similarly, if the column player plays the second column, the row player will only do worse by not playing the first row.

Such saddle points can be located by the so-called minimax principle, but this is a specialized technique that works only for saddle points, whereas the linear programming technique explained in this text will reveal the best strategy for any matrix game, including those having saddle points.

10.3 Games and Matrices

The previous section introduced you to game theory and showed how to find the value of any 2 by 2 matrix game and the optimum strategy for each player in such a game. The purpose of this section will be to deepen your understanding of these ideas by relating them to the matrix theory you learned in Chapters 1 and 2. This will prepare you for "solving" larger matrix games (that is, finding the value of the game and optimum strategy for each player) in the next section using the simplex algorithm and duality theory presented in Chapters 6 and 7. Since game theory clearly uses probability (introduced in Chapter 9), this section and the next might be considered a review and synthesis of the whole book.

10.3.1 Example

In Example 10.2.7 the matrix game $A = \begin{bmatrix} 3 & -1 \\ -4 & 2 \end{bmatrix}$ was studied. The *value* of this game was found to be $v = 0.2$; the optimum strategy for the row player was $P = [0.6, 0.4]$; and the strategy for the column player was $Q = \begin{bmatrix} 0.3 \\ 0.7 \end{bmatrix}$. Notice the following relationships, which follow from the discussion in the previous section:

$$PA = [0.6, 0.4]\begin{bmatrix} 3 & -1 \\ -4 & 2 \end{bmatrix} = [1.8 - 1.6, -0.6 + 0.8] = [0.2, 0.2]$$

This says that if the row player plays the first row 0.6 of the time in a random pattern, the expectation of the game is 0.2, whether the column player plays $Q_1 = \begin{bmatrix} 1 \\ 0 \end{bmatrix}$ or $Q_2 = \begin{bmatrix} 0 \\ 1 \end{bmatrix}$ or, therefore, any mixed strategy.

$$AQ = \begin{bmatrix} 3 & -1 \\ -4 & 2 \end{bmatrix} \begin{bmatrix} 0.3 \\ 0.7 \end{bmatrix} = \begin{bmatrix} 0.9 - 0.7 \\ -1.2 + 1.4 \end{bmatrix} = \begin{bmatrix} 0.2 \\ 0.2 \end{bmatrix}$$

This means that if the column player plays the first column 0.3 of the time in a random pattern, the expectation of the game is 0.2 whether the row player plays $P_1 = [1, 0]$ or $P_2 = [0, 1]$ or, therefore, any mixed strategy.

Furthermore, since both P and Q are probability vectors (that is, the sum of their elements is 1), we see 0.2 yet again in the expressions

$$(PA)Q = [0.2, 0.2]\begin{bmatrix} 0.3 \\ 0.7 \end{bmatrix} = 0.2 \qquad P(AQ) = [0.6, 0.4]\begin{bmatrix} 0.2 \\ 0.2 \end{bmatrix} = 0.2$$

None of the appearances of 0.2 in Example 10.3.1 are due to coincidence. For any matrix game A, if P is the row vector describing the *optimum* strategy for the row player and Q is the column vector describing the *optimum* strategy for the column player, then PAQ will be the value of the game. That is, $v = PAQ$.

Furthermore, PA will be a row vector with the same number of elements as the column vector Q (the number of columns in the game matrix A) and PA tells the expectation of the game for each pure strategy of the column player if the row player uses strategy P. Each element of PA must be at least as large as v; if any are larger, the column player should never play the corresponding pure strategy as part of a mixed strategy.

Similarly, AQ will be a column vector with the same number of elements as the row vector P (the number of rows in the game matrix A) and AQ tells the expectation of the game for each pure strategy of the row player if the column player uses strategy Q. Each element of AQ must be no larger than v; if any are smaller, then the row player should never play the corresponding pure strategy.

Techniques for finding v, P, and Q for any 2 by 2 matrix game were given in Section 10.2. These techniques do not always work for 3 by 3 matrix games. In the next example we attempt to find the value of a 3 by 3 game by trial and error.

10.3.2 Example

Study the game matrix $\begin{bmatrix} 3 & 0 & 3 \\ 3 & 1 & 1 \\ 0 & 4 & 2 \end{bmatrix}$ using the ideas above.

Solution:

We do not know the optimum strategy for either player, but suppose we see what happens when the row player plays strategy $P_1 = [\frac{1}{3}, \frac{1}{3}, \frac{1}{3}]$. (We shall use a subscript on P to denote the fact that this is not known to be the optimum strategy.) Then

$$P_1A = [\tfrac{1}{3}, \tfrac{1}{3}, \tfrac{1}{3}]\begin{bmatrix} 3 & 0 & 3 \\ 3 & 1 & 1 \\ 0 & 4 & 2 \end{bmatrix} = [2, \tfrac{5}{3}, 2]$$

From this we can conclude that the value v of this game is greater than $\frac{5}{3}$ because no matter what strategy Q_j is used by the column player, we have $\frac{5}{3} \leq P_1AQ_j$. If we let 1_{row} denote the row vector consisting of all 1's, we observe that $P_1A \geq (\frac{5}{3})1_{\text{row}}$, where the inequality sign was defined in Definition 7.1.1 (page 234).

Now let us try another strategy for the row player, namely $P_2 = [0.4, 0.2, 0.4]$. Then we can compute

$$P_2A = [0.4, 0.2, 0.4]\begin{bmatrix} 3 & 0 & 3 \\ 3 & 1 & 1 \\ 0 & 4 & 2 \end{bmatrix} = [1.8, 1.8, 2.2]$$

Since $P_2A \geq (1.8)1_{\text{row}}$, we can conclude that $v \geq 1.8$. This is because any other strategy P_3 such that $P_3A \ngeq (1.8)1_{\text{row}}$ leaves the column player some strategy that would lower the expectation below 1.8; since the row player is assumed not to want to take such a risk, P_3 would never be chosen.

In particular, we now know that the row player should not use $P_1 = [\frac{1}{3}, \frac{1}{3}, \frac{1}{3}]$ above because then the column player could obtain a payoff of $\frac{5}{3} < 1.8$ by playing strategy $Q = \begin{bmatrix} 0 \\ 1 \\ 0 \end{bmatrix}$. Thus P_1 is not an optimum strategy.

Is P_2 the optimum strategy? That depends on whether there is another strategy, P_3, such that $P_3A \geq v1_{\text{row}}$, where $v > 1.8$. It is not clear offhand whether this is so; you might try some other P_3 at random, but that gets boring.

Clearly, the trial-and-error method has its limitations. In the next section we shall show how to use the simplex algorithm to find v, P, and Q all at once. Now let us look at a slightly different example.

10.3.3 Example

Analyze the matrix game $A = \begin{bmatrix} 2 & 1 & 0 \\ 0 & 2 & 3 \end{bmatrix}$

Solution:

Since there are only two rows, we know that the optimum strategy for the row player is $P = [p, 1 - p]$, and it is possible to graph the expected payoff as we did in the previous section. When the column player plays the first column, the expected value is

$$E_1 = [p, 1 - p]\begin{bmatrix} 2 & 1 & 0 \\ 0 & 2 & 3 \end{bmatrix}\begin{bmatrix} 1 \\ 0 \\ 0 \end{bmatrix} = 2p$$

Similarly, $E_2 = p + 2(1 - p) = 2 - p$ and $E_3 = 3(1 - p) = 3 - 3p$. These are graphed for $0 \le p \le 1$ in Figure 10.3–1.

Figure 10.3–1

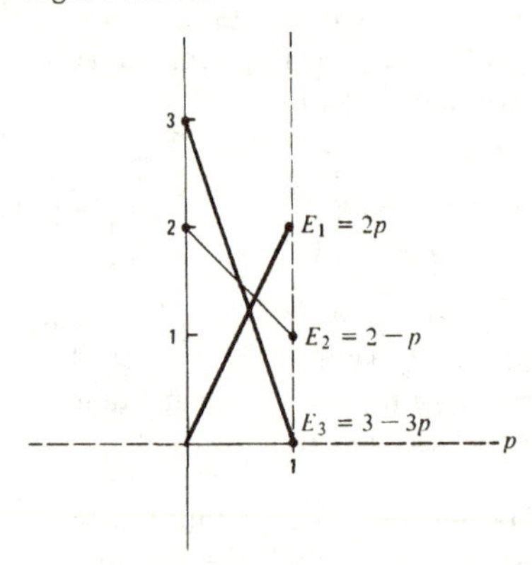

Looking at this graph we can see that the p which maximizes the least possible expected value occurs at the intersection of E_1 and E_3. Thus we solve $2p = 3 - 3p$ to get $p = \frac{3}{5}$. Since $E_1(\frac{3}{5}) = \frac{6}{5}$, the value of this game is $v = \frac{6}{5}$.

Now consider the following computation.

$$PA = [\tfrac{3}{5}, \tfrac{2}{5}]\begin{bmatrix} 2 & 1 & 0 \\ 0 & 2 & 3 \end{bmatrix} = [\tfrac{6}{5}, \tfrac{7}{5}, \tfrac{6}{5}]$$

Sure enough—$PA \ge v1_{\text{row}}$. But this time $PA \ne v1_{\text{row}}$. The fact that the middle element of PA is not v indicates that the column player should never choose the second column; if he does, the expected value will be greater than the value of the game.

This tells us that the optimum strategy for the column player is of the form $Q = \begin{bmatrix} q \\ 0 \\ 1 - q \end{bmatrix}$, and we can use the same method as above to find q. If the row player plays the first row, the expectation is $E_4 = 2q + 0 + 0$ and if she plays the second row, the expectation is $E_5 = 0 + 0 + 3(1 - q)$. Setting these equal to find the point of intersection (the minimum of the greatest possible payoffs), we have $2q = 3 - 3q$

or $q = \frac{3}{5}$. Thus the optimum strategy for the column player is of the form $Q = \begin{bmatrix} \frac{3}{5} \\ 0 \\ \frac{2}{5} \end{bmatrix}$. Again, we can confirm that the value of the game is $\frac{6}{5}$, by computing

$$AQ = \begin{bmatrix} 2 & 1 & 0 \\ 0 & 2 & 3 \end{bmatrix} \begin{bmatrix} \frac{3}{5} \\ 0 \\ \frac{2}{5} \end{bmatrix} = \begin{bmatrix} \frac{6}{5} \\ \frac{6}{5} \end{bmatrix}$$

Notice that this example verifies that, as asserted just before Example 10.3.2, PA is a row vector with the same number of elements as the column vector, Q, and AQ is a column vector with the same number of elements as the row vector, P.

Games with Positive Value

When we use the simplex method in the next section to solve matrix games, we will be able to handle only those games that are already known to have a positive value, v. [This is because we will want to divide the inequality $PA \geq (v)1_{row}$ by v and know that the sense of the inequality is unchanged.] But this is not a serious restriction. Every matrix game can be "converted" to another game with positive value having the same optimum strategies by adding some constant number to all the elements of the matrix. In Example 10.2.1, for example, we added 1 to the matrix $\begin{bmatrix} -1 & -2 & 4 \\ 3 & -1 & 2 \\ 2 & -2 & 1 \end{bmatrix}$ to get a "fair" game matrix, $\begin{bmatrix} 0 & -1 & 5 \\ 4 & 0 & 3 \\ 3 & -1 & 2 \end{bmatrix}$ with the same strategy. If we add 2 more to this second matrix, we will obtain yet another matrix with the sam strategy, but with a value that is clearly positive, since the value of he game described by a matrix must be at least as much as th smallest number in the matrix.

10.3.4 Rule

Let A and B be two matrices of the same dimension such that every element of B is exactly k more than the corresponding element of A. Then if we consider the matrix games represented by A and B:

(a) Both players have the same optimum strategy for both games.
(b) The value of the game described by B is exactly k more than the value of the game described by A.

Thus we can easily convert any matrix game to a game with all positive entries by adding a number that is 1 more than the absolute value of the negative element with the largest absolute value.

Non-Zero-Sum Games

Thus far we have discussed only those games in which the row player wins exactly the same amount that the column player loses, and vice versa. In real-world applications, this is often not the case. We then represent the game with a pair of numbers at each element of the matrix; the first tells how much the row player wins, and the second tells how much the column player wins.

10.3.5 Example

The following double matrix was obtained in 1971 by asking students at the teacher-training schools in Cyprus their opinions of the outcome to various political options. The number zero indicates a neutral reaction. Although there are many simplifications in any such mathematical analysis of political situation, this simple chart gives much insight into the fighting that broke out in Cyprus just three years after the survey was taken.

Greek Cypriot options	Turkish Cypriot options	
	Modify position	Maintain present position
Modify position	Peace (2.2, 1.2)	Partitioning island into two communities (−3.0, 3.4)
Maintain present position	Union of Cyprus with Greece (3.8, −4.7)	War (−2.0, −1.2)

In each box of the game matrix, the left number gives the average Greek rating of the given outcome and the right number gives the Turkish rating. The numbers in the upper left box indicate that both groups would consider peace to be a desirable (positive) outcome, and the numbers in the lower right box show that both groups would consider war to be an undesirable (negative) outcome. However, the numbers in the lower left box show that the Greeks would consider a union with Greece even better than peace (a 3.8 over 2.2 rating), while Turkish respondents would consider it even worse than war (−4.7 as opposed to −1.2). Similarly, the upper right box shows that the Turkish respondents would prefer partitioning of the island to peace and the Greeks find the prospect more distasteful than war.

Thus even though the greatest total satisfaction would have resulted from both groups' modifying their positions, each group alone felt it had more to win by keeping firm in its own position. War broke out in the summer of 1974.

The study reported in the previous example was administered by a Greek educational psychologist and a Turkish college principal, each of whom surveyed a majority of the students of his own background in teacher-training colleges in Nicosia, Cyprus. These schools are the most advanced on the island (for university work, students must travel to the mainland) and their students' views can be expected both to reflect those of their respective populations and to influence them. An article about this study, written by Malvern Lumsden of the Stockholm Peace Institute, appears in Vol. 17 of the *Journal of Conflict Resolution*.

Such studies are just beginning, but one can hope that eventually they may be a significant tool in bringing understanding (and, therefore, compromise) between conflicting groups and individuals.

SUMMARY

This section concluded with an example of a serious two-"person" non-zero-sum "game," and explored further the pattern of two-person zero-sum games.

Each two-person zero-sum game having a finite choice of pure strategies for each player can be expressed as a matrix A. There is an optimum strategy for the row player, written as a row vector P and an optimum strategy for the column player, written as a column vector Q. If both players use their optimum strategy, the expectation of the game is $v = PAQ$. The row vector $PA \geq v1_{\text{row}}$, where 1_{row} denotes the row vector with all 1's and the same number of elements as Q. Similarly, the column vector $AQ \leq v1_{\text{column}}$, where 1_{column} is the column vector consisting of all 1's and the same number of elements as P. If the row player tries a "wrong" strategy, P_1, then $P_1A \ngeq v1_{\text{row}}$ for $v = PAQ$. Similarly, if the column player plays a "wrong" strategy, Q_1, then $AQ_1 \nleq v1_{\text{column}}$.

If the same number k is added to every element of a game matrix A, then the payoff increases by k, but the optimum strategy for each player remains the same.

EXERCISES 10.3. A

1. Using the notation given in the summary just above, find v, P, Q, PA, and AQ for each of the matrix games given below. Use the methods of the previous section; eliminate dominated rows and columns first.

(a) $\begin{bmatrix} 0 & 1 & -2 \\ -1 & 1 & 3 \\ -2 & 0 & 1 \end{bmatrix}$ (b) $\begin{bmatrix} 0 & -1 & 2 \\ 4 & 3 & 0 \\ 2 & 2 & -3 \end{bmatrix}$

2. Write a matrix game that has the same optimum strategies as each of the following but all positive entries.

(a) $\begin{bmatrix} -2 & 0 \\ -1 & -4 \end{bmatrix}$ (b) $\begin{bmatrix} -1 & 2 & -5 \\ 0 & -3 & 5 \\ 4 & 1 & 0 \end{bmatrix}$ (c) $\begin{bmatrix} -6 & -7 & -3 \\ 0 & -2 & 5 \\ -4 & 0 & 4 \end{bmatrix}$

3. Consider the matrix game described by $A = \begin{bmatrix} 2 & 0 & 4 \\ 2 & 1 & 2 \\ 0 & 4 & 2 \end{bmatrix}$

   (a) If the row player uses strategy $P_1 = [\frac{1}{3}, \frac{1}{3}, \frac{1}{3}]$, what is the expectation for each pure strategy available to the column player?
   (b) What does the answer to part (a) tell you about the value v of the game?
   (c) If you were the column player and you knew that the row player were using strategy $[\frac{1}{3}, \frac{1}{3}, \frac{1}{3}]$, what strategy would you use?
   (d) Suppose that the column player chose each column $\frac{1}{3}$ of the time. What is the expectation for each pure strategy available to the row player?
   (e) What does this tell you about the value v of the game?

4. Example 10.3.5 is a particular case of a general type of game called "Prisoner's Dilemma" because of the following interpretation. A district attorney has two suspects, on each of whom he has some evidence, but each of whom he thinks knows enough about the activities of the other to warrant a 10-year sentence. He tells them he will interrogate each alone, and not tell the second what the first has said until all results are in. If neither confesses, both will get a year's jail sentence. If one confesses but not the other, the confessor gets off free and his cohort will get the 10-year sentence. And if both confess, each will get a 7-year prison term.
   (a) Write the situation as a matrix with a pair of numbers in each entry.
   (b) Suppose you are one of the prisoners and you know both deserve the 10 years. What would you do?

5. Two authors are writing competing textbooks that will both be published at the same time. If they work completely independently, they will split the market evenly. If one tells the other his or her best ideas but not conversely, one third of the buyers of the teller's books will instead buy from the other. But if they both share their best ideas with the other, buyers will become so enthusiastic about the books and the market will expand, each getting one-third more sales. Write this situation as a matrix whose entries are number pairs.

EXERCISES 10.3. B

1. Using the notation given in the summary of this section, find v, P, Q, PA, and AQ for each of the matrix games given below. Use the methods of the previous section: eliminate dominated rows and columns first.

(a) $\begin{bmatrix} -1 & 1 & 2 \\ 4 & 0 & 2 \\ 2 & -1 & 1 \end{bmatrix}$ (b) $\begin{bmatrix} -1 & -2 & 1 \\ 1 & 1 & -4 \\ 3 & 2 & -3 \end{bmatrix}$

2. Write a matrix game that has the same optimum strategies as each of the following matrix games, but all positive entries.

   (a) $\begin{bmatrix} -3 & 0 \\ -1 & -6 \end{bmatrix}$ (b) $\begin{bmatrix} -3 & 0 & -4 \\ 4 & -9 & 0 \\ -1 & 8 & 3 \end{bmatrix}$ (c) $\begin{bmatrix} -4 & -7 & 8 \\ 0 & 3 & -5 \\ -2 & 7 & -4 \end{bmatrix}$

3. Consider that matrix game described by

$$A = \begin{bmatrix} 0 & 2 & 3 \\ 2 & 1 & 2 \\ 4 & 0 & 2 \end{bmatrix}$$

   Answer the five questions given in problem 3, Exercises 10.3.A, for this matrix.

4. Same as problem 4, Exercises 10.3.A.
5. Suppose a shopping mall will have two different department stores. Each store has been offered the choice of two different sites, one in the middle of the mall and one on the edge. The potential edge sites are on opposite edges of the mall. Thus if they both choose the edge sites, the total sales will be one third again as much as if they both choose the center sites. But if one chooses the edge while the other chooses the center, the one on the edge will get a third less than if they are both in the center and the one in the center will get a third more. Describe this situation with a matrix whose entries are pairs of numbers.

ANSWERS 10.3. A

1. (a) $v = -\frac{1}{3}$; $P = [\frac{2}{3}, \frac{1}{3}, 0]$; $Q = \begin{bmatrix} \frac{5}{6} \\ 0 \\ \frac{1}{6} \end{bmatrix}$; $AQ = \begin{bmatrix} -\frac{1}{3} \\ -\frac{1}{3} \\ -\frac{3}{2} \end{bmatrix}$; $PA = [-\frac{1}{3}, 1, -\frac{1}{3}]$

   (b) $v = 1$; $P = [\frac{1}{2}, \frac{1}{2}, 0]$; $Q = \begin{bmatrix} 0 \\ \frac{1}{3} \\ \frac{2}{3} \end{bmatrix}$; $AQ = \begin{bmatrix} 1 \\ 1 \\ -\frac{4}{3} \end{bmatrix}$; $PA = [2, 1, 1]$

2. (a) $\begin{bmatrix} 3 & 5 \\ 4 & 1 \end{bmatrix}$ (b) $\begin{bmatrix} 5 & 8 & 1 \\ 6 & 3 & 11 \\ 10 & 7 & 6 \end{bmatrix}$ (c) $\begin{bmatrix} 2 & 1 & 5 \\ 8 & 6 & 13 \\ 4 & 8 & 12 \end{bmatrix}$

3. (a) $E_1 = \frac{4}{3}$; $E_2 = \frac{5}{3}$; $E_3 = \frac{8}{3}$ (b) $v \geq \frac{4}{3}$ (c) $\begin{bmatrix} 1 \\ 0 \\ 0 \end{bmatrix}$, because then the expectation of the game would be $\frac{4}{3}$, which is the lowest possible if the row player uses the given strategy. (d) $E_1 = 2$; $E_2 = \frac{5}{3}$; $E_3 = 2$ (e) $v \leq 2$

4. (a)

	Prisoner II	
Prisoner I	Does not confess	Confesses
Does not confess	(−1, −1)	(−10, 0)
Confesses	(0, −10)	(−7, −7)

(b) I would feel unhappy. If we both remain silent, the total punishment is least. But if I trust my friend and he squeals, I get 10 years. How trustworthy is my accomplice?

5.

First author	Second author: Helps	Second author: Does not help
Helps	(4, 4)	(2, 4)
Does not help	(4, 2)	(3, 3)

10.4 Solving Matrix Games Using the Simplex Algorithm

This section shows how to use the simplex algorithm to solve any matrix game whose value, v, is already known to be positive. This is not really a restriction on which games can be solved because, as we saw in the previous section, it is easy to add a number to all the elements of a game matrix to obtain another game matrix with the same optimum strategies.

If you have not studied dual LP problems (Chapter 7), you may want to skip the first few paragraphs of this section and begin reading at the subheading "Setup of Problem." It is not necessary to understand duality in order to read most of this section and successfully do the exercises.

In Chapter 7 we wrote dual problems with the following notation:

$$\text{Maximize } P = BX \qquad \text{Minimize } C = UD$$
$$\text{when } AX \leq D \qquad \text{when } UA \geq B$$

In this section we shall show that the solution of any two-person, zero-sum game described by a matrix A with a positive value $v > 0$ can be described by the following dual problems, where the notation is explained below.

$$\text{Maximize } M = 1_{\text{row}} X \qquad \text{Minimize } C = U 1_{\text{column}}$$
$$\text{when } AX \leq 1_{\text{column}} \qquad \text{when } UA \geq 1_{\text{row}}$$

We shall find that $M_{\max} = C_{\min} = 1/v$, the reciprocal of the value of the game.

This notational magic is made possible by a clever choice of the independent variables, X and U.

$$X = \frac{Q}{v} \qquad U = \frac{P}{v}$$

where P and Q are the strategies of the row and column players, respectively, and v is the (as yet unknown, but positive) value of the game A. As in the previous section, 1_{row} is a row vector with all 1's and 1_{column} is a column vector with all 1's.

Setup of Problem

Let us consider the game from the point of view of the column player. By the reasoning of the previous section, she or he wants to find the smallest v such that there is a probability column vector Q such that $AQ \leq v1_{\text{column}}$. For such a Q, no matter what the row player chooses, the expected payoff cannot be greater than v.

Thus the column player wants to minimize v such that $AQ \leq v1_{\text{column}}$. Since $v > 0$, this is the same as maximizing $1/v$ such that $A\frac{Q}{v} \leq 1_{\text{column}}$. Setting $X = \frac{Q}{v}$, we notice that $1/v = 1_{\text{row}}\left(\frac{Q}{v}\right) = 1_{\text{row}}X$, because Q is a probability vector and therefore the sum of its elements is 1.

10.4.1 Rule

To solve a game with matrix A, let $X = Q/v$ and use the simplex algorithm to

$$\text{Maximize } M = 1_{\text{row}}X \qquad \text{when} \qquad AX \leq 1_{\text{column}}$$

10.4.2 Example

Let $A = \begin{bmatrix} 2 & 1 & 0 \\ 0 & 2 & 3 \end{bmatrix}$ be the matrix describing a game. Find the value of the game and the optimum strategy for the column player.

Solution:

Using the form of Rule 10.4.1, we write

$$\text{Maximize } M = x_1 + x_2 + x_3 \qquad \text{when} \qquad \begin{aligned} 2x_1 + x_2 \qquad &\leq 1 \\ 2x_2 + 3x_3 &\leq 1 \end{aligned}$$

Referring back to Section 6.1, the first tableau needed for the simplex algorithm is

	x_1	x_2	x_3	s_1	s_2	Solutions
s_1	②	1	0	1	0	1
s_2	0	2	3	0	1	1
$-M$	1	1	1	0	0	0

We can choose any one of the first three columns as the pivot column; following convention, we choose the first.

	x_1	x_2	x_3	s_1	s_2	Solutions
x_1	1	$\frac{1}{2}$	0	$\frac{1}{2}$	0	$\frac{1}{2}$
s_2	0	2	③	0	1	1
$-M$	0	$\frac{1}{2}$	1	$-\frac{1}{2}$	0	$-\frac{1}{2}$

	x_1	x_2	x_3	s_1	s_2	Solutions
x_1	1	$\frac{1}{2}$	0	$\frac{1}{2}$	0	$\frac{1}{2}$
x_3	0	$\frac{2}{3}$	1	0	$\frac{1}{3}$	$\frac{1}{3}$
$-M$	0	$-\frac{1}{6}$	0	$-\frac{1}{2}$	$-\frac{1}{3}$	$-\frac{5}{6}$

Thus $M = \frac{5}{6}$ and since $M = 1/v$, it follows that $v = \frac{6}{5}$. Also, since $X = Q/v$, $Q = vX = \frac{6}{5}\begin{bmatrix}\frac{1}{2}\\0\\\frac{1}{3}\end{bmatrix} = \begin{bmatrix}\frac{3}{5}\\0\\\frac{2}{5}\end{bmatrix}$ Notice that Q is indeed a probability vector; if it were not, we would know that a mistake had been made.

Now let us consider a matrix game from the viewpoint of the row player. He or she wants to maximize v (that is, minimize $1/v$) such that there is a P with $PA \geq v1_{\text{row}}$. But this is equivalent to $(P/v)A \geq 1_{\text{row}}$ or $UA \geq 1_{\text{row}}$ when we set $U = P/v$. Using the fact that P is a probability vector (that is, its elements add to 1), we have $1/v = (P/v)1_{\text{column}} = U1_{\text{column}}$ and so the row player's goal can be stated as follows.

10.4.3 Rule

Minimize $C = U1_{\text{column}}$ when $UA \geq 1_{\text{row}}$.

This is, as predicted, the dual of the column player's problem as stated in Rule 10.4.1. Thus, using the ideas just before the conclusion to Section 7.2 (pages 243 and 244), if we use the simplex algorithm to maximize the column player's problem, the solution to the row player's problem will appear as the absolute values of the numbers under the slack variables in the last row of the final tableau.

10.4.4 Example

Find the optimum strategy for the row player in the game described in Example 10.4.2.

Solution:

Because of the discussion just above, we need only read off the numbers below the slack variables in the last row of the last tableau in Example 10.4.2. These numbers are $-\frac{1}{2}$ and $-\frac{1}{3}$. This implies that

$U = [\frac{1}{2}, \frac{1}{3}]$. Since $U = P/v$ and $v = \frac{6}{5}$, we have $P = vU = \frac{6}{5}[\frac{1}{2}, \frac{1}{3}] = [\frac{3}{5}, \frac{2}{5}]$. We note that

$$PA = [\tfrac{3}{5}, \tfrac{2}{5}]\begin{bmatrix} 2 & 1 & 0 \\ 0 & 2 & 3 \end{bmatrix} = [\tfrac{6}{5}, \tfrac{7}{5}, \tfrac{6}{5}] \geq \tfrac{6}{5}1_{\text{row}}$$

The $\frac{7}{5}$ in the row vector PA corresponds to the 0 in the column vector Q; it indicates why the column player should never play the second column.

10.4.5 Example

Solve the matrix game represented by

$$A = \begin{bmatrix} 3 & 0 & 3 \\ 3 & 1 & 1 \\ 0 & 4 & 2 \end{bmatrix}$$

[This was the matrix in Example 10.2.3(b).]

Solution:

We set up the initial tableau as follows, using Rules 10.4.1 and 10.4.3:

	x_1	x_2	x_3	s_1	s_2	s_3	Solutions
s_1	(3)	0	3	1	0	0	1
s_2	3	1	1	0	1	0	1
s_3	0	4	2	0	0	1	1
$-M$	1	1	1	0	0	0	0

The positive numbers in the last row are all the same, so we could choose any of these three columns as our pivot column; it is conventional, but not necessary, to choose the first. Then we see we have another choice! Since both the positive numbers yield the same quotient when divided into the corresponding number in the "solutions" column, we could use either 3 as a pivot. We choose the first, because there is an extra zero in that row, which makes the computations easier.

	x_1	x_2	x_3	s_1	s_2	s_3	Solutions
x_1	1	0	1	$\frac{1}{3}$	0	0	$\frac{1}{3}$
s_2	0	(1)	-2	-1	1	0	0
s_3	0	4	2	0	0	1	1
$-M$	0	1	0	$-\frac{1}{3}$	0	0	$-\frac{1}{3}$

Now the pivoting is straightforward using the techniques of Sections 6.1 and 6.2.

	x_1	x_2	x_3	s_1	s_2	s_3	Solutions
x_1	1	0	1	$\frac{1}{3}$	0	0	$\frac{1}{3}$
x_2	0	1	-2	-1	1	0	0
s_3	0	0	(10)	4	-4	1	1
$-M$	0	0	2	$\frac{2}{3}$	-1	0	$-\frac{1}{3}$

	x_1	x_2	x_3	s_1	s_2	s_3	Solutions
x_1	1	0	0	$-\frac{1}{15}$	$\frac{2}{5}$	$-\frac{1}{10}$	$\frac{7}{30}$
x_2	0	1	0	$-\frac{1}{5}$	$\frac{1}{5}$	$\frac{1}{5}$	$\frac{1}{5}$
x_3	0	0	1	$\frac{2}{5}$	$-\frac{2}{5}$	$\frac{1}{10}$	$\frac{1}{10}$
$-M$	0	0	0	$-\frac{2}{15}$	$-\frac{1}{5}$	$-\frac{1}{5}$	$-\frac{8}{15}$

Thus the value of the game is $v = 1/M = \frac{15}{8}$. The optimum strategy for the column player is $Q = vX = \frac{15}{8}\begin{bmatrix}\frac{7}{30}\\ \frac{1}{5}\\ \frac{1}{10}\end{bmatrix} = \begin{bmatrix}\frac{7}{16}\\ \frac{3}{8}\\ \frac{3}{16}\end{bmatrix}$ and the optimum strategy for the row player is $P = vU = \frac{15}{8}[\frac{2}{15}, \frac{1}{5}, \frac{1}{5}] = [\frac{1}{4}, \frac{3}{8}, \frac{3}{8}]$. It is easy to check that $PA = v1_{\text{row}}$ and $AQ = v1_{\text{column}}$.

SUMMARY

Any game matrix A with a positive value v can be solved using the simplex algorithm beginning with the following tableau:

A	I	1_{column}
1_{row}	0_{row}	0

When pivoting has been applied until the last row contains no negative numbers, the number in the lower right is the negative reciprocal of v, the elements of Q can be found by multiplying the numbers in the right column by v, and P is $-v$ times the numbers in the last row below the slack variables.

EXERCISES 10.4. A

Solve each of the following matrix games using the simplex algorithm. Also, find PA and AQ, using the notation of this chapter.

1. $A = \begin{bmatrix}0 & 2 & 3\\ 1 & 2 & 0\end{bmatrix}$ 2. $A = \begin{bmatrix}1 & 0 & 2\\ 0 & 2 & 1\\ 2 & 0 & 1\end{bmatrix}$ 3. $A = \begin{bmatrix}2 & 1 & 2\\ 2 & 0 & 4\\ 0 & 4 & 2\end{bmatrix}$

EXERCISES 10.4. B

Solve each of the following matrix games using the simplex algorithm. Also, find PA and AQ, using the notation of this chapter.

1. $A = \begin{bmatrix} 1 & 3 & 0 \\ 0 & 2 & 4 \end{bmatrix}$ 2. $A = \begin{bmatrix} 2 & 0 & 2 \\ 1 & 1 & 1 \\ 0 & 2 & 1 \end{bmatrix}$ 3. $A = \begin{bmatrix} 0 & 2 & 3 \\ 2 & 1 & 2 \\ 4 & 0 & 2 \end{bmatrix}$

EXERCISES 10.4. C

Solve each of the following matrix games using the simplex algorithm. Also, find PA and AQ, using the notation of this chapter.

1. $A = \begin{bmatrix} 0 & 1 & 2 \\ 1 & 2 & 0 \\ 2 & 0 & 1 \end{bmatrix}$
2. $A = \begin{bmatrix} 4 & 0 & 1 \\ 2 & 3 & 0 \\ 0 & 3 & 3 \end{bmatrix}$

ANSWERS 10.4. A

1. $v = \frac{3}{4}$; $P = [\frac{1}{4}, \frac{3}{4}]$; $PA = [\frac{3}{4}, 2, \frac{3}{4}]$; $Q = \begin{bmatrix} \frac{3}{4} \\ 0 \\ \frac{1}{4} \end{bmatrix}$; $AQ = \begin{bmatrix} \frac{3}{4} \\ \frac{3}{4} \end{bmatrix}$
2. $v = 1$; $P = [0, \frac{1}{2}, \frac{1}{2}]$; $PA = [1, 1, 1]$; $Q = \begin{bmatrix} \frac{1}{2} \\ \frac{1}{2} \\ 0 \end{bmatrix}$; $AQ = \begin{bmatrix} \frac{1}{2} \\ 1 \\ 1 \end{bmatrix}$
3. $v = \frac{8}{5}$; $P = [\frac{4}{5}, 0, \frac{1}{5}]$; $PA = [\frac{8}{5}, \frac{8}{5}, \frac{10}{5}]$; $Q = \begin{bmatrix} \frac{3}{5} \\ \frac{2}{5} \\ 0 \end{bmatrix}$; $AQ = \begin{bmatrix} \frac{8}{5} \\ \frac{6}{5} \\ \frac{8}{5} \end{bmatrix}$

ANSWERS 10.4. C

1. $v = 1$; $P = [\frac{1}{3}, \frac{1}{3}, \frac{1}{3}]$; $PA = [1, 1, 1]$; $Q = \begin{bmatrix} \frac{1}{3} \\ \frac{1}{3} \\ \frac{1}{3} \end{bmatrix}$; $AQ = \begin{bmatrix} 1 \\ 1 \\ 1 \end{bmatrix}$
2. $v = \frac{42}{23}$; $P = [\frac{9}{23}, \frac{3}{23}, \frac{11}{23}]$; $PQ = [\frac{42}{23}, \frac{42}{23}, \frac{42}{23}]$; $Q = \begin{bmatrix} \frac{9}{23} \\ \frac{8}{23} \\ \frac{6}{23} \end{bmatrix}$; $AQ = \begin{bmatrix} \frac{42}{23} \\ \frac{42}{23} \\ \frac{42}{23} \end{bmatrix}$

VOCABULARY

random variable, expected value or expectation, matrix game, two-person game, zero-sum game, row player, column player, payoff, saddle point, mixed strategy, optimum strategy, value of a game, pure strategy

SAMPLE TEST

Chapter 10

1. (20 pts) Suppose that the probability that a real estate agency will sell no house on a given day is 0.5, the probability that it will sell one house is 0.4, and the probability that it will sell two houses is 0.1.
   (a) What is the expected number of houses per day that it sells?
   (b) If the probability distribution for all days is the same, what is the expected number of houses that it will sell in a 6-day week?
   (c) If it sells houses in only one development, each of which yields the agency a $4000 commission, what is the expected gross commission each week?
2. Solve each of the following matrix games, using any technique you have learned in this chapter. Using the notation of this chapter, also find PA and AQ for each game.

   (a) $\begin{bmatrix} -1 & -2 & 0 \\ 3 & -4 & -3 \\ 2 & -1 & 2 \end{bmatrix}$ (b) $\begin{bmatrix} -1 & 2 \\ 3 & 0 \end{bmatrix}$ (c) $\begin{bmatrix} 1 & 0 & 3 \\ 0 & 4 & 2 \end{bmatrix}$

   (d) $\begin{bmatrix} 2 & 2 & 4 \\ 6 & 0 & 2 \\ 0 & 3 & 2 \end{bmatrix}$

The answers are at the back of the book.

Appendix 1

SIGNED NUMBERS

While adding signed numbers, it is sometimes helpful to think of arrows along the number line. To add any two numbers of the same sign, add the absolute values and keep the same sign. To add two numbers of opposite sign, subtract the absolute values and use the sign of the larger.

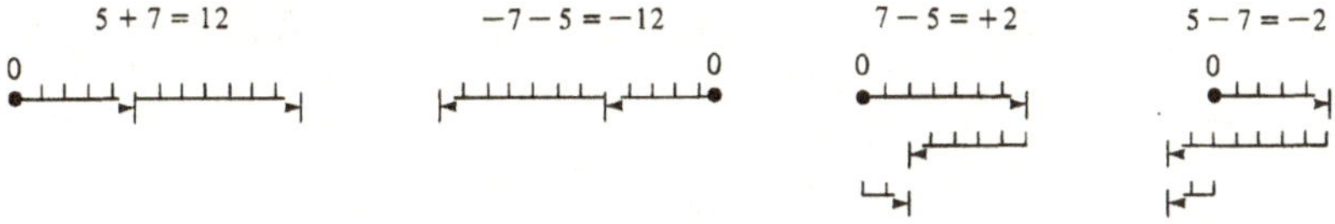

$5 + 7 = 12$ $\quad -7 - 5 = -12$ $\quad 7 - 5 = +2$ $\quad 5 - 7 = -2$

To subtract, you merely change the sign of the subtrahend (the second number) and add:

$$5 - (-7) = 5 + 7 = 12 \qquad 5 - (+7) = -2 \qquad -5 - (-7) = 2$$

Multiplication by a positive number keeps the sign of the other number being multiplied.

$(3)(4) = 12$ $\quad (3)(-4) = -12$

Multiplication by a negative number changes the sign of the other number being multiplied (as on the right just above).

$$(-3)4 = -12 \qquad (-3)(-4) = 12$$

The product of two numbers of the same sign is positive. The product of two numbers of opposite sign is negative. Division follows the same rules as multiplication, since to divide by some number (for example, 2) is the same as to multiply by its reciprocal (for example, $\frac{1}{2}$).

EXERCISES A.1. A

Perform the following computations.

1. $2 - (-6) =$ 2. $-2 - 6 =$ 3. $2(-7) + 8 =$ 4. $2(-5) - (3) =$
5. $4 + (-6)(-2) =$ 6. $(-2)(-3) - 4 =$ 7. $(-2)(-4) - (-9) =$
8. $8 - 9 =$ 9. $-8(2) + 7 =$ 10. $-2(-4) - 5 =$ 11. $-7 - 2 =$
12. $(-4)(-3) - (-2)(-4) =$ 13. $-9 + 8(-2) =$ 14. $-5 + (-4) =$
15. $(-5)(-3) + (-4)(2) =$ 16. $(-7)(-4) + (-5)(-5) =$ 17. $-2 - 3 =$
18. $(-6)3 - (-4)(-2) =$ 19. $(-2)(-3)(-4) + 4 =$ 20. $-6 - (-8) =$

EXERCISES A.1. B

1. $3 - (-7) =$ 2. $-4 - 8 =$ 3. $3(-8) + 6 =$ 4. $3(-3) - 6 =$
5. $6 + (-2)(-7) =$ 6. $(-5)(-4) - 6 =$ 7. $(-6)(-4) - (-8) =$

8. $6 - 9 =$ 9. $-7(3) + 5 =$ 10. $-3(-7) - 3 =$ 11. $-9 - 8 =$
12. $(-5)(-2) - (-3)(-7) =$ 13. $-7 + 5(-3) =$ 14. $-6 + (-2) =$
15. $(-7)(-3) + (-6)(3) =$ 16. $(-5)(-3) + (-4)(-4) =$ 17. $-5 - 7 =$
18. $(-5)7 - (-6)(-4) =$ 19. $(-3)(-4)(-5) + 5 =$ 20. $-7 - (-4) =$

ANSWERS A.1. A

1. 8 2. −8 3. −6 4. −13 5. 16 6. 2 7. 17 8. −1 9. −9
10. 3 11. −9 12. 4 13. −25 14. −9 15. 7 16. 53 17. −5
18. −26 19. −20 20. 2

Appendix 2

SLOPES AND GRAPHS OF LINEAR EQUATIONS

You probably remember from elementary algebra that equations can often be "pictured" by graphs. This section reviews such graphs for one set of equations called *linear equations*.

If an equation contains two variables, they are most often called x and y; in this case the x-axis is always the horizontal axis and the y-axis is the vertical axis. The point where they cross is called the origin.

Figure A.2–1

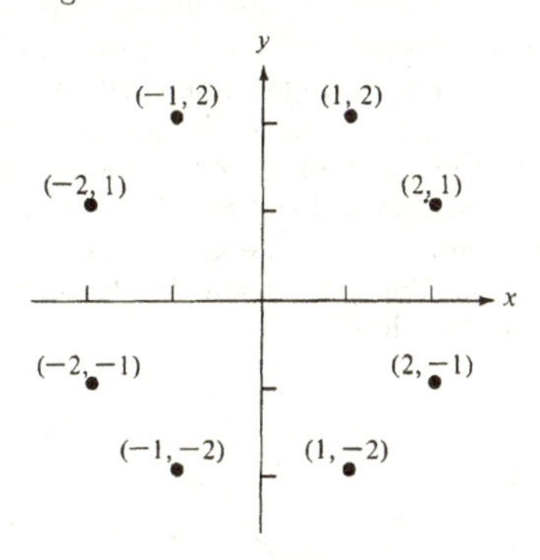

Pairs of numbers can be plotted on the plane spanned by these two axes. The first number in any pair is called the x-coordinate and the second, the y-coordinate (Figure A.2–1). To plot a number pair on the coordinate axes you move left (for negative x) or right from the origin the distance indicated by the first number (the x-coordinate) of the pair of numbers and you move up (for positive y) or down from the origin the distance indicated by the second number (or y-coordinate) of the pair.

A.2.1 Example

Plot the points $(3, 4)$, $(2, 3)$, $(1, 2)$, $(0, 1)$, $(-1, 0)$, and $(-2, -1)$.

Solution:

See Figure A.2–2.

Figure A.2–2

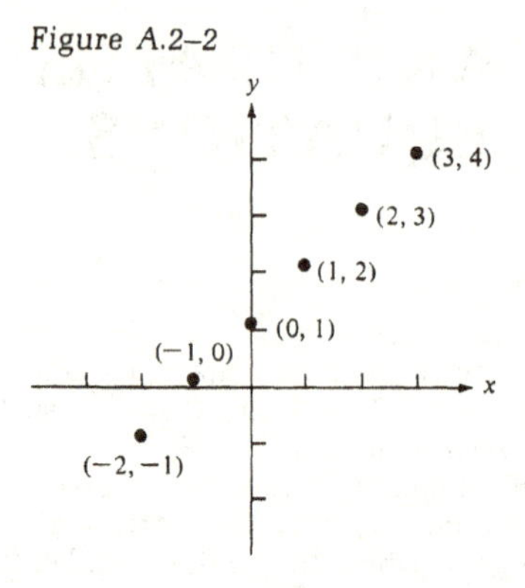

You may have noticed that in this example the second coordinate of all the pairs of points is one more than the first. Referring to the first coordinate by its most common name, x, and the second by its usual name, y, this relationship can be expressed $y = x + 1$. It is clear from the graph that the given six points lie on one straight line. Try to convince yourself that all the others such that $y = x + 1$ must lie on the same straight line. As you add some number to the value of x, the same number will be added to the value of y, and a careful consideration of the right isosceles triangle thus formed shows that it must lie on the given line. Read Figure A.2–3 counterclockwise, beginning on the left, as you try to understand why every point $(x, x + 1)$ must lie on this line.

Figure A.2–3

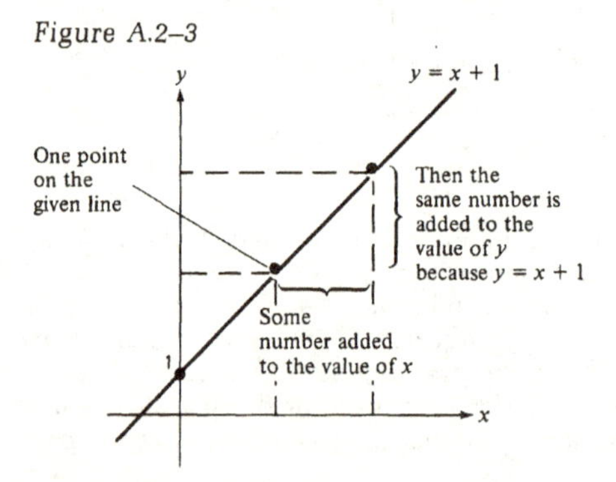

An equation in two variables whose graph is a straight line is called a linear equation. We shall show that any linear equation can be expressed in slope-intercept form,

$$y = mx + b$$

for some numbers m and b that uniquely determine both the equation and its graph. In Example A.2.1 you can see that $m = 1$ and $b = 1$. The slope-intercept form of an equation is so named because m tells its slope and b tells its y-intercept. We shall discuss each of these concepts in turn.

The number b in the expression $y = mx + b$ is called the y-intercept because it indicates where the straight line crosses (intersects) the y-axis. Substituting $x = 0$ (which is the equation that describes the y-axis) into the equation $y = mx + b$, we get $y = m \cdot 0 + b$ or simply $y = b$. Thus the point $(0, b)$ lies on the graph of $y = mx + b$ and, indeed, indicates its y-intercept. In Figure A.2–4 you can see how the y-intercept of the equations $y = x + b$ (for $m = 1$) varies as b varies.

Figure A.2–4

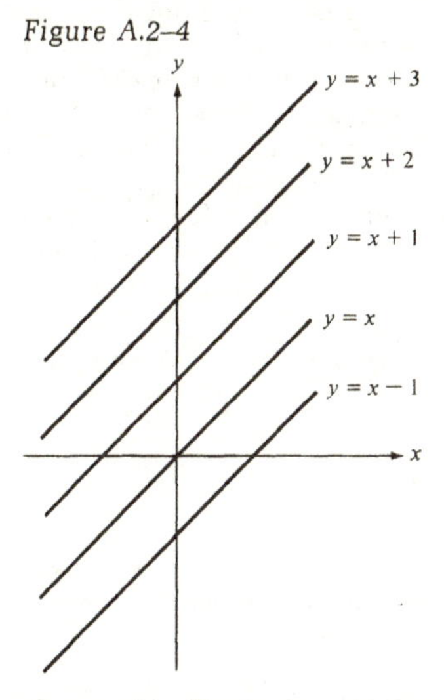

The slope m is a harder concept to understand fully than the y-intercept. As the word "slope" suggests, it indicates how fast a line is rising as we move along it from the left to the right. Figure A.2–5 indicates the slope of various straight lines going through the origin—that is, lines such that

Figure A.2–5

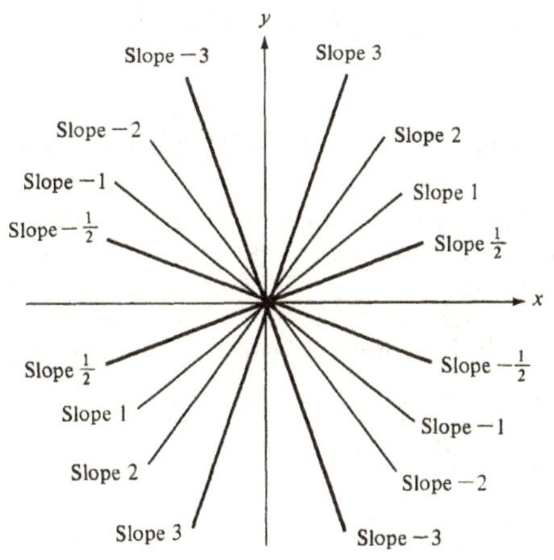

$b = 0$. Notice that we would really only have to label one side of the picture because the lines on the lower left (for example) are the same as those on the upper right.

All horizontal lines (including, of course, the x-axis) have slope zero, and the nearer a line is to being horizontal, the closer its slope is to zero. The steeper the line, the larger is the absolute value of its slope. A positive slope tells us that the line goes up as we move along it from left to right. A negative slope tells us that it goes down.

Since $b = 0$ in all the lines of Figure A.2–5, their equations are of the form $y = mx$. In each case m is the slope of the line. For example, the line labeled "slope 1" contains the points $(1, 1)$, $(2, 2)$, $(3, 3)$, $(-1, -1)$, and $(-2, -2)$ and is, in fact, the line $y = x$. The line labeled "slope 2" contains the points $(1, 2)$, $(2, 4)$, $(-1, -2)$, and $(-2, -4)$; its name is $y = 2x$.

Now we turn to more formal definitions involving the idea of slope.

A.2.2 Definition

The slope between two points (x_1, y_1) and (x_2, y_2) such that $x_1 \neq x_2$ is

$$\frac{y_2 - y_1}{x_2 - x_1}$$

A.2.3 Rule

The slope between any two points on the line $y = mx + b$ is m.

Proof:

Let x_1 and x_2 be the x-coordinates of any two different points on the line. Then the corresponding y-coordinates are $y_1 = mx_1 + b$ and $y_2 = mx_2 + b$. Thus

$$\frac{y_2 - y_1}{x_2 - x_1} = \frac{(mx_2 + b) - (mx_1 + b)}{x_2 - x_1} = \frac{mx_2 + b - mx_1 - b}{x_2 - x_1}$$

$$= \frac{mx_2 - mx_1}{x_2 - x_1} = \frac{m(x_2 - x_1)}{x_2 - x_1} = m$$

Thus it makes sense, using Definition A.2.2, to refer to m as the slope of the line, even though "slope" was originally defined for a pair of points.

In Chapter 5 you will want to graph straight lines given their equations. The equations are not always given in slope-intercept form; they are more commonly in the form $Ax + By = C$, where A, B, and C are numbers.

If $B \neq 0$, such an equation can be changed to slope-intercept form by subtracting Ax from both sides of the equation and then dividing by B:

$$By = -Ax + C$$

$$y = -\frac{A}{B}x + \frac{C}{B}$$

It is now in slope-intercept form with $m = -A/B$ and $b = C/B$.

But this algebra is not necessary merely to graph the equation. Since you know that the graph is a straight line, you can just plot two points and then draw the only straight line through them. If neither A nor B is zero, this is most easily done by setting each variable equal to zero in turn and solving for the other. This gives you the y-intercept (when $x = 0$) and the x-intercept (when $y = 0$), both of which are easy to plot.

A.2.4 Example

Graph $5x + 3y = 30$.

Solution:

Setting $x = 0$, we obtain $3y = 30$ or $y = 10$, so the y-intercept is 10 (Figure A.2–6). Setting $y = 0$, we obtain $5x = 30$ or $x = 6$, so the x-intercept is 6.

Figure A.2–6

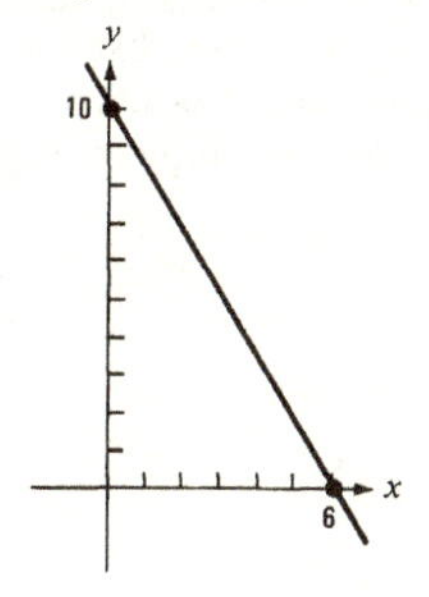

A.2.5 Example

Graph $2x + 3y = 12$.

Solution:

Setting $x = 0$, we obtain $3y = 12$ or $y = 4$, so $(0, 4)$ is on the line (Figure A.2–7). Setting $y = 0$, we obtain $2x = 12$ or $x = 6$, so $(6, 0)$ is on the line.

Figure A.2–7

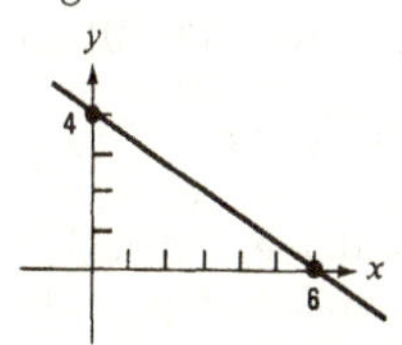

SUMMARY

Any nonvertical straight line on a coordinate plane can be expressed $y = mx + b$, where b is the y-intercept of the line and m is the slope between any two different points on the line; all such slopes are equal. To graph a linear equation given the form $Ax + By = C$, it is often easiest merely to set first $x = 0$ and then $y = 0$ to discover both intercepts, and then draw the straight line through these two intercepts.

EXERCISES A.2. A

1. Plot the points (2, 3), (3, 2), (−2, 3), (−3, 2), (2, −3), (3, −2), (−3, −2), and (−2, −3).
2. What is the slope of all the lines in Figure A.2–4?
3. Graph:
   (a) $3x + 5y = 30$ (b) $5x - 3y = 30$ (c) $-3x - 5y = 30$
   (d) $-3x + 5y = 30$ (e) $5x + 3y = 30$ (f) $3x - 5y = 30$
   (g) $3x + 2y = 12$ (h) $3x - 2y = 12$ (i) $-3x + 2y = 12$
   (j) $-3x - 2y = 12$ (k) $2x + 3y = 12$ (l) $2x - 3y = 12$
   (m) $2x + 5y = 10$ (n) $x + 3y = 6$ (o) $2x + y = 6$ (p) $5x + y = 5$
4. What is the slope and the y-intercept of each of the lines graphed here? Write the equation for each line in slope-intercept form.

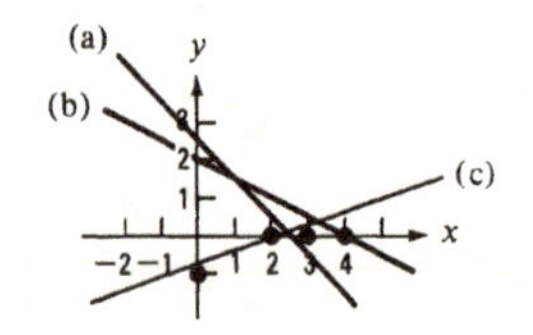

EXERCISES A.2. B

1. Plot the points (1, 3), (3, 1), (−3, 1), (−1, 3), (1, −3), (3, −1), (−3, −1), and (−1, −3).
2. What is the slope of all the lines in Figure A.2–4?
3. Graph:
   (a) $3x + 4y = 12$ (b) $3x - 4y = 12$ (c) $-3x - 4y = 12$
   (d) $-3x + 4y = 12$ (e) $4x + 3y = 12$ (f) $4x - 3y = 12$
   (g) $2x + 3y = 6$ (h) $3x - 2y = 6$ (i) $-3x + 2y = 6$
   (j) $-3x - 2y = 6$ (k) $3x + 2y = 6$ (l) $2x - 3y = 6$
   (m) $2x + 7y = 14$ (n) $x + 2y = 4$ (o) $3x + y = 6$
4. What is the slope and the y-intercept of each of the lines graphed here? Write the equation for each line in slope-intercept form.

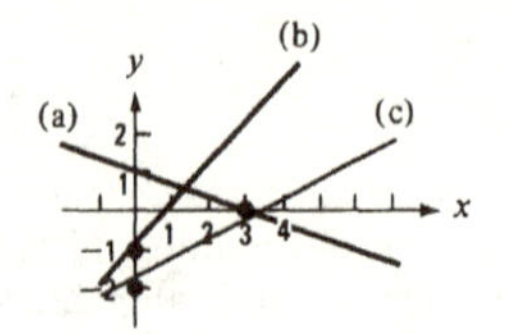

ANSWERS A.2. A

1.

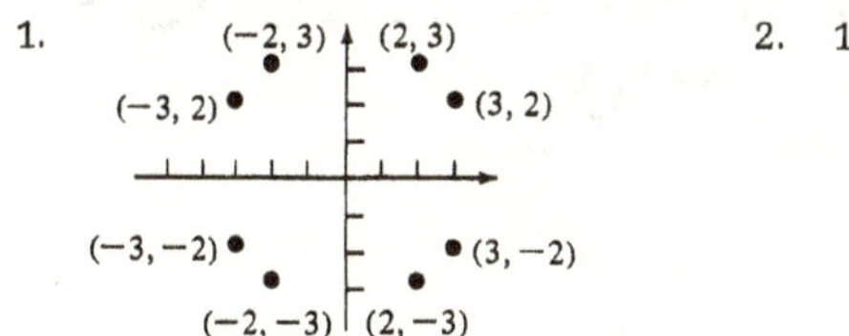

2. 1

3.

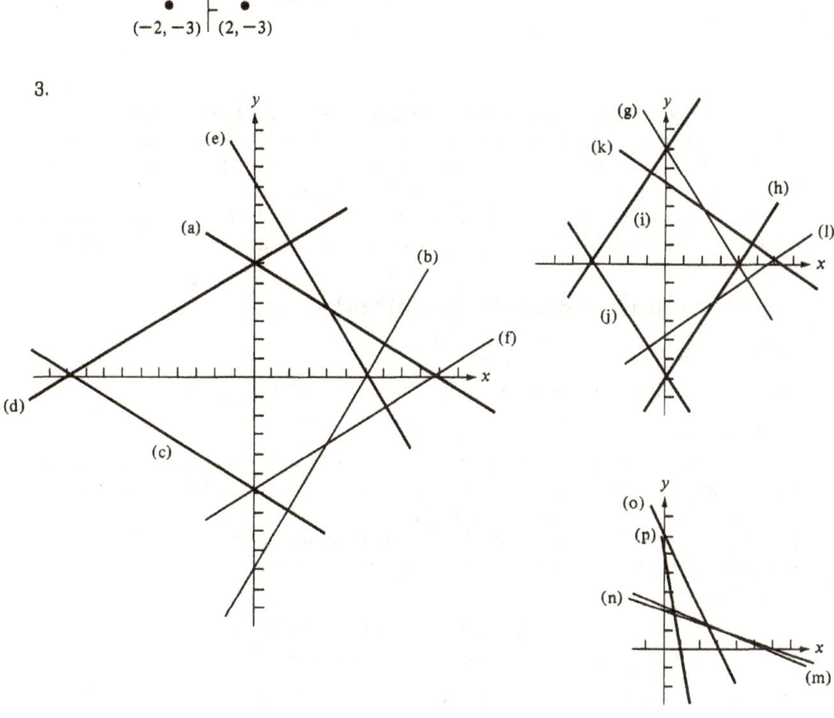

4. (a) slope -1 and y-intercept 2; $y = -x + 2$
   (b) slope $-\frac{1}{2}$ and y-intercept 2; $y = -\frac{1}{2}x + 2$
   (c) slope $\frac{1}{3}$ and y-intercept -1; $y = \frac{1}{3}x - 1$

Appendix 3

SOLVING TWO SIMULTANEOUS EQUATIONS IN TWO UNKNOWNS

In elementary algebra you studied three methods for finding the simultaneous solution of two equations in two unknowns—that is, for finding that value of x and y which makes both equations true. One of these methods was to graph both equations (using methods explained in Appendix 2) and to observe at what point they intersect. The other two methods, the addition and subtraction method and the substitution method, are the subject matter of this appendix.

Addition and Subtraction Method

A.3.1 Example

Find the simultaneous solution of $3x + 2y = 16$ and $5x + 4y = 30$.

First Solution:

In this solution we eliminate x by subtracting appropriate multiples of the two equations. Since the coefficient of x in the first equation is 3, and the coefficient of x in the second equation is 5, we can multiply each equation by the number that will change these coefficients to 15, their least common multiple. To do this, we multiply the first equation by 5 and the second by 3.

$3x + 2y = 16$ multiplied by 5 gives $15x + 10y = 80$

$5x + 4y = 30$ multiplied by 3 gives $15x + 12y = 90$

Subtracting the first equation on the right from the second:

$0 + 2y = 10$ or $2y = 10$

Dividing by 2: $y = 5$

We can now substitute $y = 5$ back into either original equation to find the value of x. If we substitute $y = 5$ into $3x + 2y = 16$, we get $3x + 2(5) = 16$, which gives $3x + 10 = 16$.

Subtracting 10 from each side: $3x = 6$

Dividing by 3: $x = 2$

Thus the solution is $x = 2$ and $y = 5$, or, briefly, $(2, 5)$. Notice that this checks in both original equations:

$3(2) + 2(5) = 16 \qquad 5(2) + 4(5) = 30$

Second Solution:

It is faster to use the addition and subtraction method to eliminate y from these equations, as you would guess if you happened to notice that the coefficient of y in the second equation is twice that of the first. Doubling the first equation, $3x + 2y = 16$, we get

$$6x + 4y = 32$$

Subtracting the second equation: $5x + 4y = 30$

We immediately obtain: $x = 2$

Substituting $x = 2$ back into $3x + 2y = 16$:

$$3(2) + 2y = 16$$

Subtracting 6 from both sides: $2y = 10$

Dividing both sides by 2: $y = 5$

Thus we get the same solution as before, but more quickly. Either method is correct.

Substitution Method

A.3.2 Example

Find the simultaneous solution of $3x + 2y = 17$ and $x + 4y = 19$.

Solution:

We could use the addition and subtraction method for this problem, but since the coefficient of x in the second equation is 1, it is easy to solve for x in this equation and substitute this value for x in the other equation to find the numerical value of y.

Subtracting $4y$ from both sides of $x + 4y = 19$: $x = 19 - 4y$

Substituting this value of x in $3x + 2y = 17$: $3(19 - 4y) + 2y = 17$

Multiplying using the distributive law: $57 - 12y + 2y = 17$

Subtracting 57 from both sides: $-12y + 2y = 17 - 57$

Performing indicated operations: $-10y = -40$

Dividing both sides by -10: $y = 4$

Then we can use the expression $x = 19 - 4y$ to solve for x:

$$x = 19 - 4(4) = 19 - 16 = 3$$

Checking (3, 4), we see that, indeed,

$$3(3) + 2(4) = 17 \quad \text{and} \quad 3 + 4(4) = 19$$

A.3.3 Example

Solve the problem given in Example A.3.1 using the substitution method.

Solution:

This is more difficult than Example A.3.2 because more effort is needed to isolate one of the variables. To isolate x in $3x + 2y = 16$,

First subtract $2y$ from both sides: $3x = 16 - 2y$

And then divide by 3: $x = \dfrac{16 - 2y}{3}$

Substituting this value for x in $5x + 4y = 30$, we obtain

$$5\left(\frac{16 - 2y}{3}\right) + 4y = 30$$

Multiplying through by 3: $5(16 - 2y) + 12y = 90$

Using the distributive law: $80 - 10y + 12y = 90$

Subtracting 80 from both sides: $2y = 10$

Dividing by 2: $y = 5$

Then, as in the first solution of Example A.3.1, we can substitute $y = 5$ back into $3x + 2y = 16$ to obtain $x = 2$.

SUMMARY

Either the addition and subtraction method or the substitution method can be used to solve any system of two linear equations in two unknowns if that system has a solution (see Example 3.1.2). But sometimes one method is easier to use for a given system than the other.

EXERCISES A.3. A

1. Solve the following systems of equations using the addition and subtraction method.
   (a) $2x + 7y = 20$, $3x + 6y = 21$ (b) $4x + 3y = 10$, $3x + 5y = 13$ (c) $5x + 2y = 23$, $4x - 3y = 0$
2. Solve the following systems of equations using the substitution method.
   (a) $x + 4y = -2$, $2x - 5y = 9$ (b) $3x + 5y = -9$, $x + 7y = -3$ (c) $3x - y = 2$, $4x + 3y = 20$

EXERCISES A.3. B

1. Solve the following systems of equations using the addition and subtraction method.
   (a) $2x + 5y = 13$, $3x + 2y = 14$ (b) $3x + 5y = 31$, $6x + 3y = 27$ (c) $5x - 2y = 4$, $3x + 4y = 18$
2. Solve the following systems of equations using the substitution method.
   (a) $x + 3y = 0$, $4x + 3y = 9$ (b) $3x - 4y = 8$, $x + 2y = -4$ (c) $5x + y = 22$, $7x - 2y = 7$

ANSWERS A.3. A

1. (a) (3, 2) (b) (1, 2) (c) (3, 4)
2. (a) (2, −1) (b) (−3, 0) (c) (2, 4)

ANSWERS TO SAMPLE TESTS

SAMPLE TEST, CHAPTER 1 (pages 46–48)

1. (a) 3 (b) $\begin{bmatrix} 0 & 8 \\ 5 & 11 \end{bmatrix}$ (c) $\begin{bmatrix} 8 & 26 \\ 10 & 38 \end{bmatrix}$ (d) $\begin{bmatrix} 3 & 9 \\ 6 & 12 \end{bmatrix}$

2. (a) $\begin{bmatrix} 0 & 0 & 0 \\ 0 & 0 & 0 \\ 0 & 0 & 0 \end{bmatrix}$ (b) $\begin{bmatrix} 1 & 0 & 0 \\ 0 & 1 & 0 \\ 0 & 0 & 1 \end{bmatrix}$

3. There are many right answers; a sample is given.

   (a) $\begin{bmatrix} 1 & 2 & 3 \\ 2 & 4 & 5 \\ 3 & 5 & 6 \end{bmatrix}$ (b) $\begin{bmatrix} 1 \\ 2 \\ 3 \end{bmatrix}$ (c) $\begin{bmatrix} 1 & 0 & 0 \\ 2 & 3 & 0 \\ 4 & 5 & 6 \end{bmatrix}$ (d) $\begin{bmatrix} 1 & 2 \\ 3 & 4 \\ 5 & 6 \end{bmatrix}$

4. Additive inverses: $\begin{bmatrix} -\frac{1}{4} & 0 \\ 0 & -3 \end{bmatrix} \begin{bmatrix} 0 & -1 \\ -1 & 0 \end{bmatrix}$

   Multiplicative inverses: $\begin{bmatrix} 4 & 0 \\ 0 & \frac{1}{3} \end{bmatrix} \begin{bmatrix} 0 & 1 \\ 1 & 0 \end{bmatrix}$

5. (a) $\begin{bmatrix} 1 & 3 \\ 4 & -6 \end{bmatrix} \begin{bmatrix} x_1 \\ x_2 \end{bmatrix} = \begin{bmatrix} 7 \\ 8 \end{bmatrix}$ (b) $\begin{bmatrix} 1 & 0 & -1 \\ 0 & 2 & 3 \end{bmatrix} \begin{bmatrix} x_1 \\ x_2 \\ x_3 \end{bmatrix} = \begin{bmatrix} 7 \\ 4 \end{bmatrix}$

6. (a) upper triangular (b) symmetric

7. There are many right answers; a sample is given.

   (a) $\begin{bmatrix} 1 & 1 \\ -1 & -1 \end{bmatrix}$ (b) $\begin{bmatrix} 1 & 0 \\ 0 & 0 \end{bmatrix} \begin{bmatrix} 0 & 0 \\ 0 & 1 \end{bmatrix}$

8. $\begin{bmatrix} 0.3 & 0.5 \\ 0.5 & 0.25 \end{bmatrix} \begin{bmatrix} 10 \\ 12 \end{bmatrix} + \begin{bmatrix} 1 \\ 4 \end{bmatrix} = \begin{bmatrix} 10 \\ 12 \end{bmatrix}$

9. (a)

	S	F	B	R	L
S	0	0	0	0	0
F	1	0	0	0	0
B	8	4	0	0	0
R	4	1	0	0	0
L	0	4	0	0	0

   (b) $\begin{bmatrix} 0 & 0 & 0 & 0 & 0 \\ 1 & 0 & 0 & 0 & 0 \\ 8 & 4 & 0 & 0 & 0 \\ 4 & 1 & 0 & 0 & 0 \\ 0 & 4 & 0 & 0 & 0 \end{bmatrix} \begin{bmatrix} 5 \\ 3 \\ 0 \\ 0 \\ 0 \end{bmatrix} = \begin{bmatrix} 0 \\ 5 \\ 52 \\ 23 \\ 12 \end{bmatrix}$

10. $[7, 10, 15] \begin{bmatrix} 8 \\ 3 \\ 2 \end{bmatrix} = 116$ cents 11. I am tired.

SAMPLE TEST, CHAPTER 2 (pages 88–89)

1. 5 large assortment packages, 10 Mix Variety packages, and 7 small packages

2. (a) $\begin{bmatrix} 1 & 0 & 0 \\ 0 & 3 & 0 \\ 0 & 0 & 1 \end{bmatrix}$ (b) $\begin{bmatrix} 1 & 0 & 0 \\ 0 & 0 & 1 \\ 0 & 1 & 0 \end{bmatrix}$ (c) $\begin{bmatrix} 1 & 0 & 0 \\ 0 & 1 & 4 \\ 0 & 0 & 1 \end{bmatrix}$

(d) $\begin{bmatrix} 1 & 0 & 0 \\ 0 & \frac{1}{3} & 0 \\ 0 & 0 & 1 \end{bmatrix}$ (e) $\begin{bmatrix} 1 & 0 & 0 \\ 0 & 0 & 1 \\ 0 & 1 & 0 \end{bmatrix}$ (f) $\begin{bmatrix} 1 & 0 & 0 \\ 0 & 1 & -4 \\ 0 & 0 & 1 \end{bmatrix}$

3. (a) $C^{-1}B^{-1}A^{-1}$ (b) elementary (c) A (d) $CDEF$

4. $\begin{bmatrix} 0 & 0 & \frac{1}{2} \\ -\frac{1}{20} & \frac{1}{5} & \frac{3}{40} \\ \frac{1}{4} & 0 & -\frac{3}{8} \end{bmatrix}$

5. $\begin{bmatrix} 0.4 & 0.4 \\ 0.6 & 0.3 \end{bmatrix} \begin{bmatrix} 185 \\ 210 \end{bmatrix} + \begin{bmatrix} 27 \\ 36 \end{bmatrix} = \begin{bmatrix} 185 \\ 210 \end{bmatrix}$ 6. $X = \begin{bmatrix} 3 \\ 10 \\ 38 \end{bmatrix}$

SAMPLE TEST, CHAPTER 3 (pages 129–130)

1. (a) (10 pts) inconsistent (b) (5 pts) null set
2. (a) (10 pts) redundant (b) (5 pts)

$$X = \begin{bmatrix} 0 \\ -1 \\ 0 \\ 0 \end{bmatrix} + a \begin{bmatrix} -1 \\ -2 \\ 1 \\ 0 \end{bmatrix} + b \begin{bmatrix} -2 \\ 0 \\ 0 \\ 1 \end{bmatrix}$$

(c) (5 pts) There are many right answers; here are three:

$$\begin{bmatrix} 0 \\ -1 \\ 0 \\ 0 \end{bmatrix} \begin{bmatrix} -1 \\ -3 \\ 1 \\ 0 \end{bmatrix} \begin{bmatrix} -3 \\ -3 \\ 1 \\ 1 \end{bmatrix}$$

(d) (5 pts) A plane in 4-space.

3. There are many correct answers to each part; we give here one for each part.

(a) $\begin{aligned} x + y + z &= 3 \\ x + y + z &= 4 \\ 2x + 3y + 17z &= 21 \end{aligned}$ (b) $\begin{aligned} x + y + z &= 3 \\ x + y - z &= 1 \\ 2x - 3y + 2z &= 1 \end{aligned}$

(c) $\begin{aligned} x + y + z &= 3 \\ 2x + 2y + 2z &= 6 \\ x + y - z &= 2 \end{aligned}$ (d) $\begin{aligned} x + y + z &= 3 \\ 2x + 2y + 2z &= 6 \\ 3x + 3y + 3z &= 9 \end{aligned}$

4. $v_1 + v_2 + 5v_3 = O$
5. Linearly dependent—use Rule 3.5.4.
6. 2
7. (a)

A: $500 + x_1 = 300 + x_2$
B: $400 + x_2 = 100 + x_3$
C: $200 + x_3 = x_4 + x_7$
D: $100 + x_4 = x_5 + 500$
E: $300 + x_5 = x_6 + 200$
F: $x_6 + x_7 = 400 + x_1$

(b)

$x_1 = x_6 + x_7 - 400$
$x_2 = x_6 + x_7 - 200$
$x_3 = x_6 + x_7 + 100$
$x_4 = x_6 + 300$
$x_5 = x_6 - 100$

(c) 100 is the least possible value for x_6 that will make x_5 nonnegative.

SAMPLE TEST, CHAPTER 4 (page 150)

1. (a) -8 (b) Yes, because its determinant is not zero.
   (c) No, because the coefficient matrix has a nonzero determinant.
2. Linearly dependent, because the determinant is zero.
3. See Rule 4.2.1, page 139.
4. Yes, because $\begin{vmatrix} 2 & 6 \\ 1 & 3 \end{vmatrix} = 0$; $x = -3$, $y = 1$
5. 142

SAMPLE TEST, CHAPTER 5 (pages 192–193)

1. x = number of bracelets, y = number of necklaces; constraints: $x \geq 0$, $y \geq 0$, $40x + 120y \leq 360$, $30x + 60y \leq 240$; objective equation: $P = 3x + 5y$.
2. x = number of Morgan 55, y = number of Columbia 42, z = number of Trident 36; constraints: $19x + 10y + 5z \leq 1960$, $16x + 20y + 50z \leq 1800$, $x \geq 0$, $y \geq 0$, $z \geq 0$; objective equation: $P = 5000x + 4600y + 3900z$.

3. 4.

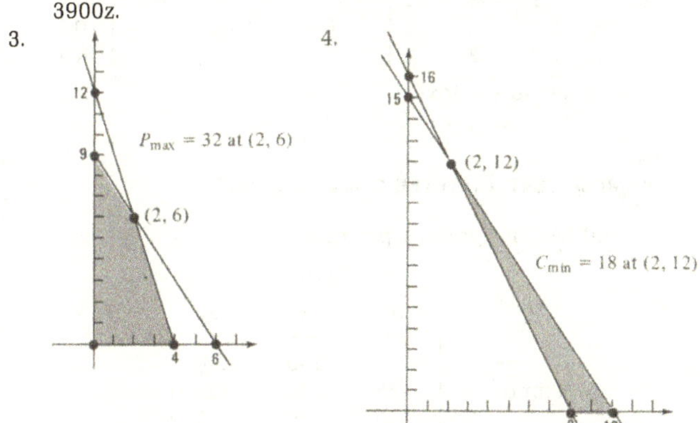

5. $C_{min} = 48.5$ at $(0, 6, 2.5)$ (Equations 1, 4, and 5)
6. $P_{max} = 54.5$ at $(0, 13, 0.5)$ (Equations 1, 4, and 6)

SAMPLE TEST, CHAPTER 6 (page 233)

1. $P_{max} = 19$ at $(2, 0, 5)$ 2. Counterclockwise
3. Multiply the objective equation by -1 and maximize $-C$.
4. (a) 2 (b) 3

	x	y	z	s_1	s_2	a_1	a_2	a_3	Solution
s_1	4	2	1	1	0	0	0	0	12
a_1	2	3	0	0	0	1	0	0	8
a_2	0	1	4	0	-1	0	1	0	6
a_3	1	0	1	0	0	0	0	1	3
$-A$	3	4	5	0	-1	0	0	0	17

5.

	x_1	x_2	x_3	x_4	Solution
x_2	0	1	0	0.4	0.2
x_1	1	0	0	−0.4	0.8
x_3	0	0	1	1	1
$-C$	0	0	0	−0.6	3.2

SAMPLE TEST, CHAPTER 7 (page 249)

1. Minimize $C = 14u_1 + 10u_2 + 4u_3$ when $u_1 + u_2 \geq 4$, $3u_1 + 2u_3 \geq 2$, and $u_1 + u_2 + u_3 \geq 6$. $C_{min} = 48$ when $u_1 = 0$, $u_2 = 4$, and $u_3 = 2$.
2. Maximize $P = 6x_1 + 8x_2 + 2x_3$ when $2x_1 + 4x_2 \leq 24$, $6x_2 + 4x_3 \leq 20$, and $2x_1 + 4x_2 + 2x_3 \leq 12$. $P_{max} = 36$ when $x_1 = 6$, $x_2 = x_3 = 0$.
3. Maximize $P = BX$ when $AX \leq D$.
4. $$A = \begin{bmatrix} 1 & 3 & 1 \\ 1 & 0 & 1 \\ 0 & 2 & 1 \end{bmatrix} \qquad B = [4, 2, 6] \qquad D = \begin{bmatrix} 14 \\ 10 \\ 4 \end{bmatrix}$$

5–6. Refer to the text.

SAMPLE TEST, CHAPTER 8 (pages 291–292)

1. (a) An assignment problem

		Customer 1	Customer 2	Customer 3
Piano	1	130	145	140
	2	125	145	135
	3	150	170	160

or

		Piano 1	Piano 2	Piano 3
Customer	1	130	125	150
	2	145	145	170
	3	140	135	160

(b) A transportation problem

		Product A	Product B	Product C	Product D	
Machine	I	25	45	40	0	35
	II	30	35	35	0	55
		25	25	25	15	

2. 1st team gets 4th job; 2nd team gets 3rd job; 3rd team gets 1st job; 4th team gets 2nd job

3. (a)

40	15		
	30	25	
		25	25

(b)

		50	5
35			20
5	45		

(c)

		30	25
10	45		
30		20	

(d) $1325

4.

	20	80	
90			20
	80		30

or

	40	60	
90		20	
	60		50

SAMPLE TEST, CHAPTER 9 (page 335)

1. (3)(20)(20) = 1200
2. (a) $\frac{1}{4}$ (b) $\frac{1}{2}$ (c) $\frac{2}{5}$ (d) 0.8
   (e) [0.1, 0.2, 0.3, 0.25, 0.1, 0.05]
3. $1/5! = 1/120$
4. $1/{}_7C_2 = \frac{1}{21}$
5. (a)

	Auto deaths	Other causes	
Age 15–24	0.008	0.016	0.024
Other ages	0.015	0.961	0.976
	0.023	0.977	1.000

   (b) No; $(0.024)(0.023) = 0.000552 \neq 0.008$
6. (a) A regular Markov matrix
   (b) (0.2, 0.8)

SAMPLE TEST, CHAPTER 10 (page 370)

1. (a) 0.6 (b) 3.6 (c) $14,400

2. (a) $v = -1$; $P = [0, 0, 1]$; $Q = \begin{bmatrix} 0 \\ 1 \\ 0 \end{bmatrix}$; $PA = [2, -1, 2]$; $AQ = \begin{bmatrix} -2 \\ -4 \\ -1 \end{bmatrix}$

   (b) $v = 1$; $P = [\frac{1}{2}, \frac{1}{2}]$; $Q = \begin{bmatrix} \frac{1}{3} \\ \frac{2}{3} \end{bmatrix}$; $PA = [1, 1]$; $AQ = \begin{bmatrix} 1 \\ 1 \end{bmatrix}$

   (c) $v = \frac{4}{5}$; $P = [\frac{4}{5}, \frac{1}{5}]$; $PA = [\frac{4}{5}, \frac{4}{5}, \frac{14}{5}]$; $Q = \begin{bmatrix} \frac{4}{5} \\ \frac{1}{5} \\ 0 \end{bmatrix}$; $AQ = \begin{bmatrix} \frac{4}{5} \\ \frac{4}{5} \end{bmatrix}$

   (d) $v = 2$; $P = [0, \frac{1}{3}, \frac{2}{3}]$; $Q = \begin{bmatrix} \frac{1}{3} \\ \frac{2}{3} \\ 0 \end{bmatrix}$; $PA = [2, 2, 2]$; $AQ = \begin{bmatrix} 2 \\ 2 \\ 2 \end{bmatrix}$

GLOSSARY

Artificial variables: variables inserted into constraints of a nonstandard linear programming problem to find an initial feasible solution

Assembly graph: a diagram showing which parts are needed for making more complex items

Basic outcomes: the elements of a set of possible outcomes in a probability experiment that is both exhaustive and exclusive

Basic variables (of a simplex tableau): variables that are not necessarily zero

Constraints: inequalities and equalities stating the restrictions on the independent variables in a linear programming problem

Diagonal matrix: a square matrix with zeros except perhaps on the major diagonal

Elementary matrix: a matrix that performs a Gauss–Jordan row operation on any other matrix A when it premultiplies A

Event (in a probability experiment): a subset of the sample space

Feasible region: the set of points satisfying all the constraints in a linear programming problem

Fixed probability vector (of a matrix M): a probability vector X such that $XM = X$

Fixed vector (of a matrix M): a vector X such that $XM = X$

Game theory: the mathematical analysis of conflict situations between rational beings

Gauss–Jordan row operations (on a matrix): interchanging two rows, or multiplying a row by a nonzero number, or adding a multiple of one row to another row

General solution: a vector that symbolically represents *all* the solutions of a given system of equations

Inconsistency: a system of equations with no solution

Independent event: two events, A and B, such that $P(A) = P(A|B)$

Inverse (multiplicative inverse of a matrix A): a matrix A^{-1} such that $A \cdot A^{-1} = I$, where I is the multiplicative identity

Leontief inverse: $(I - A)^{-1}$ if A is a Leontief matrix and all elements of $(I - A)^{-1}$ are nonnegative

Leontief matrix: a square matrix A such that all elements are positive or zero and the sum of the elements in each column does not exceed 1

Leontief model: (open) $AX + D = X$; (closed) $AX = X$

Linear equation: an equation of the form $a_1x_1 + a_2x_2 + \cdots + a_nx_n = b$

Linear programming (or LP): the study of maximizing and/or minimizing a linear function subject to linear constraints

Lower triangular matrix: a square matrix with all elements zero above the major diagonal

Major diagonal: the numbers in a square matrix from the upper left to the lower right

Markov matrix: a square matrix such that all elements are nonnegative and the sum of each row is 1

Matrix: a rectangular array of numbers

Model (mathematical): is formed when a mathematical object such as a matrix, probability distribution, or graph is used to describe a real-world relationship

Mutually exclusive (events in a probability experiment): two events that have no basic outcomes in common

Nonnegativity constraints: inequalities stating that certain variables cannot be negative

Objective function: an equation giving the value of a variable that is to be maximized or minimized in terms of other variables

Particular solution: a vector containing particular numbers that satisfies a given system of equations

Phase 1: a step used in a nonstandard linear programming problem to find an initial feasible solution

Postmultiply: to multiply after (in AB, B postmultiplies A)

Premultiply: to multiply before (in AB, A premultiplies B)

Probability distribution: assigns to each basic outcome of a probability experiment some positive number such that the total is 1

Probability vector: a vector $(p_1, p_2, \ldots, p_n)$ such that $p_i \geq 0$ for all i and $p_1 + p_2 + \cdots + p_n = 1$

Quantity matrix: a matrix showing which parts are used to make more complex items

Redundancy: a system of linear equations with more than one solution

Regular (Markov) matrix: is such that there is some positive integer n such that all elements in M^n are positive

Saddle point: a payoff of a game that will always be chosen by both players if they are rational

Sample space: an exhaustive and exclusive set of basic outcomes in a probability experiment

Scalar multiplication: the process of multiplying each element of a matrix by the same number, called a scalar

Significant constraints: constraints in a linear programming problem other than the nonnegativity constraints

Simplex algorithm: a method for solving linear programming problems involving writing a series of tableaux, each describing a vertex of the feasible region, so that the value of the objective function in each successive tableau is nearer the optimum value

Slack variables: variables inserted into an inequality to change it to an equality

Standard linear programming problem: a linear programming problem that maximizes the objective function and such that all significant constraints are of the form $a_1x_1 + a_2x_2 + \cdots + a_nx_n \leq b$ with $b \geq 0$

Symmetric matrix: a square matrix such that $a_{ij} = a_{ji}$

Vector: a matrix having only one row or only one column

INDEX